*This book is dedicated
to the memory of
Mary Alan Hokanson
Nels Magnus Hokanson
and
Magdalena Mora*

Donde sembraba algodón
que el mundo llamó
erróneamente el Oro Blanco

Tal vez porque él ignoró
el negro sudor
que me costó cultivarlo.

(Where he grew cotton
which the world called,
mistakenly, White Gold,

perhaps because he ignored
the dark sweat
it cost me to cultivate it.)

from "La Canción del Oro Blanco"
("The Ballad of White Gold"),
a Mexican corrido

El negro sudor is a metaphorical expression indicating not only dark sweat but also the implied enslavement of workers to cotton cultivation. While the corrido refers to Mexicans, the image plays off the history of African-American slaves who picked cotton. My thanks to Alicia Arrizón for this comment.

Dark Sweat,
White Gold

California Farm Workers,
Cotton, and the New Deal

Devra Weber

UNIVERSITY OF CALIFORNIA PRESS
Berkeley · *Los Angeles* · *London*

This book is a print-on-demand volume. It is manufactured using toner is place of ink. Type and images may be less sharp than the same material seen in traditionally printed University of California Press editions.

University of California Press
Berkeley and Los Angeles, California

University of California Press, Ltd.
London, England

© 1994 by
The Regents of the University of California

First Paperback Printing, 1996

Library of Congress Cataloging-in-Publication Data

Weber, Devra, 1946–.
    Dark sweat, white gold : California farm workers, cotton, and the New Deal / Devra Weber.

      p.   cm.
  Includes bibliographical references and index.
  ISBN 0-520-20710-6 (alk. paper)
  1. Cotton farmers—California—History. 2. Migrant agricultural laborers—California—History. 3. Alien labor, Mexican—California—History. I. Title.
HD8039.C662U68   1994
331.6'2720794–dc20

                93-36933
                CIP

Printed in the United States of America

The paper used in this publication meets the minimum requirements of ANSI/NISO Z39.48-1992(R 1997)(Permanence of Paper)

Contents

Illustrations and Maps

Acknowledgments

This book was not a solitary undertaking. It has developed with the support of friends, fellow scholars, and, especially, the people who made this history. This is their book. Gil Flores, director of the Delano office of the California Rural Legal Assistance (CRLA), and his mother, Belén Flores, of Hanford, California, introduced me to many people and shared their own lives with me. Gil drove me around the Valley for hours, explaining the ranches, the work, the camps, and the stories of multiple generations of workers and unionists. The graceful eloquence of Roberto Castro gave me a feeling of the period and his family and exuded a warmth he has passed down to his daughters, Anna Castro and Sally Castro Collins. He was a man who cared intensely about history, knowledge, and what they meant to people. I only wish he could have known how deeply he touched those who have read his words. Edward Bañales shared time, laughter, photographs, and insights about Corcoran, his family, and his Purépecha heritage. His sister, Virginia Bañales Godina, generously provided photographs of the Bañales family. Juana Padilla, Macario Pacheco, and Luis Lima opened their homes and gave of themselves and their time. Narciso Vidaurri provided personal insight into contracting and the strains contractors feel. Special thanks are due also to Luis Sálazar, Ray Magaña, Sabina Cortez, and Soledad Regelado.

I owe a debt to the generations of Mexican and Chicano organizers. Guillermo Martínez, whom I met in 1968, eloquently taught me the meaning of "sin fronteras," and he continues to teach history and its meaning to new generations of immigrants and organizers. Leroy Parra shared with me his memories of his family and progressive traditions among Mexican activists north of the border, as did Eduardo and Lillie Gasca-Cuéllar and Ed Cuéllar Jr. Josefina Arancibia, active in the Partido Liberal Mexicano (PLM) and mother of 1940s organizer Josefina Fiero de Bright, sang the songs of the PLM, described the meetings in the Placita in downtown Los Angeles between the Mexican anarchists and their Industrial Workers of the World (IWW) allies, and talked to me long and hard of the still-crucial lessons of internationalism. Gene and Grace Luna shared with me their memories of organizing with the Congress of Industrial Organizations (CIO) in Madera. Jessie de la Cruz remembered the 1933 strike, which took place when she was a young teenager; later she became an organizer with the United Farm Workers. Jessie, her husband Arnold, and her brother Eduardo López gave me a multifaceted view of a family of *huelgistas*.

My thanks to Cannery and Agricultural Workers Industrial Union (CAWIU) organizers Pat Chambers, Caroline Decker Gladstein, and Dorothy Ray Healey. Over several years Pat Chambers offered insights about the legacy of the Anglo Wobbly alliance with Mexican workers and about the interstices of grass-roots organizing in the 1930s. Caroline Decker Gladstein shared her memories of organizing. Dorothy Ray Healey generously provided me with an eloquent analysis of Mexican organizing, organizers, and their alliances with young CAWIU organizers in the 1930s. Her insights into the friendships and alliances with Mexican activists of the 1930s helped flesh out the importance of Mexican workers and Mexican communists to the success of CAWIU. And a special thanks is due to Fred Ross Sr., manager of the Arvin Farm Security Administration camp, a tireless organizer who worked with farm workers and taught the tools of organizing to several generations. I grew up listening to my father's nostalgic stories about Fred Ross; it was my good fortune to have gotten to know him in the course of writing this book.

In Mexico, the Bañales family of San Francisco Angumacutiro, Michoacán, gave me a perspective on the Bañales family "del otro lado" ("of the other side"). The elder Bañales' rich memory of their family who migrated north gives eloquent testimony to the social networks and connections of Mexican families across the border. Trina Medina and her daughter, Pati Gutiérrez-Medina, extended their hospitality and friendship in rancho Ancihuácuaro, Michoacán. Special thanks must also be extended to Vicente Trinidad Navarro, who dropped by one day and stayed five hours to entrance me with the history of the background of the rancho and of his family, who are only touched upon here, and to Juan García Gutiérrez, his wife, and other people of rancho Ancihuácuaro. My visit there gave me a visceral understanding of migration in a way no book could. Their warmth and hospitality left an indelible impression.

Along the way many people gave unstinting hospitality, friendship, good humor, and kindness, as well as their knowledge. Through them, the project has been personally enriching. In Mexico, Francisco Cervantes, Blas Manriquez, Arnoldo Martínez Verdugo, and the staff of El Centro de Estudios del Movimiento Obrero y Socialista (CEMOS) provided unflinching help and a measure of generosity I hope someday to be able to repay. José Luis Peréz-Canchola graciously introduced me to activists from earlier generations in the border area. Alvaro Ochoa of Zamora, Michoacán, shared his work and notes with me. I met Cuca Cornoa outside a church in San Francisco Angumacutiro: she spent time walking the town with me to find remnants of the history connecting Angumacutiro to the people of the San Joaquin Valley. Marcia Vanderlip, a reporter at the *Corcoran Journal*, offered me hospitality and helped me better understand the town. Elsa Malvido gave me intellectual sustenance and friendship, and generously introduced me to scholars over several visits to Mexico City. A special thanks to Lourdes Hernández Alcalá and Victor Díaz for their hospitality. The busy house of Leticia López-Serna in the Mexico City suburb of Coyoacán was my home for several months over the years, and where I joined with the self-proclaimed "chusma" of Letty, Adelaida del Castillo, Elsa Medina, and Adela Felix-Niebles in friendly and usually joyful cacophony. Coyoacán, this house, and this group of friends are among my warmest memories.

The input of my fellow academics has been more than generous. My dissertation advisers, Alexander Saxton and John Laslett, spent hours poring over drafts of the manuscript and helped sharpen both arguments and prose. I thank Alex for insisting I write this as a book, not as a dissertation, and for his constant but kind nudgings. Both Alex and John remain friends and intellectual mentors and are examples of committed scholarship many of us hope to emulate. My readers, James Gregory and Sarah Deutsch, provided helpful comments and encouragement along the way. I learned much from each of them, both intellectually and of generosity within academia.

Juan Gómez-Quiñones's work, friendship, and our amicably contentious discussions have been a help and inspiration over the years. Emilio Zamora read parts of the draft and gave valuable input. Discussions with

Adelaida del Castillo have sharpened my understanding of Mexican women, families, and social networks. Teresa McKenna's friendship, support, and incisive comments have been vital to this project. Judith Grant's suggestions, friendship, and encouragement buoyed me. Beverly Olevin patiently went through the manuscript. Beverly's sense of humor and warmth lightened this long process. Chuck Noble provided valuable comments about the federal government and class relations. Steve Brier, Manuel García y Griego, Miriam Silverberg, Andor Skotnes, and Ronnie Fraser generously gave their support, time, and useful comments on parts of the manuscript. Portions of the book, at an earlier stage, were shown to members of the Los Angeles Labor Study Group, who contributed their gastronomic talents and intellectual acumen to this undertaking: David Brundage, Roberto Calderón, Nancy Fitch, Jacklyn Greenberg, John Laslett, Gary Nash, Steve Ross, Alexander Saxton, Cindy Shelton, and Frank Stricker. A special note of thanks to my editor at the University of California Press, Eileen McWilliam. My copy editor, Jane-Ellen Long, skillfully went over the manuscript and helped me through the final stages. Project editor Erika Büky helped shepherd this book through to its completion. My deepest thanks to all of the University of California Press. Irene Morris Vasquez helped me gather material, sorted out release forms, brought me tamales, and otherwise helped to get this project off the ground. Leticia López and Ivelise Estrada assisted in transcribing portions of the oral history tapes. Alicia Arrizón went over the Spanish carefully, bringing to the translation a richness which more fully conveys the meaning of the voices in the text.

My thanks to the many librarians who helped sift out valuable documents at the National Archives, the Bancroft Library at Berkeley, the Bank of America Archives, the Bakersfield Public Library, UCLA libraries and special collections, Stanford special collections, and the National Archives at San Bruno. In Mexico, I received help at the historical archives of the Secretaría de Relaciones Exteriores and from the staff at CEMOS. My thanks to the following for use of their oral histories: Rudolfo Acuña, Anne Loftis, Lea Ybarra, the California Odyssey Project of Bakersfield, and the University of California's Regional Oral History Office.

The Chicano Studies Research Center at UCLA provided intellectual and financial support. The Institute of American Cultures Graduate Fellowships and Research Grants at UCLA aided in initial research. An Educational Foundation Fellowship from the American Association of University Women helped me complete the dissertation. An American

Council of Learned Societies Fellowship enabled me to take time off to write. Release time while teaching at California State University, Long Beach, and the University of California at Riverside permitted me to conduct additional research and gave me time to complete the manuscript.

And finally some personal thanks. To Walter Karplus, for his constant support and the hikes we shared over the years. To Onnie and William Darrow, for their unflinching support. And to my mother, Mary Alan Hokanson, for her spunkiness, wit, and independence, and for the support and love she has given me over the years. I am especially grateful for the introduction to the Mexican community when I was about four, my times at the Placita, celebrations, and happy afternoons spent in kitchens in the San Fernando Valley. Being too young to understand the harsh distinction I later learned I was supposed to make between "my" people and "their" people, I comfortably assumed this Mexican world was mine as well. The idea stuck. And, finally, to all my friends, who have heard of this book much too long.

This book is for all of you, and for the next generation, who will, I hope, combine their intellectual interests and desires with the needs of the community we all live in.

Los Angeles, California
29 April 1993

Introduction

Sixty miles north of downtown Los Angeles, California's interstate highway 5 climbs into the low-lying Tehachapi mountains, which mark the southernmost border of the San Joaquin Valley. Before the descent into the Great Central Valley, of which the San Joaquin is a part, the impressive vista of this agricultural cornucopia becomes visible. Averaging fifty miles wide and four hundred and fifty miles in length, the expanse of land is quilted in muted colors, created by patches of cultivated cropland.[1] This is the heartland of California's capitalist agriculture, which has dominated the Valley economy since the 1870s. In the 1990s cotton is a major, though dwindling, crop in the Valley, grown in the fields of Kern County and northward, in vast tracts in Kings, Tulare, Fresno, and Madera counties. On the Valley's east side, cotton fields mingle with acres of potatoes, vineyards, and other crops in an area dotted by small towns. On the west side the balance shifts, and the small towns are overshadowed by vast expanses of crop land.

Corcoran, a center of cotton production on the west side, is symbolic of both the expansiveness of the cotton industry here and the fate of its workers. Corcoran is a small town, dwarfed by the Salyer and Boswell ranches with their private airports, ginning and processing plants, and large private homes which loom over the more modest homes typical of Corcoran. Cotton picking is now mechanized, and plants are cleared by a flame-throwing dragon of a machine that burns the naked cotton stalks as it lumbers along the rows. The same shacks that once housed thousands of workers on labor camps have been moved to the

1

town and now house the families of cotton workers displaced by mechanization.

The history of the cotton industry in this Valley, and of its workers,
emerged from a changing interrelationship among growers, processors
and investors, workers, unions and their sympathizers, and agents of the
local, federal, and state governments. This is a study of these groups in
the years from 1919, when cotton became a major crop, through the
introduction of the New Deal in 1933, to 1939 when the United States'
entry into the war directly altered conditions in cotton and paved the way
for a contract labor program.

Migrant agricultural workers have until recently been largely ignored as
active participants in United States history. Most studies have focused
on the growth of capitalist agriculture, the related decline of the family
farm, and the critical and foreboding implications this held for the
Jeffersonian vision of a yeoman farmer–based democracy. If freeholding
family farmers were the basis of a democratic society, it was argued,
capitalist (or slave) agriculture was its antithesis. Studies of agricultural
development in the United States became locked into the broader questions of American democracy and a Turnerian view of the West, which
measured change against a mythologized past of conflict-free small farming thriving on a classless frontier. The focus of these studies was the
family farmer and, by extension, the nature of American character and
society.[2]

When noticed at all, field workers were usually considered only in
relation to questions framed by these assumptions. Agricultural laborers
were not conceptually included as part of the working class, but were
viewed as a frightful and degrading result of the demise of the family
farm. The most thoughtful studies of farm workers were exposés, written
to sway public opinion on the complex arrangement of social, economic,
and political power that perpetuated the conditions of farm workers:
wages below the poverty level; abysmal housing and working conditions;
the painful human toll recorded in the high rates of sickness and death
and the short life expectancy; child labor and the attendant low education rates. Some authors wrote with perception and sensitivity of
agricultural workers. Yet much like the history of workers in unskilled
industries, the written history of farm workers became molded by the
pressing conditions of their lives and thus obscured their long-term
struggles. Pictured as victims of a brutal system, they emerged from these
studies as faceless, powerless, passive, and, ultimately, outside the flow

of history. Racial, cultural, and ethnic stereotypes and systematic exclusion from the broader labor movement perpetuated this image. They were viewed as objects, not subjects, of history.[3]

Until the 1960s, historians had often viewed culture as "cultural baggage," comprised of allegedly negative attributes that precluded immigrants' successful participation in United States society. But then the social and political movements of the 1960s raised new questions for historians and prompted the examination of the history of those previously excluded. Historians began to examine culture in its broadest sense, and discussions of culture increasingly focused on a concern for the knowledge and representation of a person's values, universe, community, and social relations as they occurred in practice.[4]

E. P. Thompson, Herbert Gutman, and others paved the way for a generation of social historians who focused on workers as agents in history and sifted through previously ignored areas of family life, culture, the work place, and the community.[5] Women, Chicanos, African-Americans, and Native Americans—the others "not seen" in traditional history—slowly came to be viewed as important subjects in a richer historical mosaic. Yet the questions raised elsewhere about the interplay of workers and capital growth were not addressed in historical analyses of agriculture. Historians largely ignored the creative ways agricultural workers dealt with the conditions they faced and how they formed communities and social ties within a fragmenting system. By omission, historians relegated field workers to the condition of poor relations of the industrial working class. This omission reflected the long-standing neglect, particularly in agriculture, of unskilled, nonwhite workers by the American Federation of Labor (AFL).

This book attempts to rectify that neglect. Agricultural workers themselves, as producers of basic commodities in an economically strategic industry, were and are a vital part of the United States working class, and their history is an essential component of working-class history. This project began as a dissertation and has over the last decade developed into a book. At its inception I envisioned it as a contribution to debates in several historiographical areas: the United States working class, Mexicans and Chicanos, capitalist agriculture and the New Deal. My analysis was antithetical to the reigning discourse in the history of the trans-Mississippi West, in which the Turnerian view served as a key ideological underpinning in explaining American "exceptionalism" and as a rationalization for United States expansionism. I had, and still have, problems with the conceptualization of "the West" as a separate field. Such a concept is Eurocentric, ignores the perspective of Native Americans and Mexicans, who used different geographic nomenclatures, and limits a long and rich history to the post-1848 era that followed United States acquisition of the territory. Yet this book does address questions that have been raised by what is being called the new Western history. Many new Western historians call Turner into question; others refuse to take Turner as the starting point for their work. Drawing heavily on decades of work by Chicano, Native American, Asian American, and women scholars of all groups, historians are creatively coming to grips with the intersection of class, gender, race, and nationality to develop a new concept of the region. As Pegge Pascoe points out, this area forms a unique laboratory for exploring these questions and linking them to larger trends within the United States as a whole. While my project was conceived well before the recent wave of new historiography on the trans-Mississippi West, it fits squarely into many of the questions being raised.[6]

While this study focuses on the California cotton industry and its workers, it is also driven by the underlying question of the relationship between structure and human agency, that is, to what degree were workers shaped by the economic, social, and political conditions they labored and lived within, and to what degree were they able, within this system, to shape their own lives.[7] Authors have debated for more than a century how much relative weight should be assigned to external constraints (structure) and how much to individual motivation (agency), and what the relation is between these factors. Karl Marx raised the question of the extent to which people make their own history and the extent to which they are molded by the historical and economic parameters inherited from the past.[8] Jean Paul Sartre contributed to the debate in 1963 in *Search for a Method*, where he argued that "man in a period of exploitation is *at once both* the product of his own product and a historical agent who can under no circumstances be taken as a product. The contradiction is not fixed; it must be grasped in the very moment of *praxis*. . . . [While] men make their history on the basis of real, prior conditions . . . it is *the men* who make it and not the prior conditions."[9] Antonio Gramsci, Georg Lukács, and writers of the Frankfurt School also stressed the role of human consciousness and agency in history.[10] Louis Althusser claimed that, whereas Marx's earlier writings emphasized human agency, his later theory of historical materialism was shaped by the belief that a society's structural economic base determines its superstructure. Althusser argued that the structural governed historical development, and he denied that human beings were the authors of this process.

In 1963 E. P. Thompson's *The Making of the English Working Class* rejected the institutional framework that had long dominated labor history and focused instead on working-class culture and the notion of human agency. Thompson clearly emphasized culture and human agency over structure. His very definition of class as a "cultural as much as an economic formulation" rejected the notion of economic determinism. Thompson defined class and class consciousness thus: "Class happens when some men, as a result of common experiences (inherited or shared), feel and articulate the identity of their interests as between themselves, and as against other men whose interests are different from (and usually opposed to) theirs. The class experience is largely determined by the productive relations into which men are born—or enter involuntarily. Class consciousness is the way in which these experiences are handled in cultural terms: embodied in traditions, value-systems, ideas and institutional forms. If the experience appears as determined, class consciousness does not."[11] In 1980, Perry Anderson challenged Thompson's conflation of class membership with class consciousness and reasserted the need for a Marxian conception: "Social classes may not become conscious of themselves, may fail to act or behave in common, but they still remain—materially, historically—classes."[12]

The long-standing debate about agency is important for a number of reasons. Questions about agency and structure have been raised about immigrant, especially undocumented, workers, which are of course pertinent to immigrant agricultural workers.[13] Some scholars, such as Manuel Castells, have argued that the vulnerable position of immigrant workers in relation to the economy and, moreover, the relationship of undocumented immigrants to the government that has defined them as "illegal" limit their "capacity for organization."[14] Yet Castells's inattention to workers distorts his analysis. Immigrants' vulnerable position has not, as Castells suggests, induced a numbing paralysis. Their relationships to unions, the state, capital, and other workers do differ from those of citizens. Yet the *particular* nature of that relation varies with the economic sector they enter, the economic climate, the level of working-class organization, and their particular relationship to the state.[15] Moreover, it depends on gender, culture, family structure, community, ethnic relations, and the subjective intangibles of human beings, their experience and consciousness: in short, it depends on human agency.[16]

Concerns about human agency and economic and political structure have been framed in other ways by U.S. labor historians. In the United States, the question of the relationship among workers, their culture, and economic and political factors has been raised in calls for a synthesis that would integrate social histories within the larger framework of economic, political, and social structures. David Brody's *Steelworkers in America* provided a model for this approach with its skillful elucidation of the dialectical interrelationship between Slavic steel workers and the industry in which they worked. Brody directly questioned the emphasis on cultural studies to the exclusion of the broader economic framework and questions.[17] As J. Carroll Moody points out, "unlike the many historians who studied class formation through a culturalist approach, Brody focuses on class development through the prism of economic structure, managerial policy, trade union practices, and the role of the state."[18] In *Beyond Equality* David Montgomery placed working-class culture within a political framework; in *Workers' Control in America* he addressed the New Deal in relation to workers; and in *The Fall of the House of Labor*, Montgomery placed that culture within the context of the economic structure of the industry, the actions of management, and the dialectical relationship between business and workers in fights over workers' control and the expropriation of "the manager under the worker's hat."[19] All of these studies presented a broader, more complex and dialectical framework within which to view working people.

My study examines the interrelation between human agency, economic structure, and the political forces of the state within the context of California agriculture. It focuses on several questions: How did the social and economic structure of specialized capitalist agriculture influence the formation of the labor force and relations between workers and the industry? How did the experiences and culture of workers shape conditions and affect their responses to those conditions? How did they differ between Mexican and Anglo workers? And, finally, what effects did the state have on the industry and the workers, and how did it alter class relations?

I chose California's cotton industry for several reasons. Cotton is a basic commodity, pivotal to the local, national, and international economy. Cotton has a rich history and, as an extremely labor-intensive crop until the mid-twentieth century, was critical in defining agricultural labor relations.[20] During the industrial revolution, the growing demand for cotton helped solidify and expand the slave system and, after 1865, the sharecropping system. By the early twentieth century, California agriculture and particularly cotton represented one of the most developed forms of capitalist agriculture in the world. In part this was a matter of sheer size. By 1929, California cotton ranches were the largest in the nation, whether measured by acreage, production, or number of workers employed. The introduction of cotton dramatically increased the demand for labor in California agriculture (thus changing the relation between the supply and availability of workers in other crops as well). Cotton workers became the largest labor force in the agricultural industry. The cotton industry proved pivotal in helping agricultural interests establish mechanisms to control the labor force. Partly as a result, the largest strikes of the 1930s erupted among cotton workers.

The question of the state's relation to capital and labor can also be examined within the California cotton industry. Cotton's economic importance to the national economy was reflected in the New Deal's programmatic focus on cotton as pivotal to agricultural recovery. New Deal intervention was pronounced in the cotton industry, through relief payments, federal mediation of labor disputes, and the Agricultural Adjustment Act, and had a decisive, if multilayered, effect on class relations.

The interlocking questions raised earlier about the relation between structure and agency are particularly apt in California agriculture, where the power imbalance between the agricultural industry and its workers has been so profound. In the cotton industry, affected though it was by the limitations inherent in agricultural production, steps were taken to standardize production and control labor. These steps facilitated a concentration of centralized economic control which in turn contributed to the development of intra-industrial organization. Agricultural interests used this organization both to deal with workers and to exert pressure on the government. The cotton industry—where the economic structure was powerful and well organized, the work force was relatively powerless, and the state system worked in conjunction with capital—provides a choice area to look at the interrelated questions of culture and agency, economic structure, and state intervention. Despite the strength of the industry and the relative weakness of its workers, workers did affect the nature of their labor. Workers' responses to conditions varied widely, dependent on experience, legal status, gender, historical consciousness, family and social structures, and economic conditions.

For material on Mexican workers, I draw heavily on Chicano historians such as Juan Gómez-Quiñones, Emilio Zamora, David Montejano, Ricardo Romo, and others who have explored the development of Mexican communities in the United States and have studied the questions of Mexican migration and labor organizing.[21] Historians of Chicana history such as Vicki Ruiz, Antonia Castañeda, Deena González, and Sarah Deutsch have elucidated the interaction among gender, class, nationality, and culture.[22] Studies of migration, communities, and cross-border organizing between Mexico and the United States by scholars such as Paco Ignacio Taibo II, Javier Torres Parés, Rafael Alarcón, Jorge Durand, and Roger Rouse have contributed to a broader understanding of the transnational migration of people, ideas, networks, and culture, which is crucial to developing a valid history of Mexican workers of this period.[23]

Through extensive use of oral histories, this study elucidates the central roles of experience, consciousness, and culture in adaptation, response, and the changing relation between the industry and workers. Mexicans, who were the backbone of the labor force until the mid 1930s, had been shaped by their historical experience as transnational workers in Mexico and the United States. Until recently, they were often depicted as malleable peasant sojourners unlikely to join in effective collective action. This misperception came to be used as an ideological justification: the American Federation of Labor trundled it out to justify its inattention to Mexicans; growers used it to dismiss unrest as the product of fevered manipulation by outside agitators. The question of whether or not or to what extent Mexicans adapted and/or resisted remains part of a historical debate. Mario García has argued that because Mexicans "did not see themselves as members of a proletariat class but as Mexicans temporarily in a foreign land . . . they organized and protected themselves along ethnic lines [and] adjustment, not resistance, characterized their stay in the United States."[24] Others, such as Juan Gómez-Quiñones and Emilio Zamora, disagree. Pointing to a long, intense, and extensive participation by Mexicans in labor and social struggles on both sides of the border, they argue that Mexicans in the United States fought to preserve their culture, language, social institutions, and communities within a hostile environment. Far from being unconcerned about conditions, Mexican workers organized to obtain and protect their rights.[25]

This book is part of that debate. I argue that Mexican cotton workers were molded in the crucible of displacement and proletarianization in Mexico and the United States in the late nineteenth and early twentieth centuries. For several generations, both capital and workers spanned the border. The consciousness of this transnational work force could be described best by the term *sin fronteras* ("without borders").[26] Envisioning Mexicans as transnational workers clarifies their responses in the

California cotton fields. Mexicans who picked cotton were not only agricultural workers in California. They also had been (and many remained) miners, railroad workers, teachers, artisans, and industrial workers who labored on both sides of the border.

A transnational perspective elucidates Mexicans' experience in social conflicts that dramatically shaped their responses in the cotton fields. In Mexico, preindustrial peasants had hardly been consistently malleable agrarians. A succession of Indian and peasant revolts had punctuated Mexican history. With the proletarianization of workers in the late nineteenth century, Mexicans formed unions and participated in both rural and urban conflicts. These culminated in the Mexican Revolution of 1910 to 1920, the first major social upheaval of the twentieth century. Many who picked cotton were witnesses to and participants in the military conflict of these years, as well as in the labor organizations that crisscrossed the border. As I will argue, alliances in the teens between Mexican and Anglo radicals on both sides of the border in strikes, labor organizing, and even the ill-fated 1911 expedition by the IWW and the PLM to retake Baja California for insurgent forces formed a basis for later cooperation between progressive Anglos and Mexicans in California's agricultural fields in the 1930s. The myths and ideologies of Mexico and Mexican social upheavals, although still in the process of transformation, provided a vital historical reference point for these later conflicts. The interpenetration of organization and migration facilitated the transmission of ideas and organizations that would form a model for workers well into the 1930s.

Familial and social networks were also crucial to Mexicans' responses. Until recently, familial and social networks have often been considered unrelated to larger social conflicts defined by political parties, union structures, or male political and social formations. Activities outside these traditional spheres have at best been seen as auxiliary, or at worst been ignored as irrelevant. A broader vision of working-class life challenges the preeminence of unions and formal organizations as the essential forms of working-class organization and links women's networks, neighborhoods, and daily life to the development of social structures. This approach changes the perception of strikes, community organization, and social struggles. An analysis of families, neighborhoods, networks, and alliances reveals a more complex working-class response and includes community members who, while **not** directly involved in capitalist work relations, were still affected by **them** and participated in protests against them.

Recent scholarship has called into question the notion that capitalism undermined the family. Far from being destroyed, working-class families were as essential to capitalism as they were to workers' adaptation to the new economic order.[27] As John Bodnar pointed out, the ability of these networks to respond to the demands of work, the individual, and the group created a relationship between workers' families and capitalism that was "almost symbiotic."[28] Hardships imposed by industrialization, displacement, and migration certainly created tensions within these networks. Yet these same hardships simultaneously reinforced the individuals' dependency on human ties for survival. Overlapping familial, social, and community networks formed the structural basis for Mexicans' lives in California. Mexican communities in the United States were formed from these networks. They were also the basis for work crews, the basic unit of production in cotton. Through these networks Mexicans partially transferred their social relations from Mexico to the fields of California. These networks helped define relations among workers and between workers and contractors, and they helped workers adapt to larger transformations in social relations.

An ethos of mutuality infused these networks. As Emilio Zamora explained, "Mutualism incorporated such values as fraternalism, reciprocity, and altruism into a moral prescription for human behavior, a cultural basis for moralistic and nationalistic political action that was intended to set things right."[29] Within families and social networks, the belief in mutuality and reciprocity inspired attempts to subsume individual desires to the needs of the group and mediated tensions between individual impulses and collective needs. This mutual aid was crucial to workers' survival both as a group and individually. Yet mutualism, whether expressed in community organizations or within the families and work crews, was double-edged. At times it helped enforce work discipline. At other times it laid the basis for collective action and reformed some labor crews into de facto labor organizations.

Social networks, a sense of mutualism, past experience, and consciousness were crucial to the strike wave that tore through California's fields in 1933. Under the leadership of the progressive Cannery and Agricultural Workers Industrial Union (CAWIU), it culminated in a strike by 18,000 cotton workers. Historians have focused primarily on the CAWIU,[30] yet oral histories from leaders and participants elucidates the complexity of workers' responses. The union, while acting as an umbrella organization, was small, poor, and lacked the resources to wage a major strike. Strike organization was based in informal structures within worker networks which were effectively utilized by the CAWIU. Day-to-day leadership came from Mexican communists, veterans of earlier strikes, labor contractors, crew leaders, and workers. Women activated female networks which became the organizational bases for gender-specific collective action.[31] Working with union organizers, these social networks provided an organizational framework that facilitated the creation of effective strategy and tactics.

By 1935 there was a dramatic shift in the work force from Mexican to Anglo-American migrants from the Southeast and Southwest. This shift raised questions about how this new group, with its own distinct historical, economic, and cultural identity, would respond to work in cotton. The response by these Anglo migrants emphasizes the need to look at the intersection of economic, cultural, and political factors. The Anglo migrants entered the work force during the Depression, the latest in a long line of refugees out of the depressed cotton areas of the Southeast and Southwest. Contemporaries viewed this migration as a direct result of the family farm's decline and argued that as citizens these migrants would challenge and change the agricultural system. But economic hardships sapped their material basis for security and tore the fabric of families and social networks. The problem was one of circumstance. The Depression enlarged the labor pool and depressed wages. Grinding poverty intensified a disintegration of their world which, while temporary, affected their sojourn as cotton workers in California. Their ability to rely on extended social networks was weakened by their tendency to migrate in smaller social groupings than Mexicans. And those who were new arrivals simply had not had the time to develop networks as extensive as those of the Mexicans who had preceded them.

Historians have debated the responses of these new Anglo migrants to unionization. Walter Stein persuasively argues that Anglo workers were less responsive to unions than were Mexicans. The qualities that contemporaries viewed as favorable to Anglo organization (that they were whites and citizens) undercut their solidarity with other workers. Anglo workers shared an ambivalent political heritage which intermingled populist hostility toward the rich with longings to belong to the propertied farmer class and be recognized as members of the white community. This undermined class cohesion and made it less likely that they would join unions. Ostracized as poor whites, they found an identification with the Anglo community as voting citizens.[32]

Yet, as James Gregory points out, Anglo responses were heterogeneous. Some had been influenced by populism or socialism and sup-

ported unions.[33] Examination of the *Bakersfield Californian* and other
sources suggests that Anglo participation in unions rested on an established base of residents whose presence predated the Depression. Their
longer residency and firmer roots facilitated the growth of progressive
tendencies and the development of political and labor organizations that
supported small farmers and farm workers.

The extent to which workers accepted the state as a vehicle for viable
political expression affected conflicts in cotton. Overall, Mexicans had
an ambivalent relation to the state. For non-U.S. citizens, the electoral
process was simply not an option. Among citizens, while some looked
to the New Deal for help, others found that personal experiences of
exploitation, deportation, and racism undermined their expectations
that the government would treat them benevolently. Anglo-American
workers, however, came from areas where populist and socialist traditions had encouraged the transformation of the government through the
ballot and had emphasized electoral politics over syndicalism. When and
if they participated in attempts to change class relations, they did so
increasingly within the arena of electoral politics. Anglo farmers and
workers fought out the meaning of the New Deal, demanded protection
from agricultural monopolists, urged relief reform, and took to the polls
to cast their votes, first for Upton Sinclair and then for Culbert Olson.
Their support strengthened labor's vote. Yet in relying on electoral
politics they were placing their trust in a process that did not support
farm workers' interests. Citizenship and white skin gave little protection
from the deplorable conditions in the agricultural system and were
ultimately more important in propelling them out of the fields than in
changing the conditions in the fields.

This raises the question of the state's role in shaping economic and
political conditions, and the nature of the interrelationship among the
state, the cotton industry, and workers.[34] Scholars first explained the
New Deal within the framework of democratic pluralism, in which interest groups competed within the government to form policy and programs that benefited the public good. Arthur Schlesinger, for example,
lauded the New Deal for inhibiting employer repression of labor, improving working conditions, and welcoming workers and unions into the
society and government.[35] Historians writing in the less benign years of
the 1980s and 1990s have been more critical.[36] While David Montgomery, for example, agreed that New Deal policies and limited support
for collective bargaining initially benefited industrial workers, ultimately,
he argued, the relation between organized labor and government became

co-optive and evolved into a "restrictive quagmire" that curtailed workers' ability to organize and limited working-class participation to unions whose actions were sharply defined by the law.[37] Christopher Tomlins compellingly argued that by defining collective bargaining as an expression of public interest, the government denied federal support for unions to determine their own structure and activities. Although allowed to engage in collective bargaining, unions' actual ability to do so was subject to the "state's determination of how the public interest might be best served . . . [and] this would eventually come to mean in practice that the right to organize and bargain could be maintained only so far as the state conceived it to serve an overriding goal of industrial peace."[38]

But these writers ignore agricultural workers. As Theo Majka, Linda Majka, and Theda Skocpol have pointed out, for agricultural workers the effect of the New Deal was dramatically different. Excluded from protective legislation, their position declined precipitously, both absolutely and in relation to nonagricultural workers, in large part because of New Deal policies.[39] The problem was not only their omission from federal legislation. It was the constellation of New Deal programs and policies that institutionalized the position of agricultural workers.

Without a lengthy digression, a few words are in order about how this study defines the state and its relation to social classes. What I call the *state* is composed of branches of government at the local, state, and federal levels, including all government programs, agencies, and projects.[40] The state controls coercive mechanisms (such as the army and police), administers a given geographical territory, and finances its activities through taxes or loans.

I will argue that the state is shaped by class relations that emerged from industrial capitalism. The state apparatus reflects the heterogeneous nature of classes and the varying interests of capital and workers. Government agencies, programs, and policies become contested terrain between classes and often result in seemingly conflicting aims, programs, agencies, and personnel.[41] The state is neither a mechanistic instrument of the capitalist class nor completely autonomous.

The New Deal was the most important expansion of the role of the state in economic affairs in United States history to that point in time, and since then, inter- and intra-class relations have focused increasingly on the state. During the 1930s, federal programs were shaped by the need to facilitate capital accumulation, reproduce class relations, and maintain or reestablish hegemony. Reflecting conflicts of the period, government priorities were shaped by class alliances and conflicts as much as by federal administration. Government policies and administration, agencies, and programs became battlefields for different sectors of society. Yet in responding to class pressures, the state redefined the parameters of relations between workers and capital. In so doing, the New Deal affected the overall balance of power between classes.

How was this played out in cotton? For a brief period the New Deal appeared to extend the possibility of restructuring class relations. Workers and their supporters fought over the definition and implementation of New Deal programs, legislation, and the Agricultural Adjustment Administration (AAA). Increasing government intervention spurred organizing by cotton interests, such as the Farm Bureau and the Associated Farmers, in order to form, influence, and administer policy and undermine unions. Small cotton farmers struggled with larger farmers over the policies of the AAA. Unions in agriculture, as in industry, shaped strategy around government programs and actions. And councils of the unemployed fought for their right to relief. Many factors influenced these battles: Roosevelt's reliance on the political clout of agricultural capital, especially Southern Democrats; the perennial weaknesses of agricultural labor and obstacles to labor organizing; and the effect of liberal reformers who refused to see the situation in class terms and envisioned the government acting as a form of neutral broker.

Previous works have focused on agricultural workers' exclusion from New Deal labor legislation. But a broader scope is needed to assess the overall impact of the New Deal on class relations in cotton. This book argues that the New Deal accentuated and institutionalized the power balance within the industry. The AAA enhanced the industry's economic control, indirectly subsidized anti-union elements, and, by helping propel southern tenants and sharecroppers into the wage-labor pool, contributed to the further disorganization of the California work force. The relative powerlessness of workers and the cotton industry's strength ultimately determined the direction of government programs. What respite the government offered came from relief and its housing programs. Relief to destitute migrants briefly threatened to establish a minimum wage and to replace collective bargaining as a mechanism to raise wages. Federal housing provided a potential base for unions and shelter free from the control of growers. Yet agricultural interests ultimately expropriated and utilized federal programs to their own advantage, influenced legislation, and stymied attempts to assist agricultural workers who were excluded from concessions to labor and social security in the 1930s and the National Labor Relations Act (NLRA). By 1939, the New

Deal had promoted the centralization of power and concentration of control within the cotton industry, fueled the proletarianization of smaller farmers, and contributed to institutionalizing the relative powerlessness of farm workers.

This study focuses first on the development of the cotton industry, the growth of the work force, and the dialectical relationship between them. The first chapter shows that although the foundations of California's labor system were laid in the nineteenth century, the introduction of cotton changed labor demands sufficiently to propel the agricultural industry to increase control over workers. As cotton growing became increasingly specialized, that specialization increased the industry's vulnerability to market fluctuations and problems of labor distribution, leading growers to attempt to tighten control over workers by developing company towns, increasing supervision, and forming managerial hierarchies. But it also led to a transformation of employer-employee relations as recruitment and wages were centralized and standardized in the Agricultural Labor Bureau. The second chapter focuses on efforts by Mexican migrants, who formed the bulk of the work force, to create stability for themselves within an unstable labor system. It explores how the particularities of Mexican culture and family life and the experience of Mexican workers as laborers and participants in labor and social struggles affected labor relations on the cotton ranches. The third chapter deals with the 1933 cotton strike under the leadership of the Cannery and Agricultural Workers Industrial Union. The strike, shaped by the social structure of the cotton industry and the effects of changing class relations, marked a major transition. After 1933 federal intervention became increasingly decisive in class relations. The promises of the New Deal precipitated the strike, federal relief aided workers in maintaining it, and a federal mediator helped settle it. Yet despite hopes that the New Deal would extend its intervention in agriculture, the strike marked the furthest extension of federal power in support of agricultural workers.

The study then moves to examine the balance between workers and the industry established from 1933 to 1942. The fourth chapter examines the ways in which relief and housing programs, federal labor legislation, and the Agricultural Adjustment Act influenced class relations in cotton. The fifth chapter, focusing on the impact of new migrants on the cotton industry and its work force, explores how the background and experience of the white migrants who replaced Mexican workers by

1936 affected unionization efforts. Chapters six and seven examine the growing conflicts of 1937, 1938, and 1939 as the impact of federal policy in cotton became clear and class conflict became increasingly mediated through state programs. This process culminated in the cotton strike of 1938 and 1939, led by the United Cannery, Agricultural, Packing and Allied Workers of America (UCAPAWA), which is the subject of chapter seven.

This study traces the development of the cotton industry, its work force, and its relations with the state in the years from 1919 to 1939. In doing so it brings agricultural workers within a larger historical paradigm and addresses issues of relationship among economic structure, human agency, and the state. The discussions about Mexican and Anglo workers and the role of communities, families, and women underscore the need to look at the broad spectrum of working-class life. In so doing, this book touches on issues that are historically relevant to the history of the United States' Southwest and of the Mexican community of this area. It also has implications for future relations among the growing Mexican population, the labor force, and the state.

CHAPTER ONE

"We are producing a product to sell . . ."

The Business of Cotton

Sheltered in the middle of the state of California lies a two-hundred-mile valley called the San Joaquin. Surrounded by mountain ranges separating it from the Pacific Ocean on the west, the Mojave Desert on the east, and the Los Angeles basin to the south, its dry, hot days, cool nights, and long growing seasons provide an ideal climate for growing crops (see map 1). By the 1920s the Valley had been the heartland of California's agricultural production for sixty years. This was highly specialized, capitalist agriculture, and in the 1920s cotton was its most rapidly growing sector.

Cotton growing was a business. Its development was part of the expansion of California's agricultural industry. Market demands to make the crop profitable and competitive, both with other cotton sources and for its space on Valley land, led cotton interests to form an industrial structure to help protect their position. The structure the industry created to remain competitive was instrumental in its relations with workers.

THE INTRODUCTION OF COTTON

Cotton was first introduced into California in the 1850s and 1860s, when the unavailability of southern cotton and the soaring prices caused by the Civil War made it an attractive investment. Planters marveled at the large stretches of land and the hot temperatures ideal for cotton growing. The state legislature, anxious to encourage viable crops, offered bounties to those who planted cotton. Yet the vision that "California is destined to become a large grower of cotton, rice, tobacco, sugar, tea and coffee" failed to materialize.[1] Early cotton growers faced several problems. Until the 1880s they, along with other growers, lacked enough workers to pick the labor-intensive crop. The absence of extensive irrigation limited the area in which cotton could be grown. And there were few easily accessible markets: there were no cotton-related industries in California, and transportation to Eastern markets was prohibitively expensive.

Yet the major obstacle was cotton's inability to compete for land use or capital in California. By 1877, grower Col. J. W. Strong recognized that cotton, "while moderately successful, [has] been attended by loss when compared with the results which could have been attained in the culture of cereals. . . . A hundred acres devoted to wheat raising will yield more actual return than the same acres devoted to . . . cotton. . . . Every acre should be devoted to the production of that commodity which will bring the largest reward."[2] By the 1880s, the introduction of refrigerated railroad cars and an expanding urban market made fruit-growing more profitable, and growers invested in grapevines and fruit trees, not in cotton. Cotton production reached no more than a few thousand acres and, by the 1880s, disappeared from public record.

Cotton was reintroduced to California in the early twentieth century, when the development of specialized cotton made it profitable. In 1909 the United States Department of Agriculture (USDA), aware that "cotton of high quality must be grown if cotton is to be profitable," tested a variety of long-staple cotton strains, such as Pima, Yuma, Egyptian, Durango, and Mebane.[3] Cotton was grown in the southern part of the state from Riverside County to the Imperial Valley. Growers in the Imperial Valley, on the border with Mexico, by 1920 had planted 104,000 acres of cotton.[4] Finally, during World War I, the USDA sent W. B. Camp to the San Joaquin Valley to test new strains of long-staple cotton in the San Joaquin Valley in an effort to develop new cotton sources for manufacturing wartime airplane wings and tires.[5] By 1923 Camp had settled on a cotton strain called the Acala, a superior grade of cotton which commanded a market price two to five cents higher per pound than the market average and could, under optimum conditions in the San Joaquin Valley, produce more lint per plant, with higher overall yields.

The development of Acala coincided with an agricultural depression that eliminated Southern sources of long-staple cotton and opened up the market. The boll weevil, the postwar economic crash, deflation, and plummeting cotton prices reduced southern production of long-staple

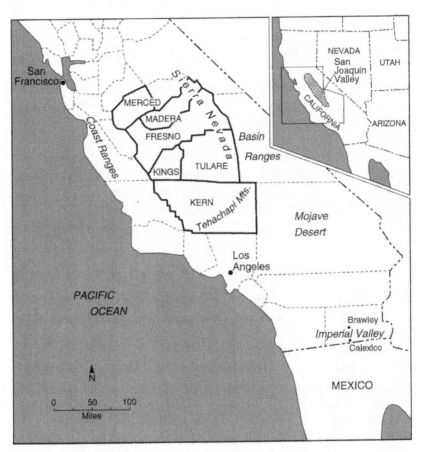

Map 1. Cotton Counties of the San Joaquin Valley

cotton from 117,000 bales in 1916 to only 2,000 bales in 1920 and led to an increasing demand for other sources of the long-stapled crop.[6] At the same time, the agricultural crisis stimulated California growers to switch to more profitable crops. Cotton declined in the Imperial Valley, as growers switched to the more lucrative melons and lettuce. But in the San Joaquin Valley, where Acala had been developed, growers sought out cotton. By 1924 the average cotton yield for Acala in the San Joaquin Valley was 459 pounds per plant, almost three times the national average. Combined with high prices, cotton reaped substantial profits. While the average United States cotton farm netted only $35 an acre, profits of $64 an acre or higher were not uncommon in the Valley.[7] Some growers netted even more; J. W. Guiberson, for example, made a net profit of $95 an acre.[8] Stanley Pratt of the San Joaquin Cotton Oil Company wrote, "While the cost per acre is higher, the cost per pound is lower due to the splendid yield."[9] By 1925, cotton was called "the biggest money crop per acre in the San Joaquin Valley," and the center of California cotton growing shifted to the San Joaquin Valley.[10]

THE COTTON BOOM OF 1924

In the 1920s cotton became a boom crop, promoted by government, utilities, and business interests. Land speculators and landlords urged tenants to grow cotton. Public utilities, aware that cotton was a heavy user of electrically powered irrigation, sent agents to valley towns to stress the advantages of cotton-growing. USDA agents worked closely with local businesses to demonstrate the latest growing techniques. The California Development Association, the statewide organization of local Chambers of Commerce, sponsored meetings to educate growers about cotton-growing. Local Chambers of Commerce enlisted growers to plant cotton and contracted ginning companies to erect plants in potential cotton communities. Advertisements were placed in southern and midwestern newspapers to entice potential growers to California; this 1924 advertisement was published in twenty-eight southern newspapers by local Chambers of Commerce in cooperation with the San Joaquin Light and Power Company:

Cotton Growers Wanted in California!

Practical men with some capital can make money. Millions of good lands available; irrigation guarantees steady heavy yield; no boll weevil; 1923 production averages a bale to the acre; living conditions ideal. For information write San Joaquin Light and Power Company, Fresno, California, Chambers of Commerce at Fresno, Madera, Merced, Selma, Dinuba, Corcoran or Bakersfield, California.[11]

A similar ad, run in midwestern newspapers by the Corcoran Chamber of Commerce, received 3,527 responses.[12]

Financial interests pushed cotton. Landlords pressured tenants to switch to cotton. Cotton merchants urged growers to plant Acala cotton and even purchased land to facilitate cotton production. The merchant firm of McFadden and Brothers, for example, leased 2,000 acres to small cotton farmers and offered financing at the munificent rate of $22 per acre.[13] By 1924 cotton had become a boom crop. Whole towns had switched to cotton-growing. In 1924 the Chamber of Commerce of Porterville, a small town in Tulare County heretofore known for its peaches and vineyards, urged growers to plant cotton. Encouraged by the merchant firm of Claggett, Gooch, and Harris, Porterville growers planted over 6,000 acres to cotton in 1925.[14]

In 1925 the *Pacific Rural Press* reported: "Men are picking cotton, hauling cotton, ginning cotton, thinking cotton, breathing cotton, dreaming cotton."[15] Cotton acreage expanded from only 9,000 acres in 1923 to 95,000 in 1925, and was to expand to 106,000 in 1926 and then to 247,000 by 1929. "White gold," as it was called in the local press, became increasingly valuable to the state economy, ranking eleventh in importance in 1924. By 1929 cotton had become the fourth most valuable crop in the state, accounting for an annual return of $24 million.[16]

COTTON PRODUCTION

How cotton was grown and processed had remained virtually unchanged since the invention of the cotton gin revolutionized the cotton industry in 1793. Agriculture, unlike other industries, was tied to natural cycles of gestation. Cotton was planted in the spring, flowered in the summer, and came to maturity in the fall. Conditions in the San Joaquin Valley gave cotton grown there an edge over southern cotton. The fecund soils, mostly virgin land unravaged by years of crop production, and the weather—blistering hot in the day, cool at night, with a long growing season—were ideal for cotton. An extensive irrigation system, developed in the 1910s and 1920s, provided water and eliminated the hazards of rain water, which could damage the crop and provide a comfortable home for cotton's ravaging pest, the boll weevil.

In April or May, when the last danger of frost had left the fields, farmers began to plant seed for the next season's crop. Plows, pulled by horses, mules, or, on the largest farms, tractors, carefully prepared the soil for the seeds. By June or July the plants flowered. Red, hard flowers burst forth where the bolls would later emerge. Workers began to trim or "chop" the plants, making room for the sturdiest. The plants matured first in the southern Valley counties of Kern, Tulare, and Kings, and then the white bolls spread northward into Fresno, Madera, and Merced. Picking began as early as August in the more southerly areas; in the north, the harvest season peaked in September or October but often lasted as late as January. There were two to three pickings. The first gathered the bulk of the crop, and the second and third caught late-developing bolls. Cotton was perishable: it had to be picked before the frosts began and after the last dew of night had evaporated, leaving the cotton bolls dry enough to pick without damage. Hired migrant men, women, and children moved along the mile-long rows of cotton, each with a twelve-foot tubular bag that held one hundred pounds of cotton, the bag dragging between their legs or slung over their backs. They worked rapidly, pulling off the bolls, cleaning them of leaves and twigs, stuffing them into their bags, and then emptying the bags onto wagons or trucks to be carried to the cotton gin. At the cotton gin the lint was separated from the seed. The seed was sold to cotton oil mills.[17] The lint was formed into five-hundred-pound bales, tagged, and sold to merchants. The cotton bales were then transported by truck over the Tehachapi mountains to Los Angeles and its new port in San Pedro, where they were compressed and shipped to markets in Asia and Europe and, in small amounts, to New England.[18]

In the early 1920s, California's cotton industry was dominated by large merchants and processors, initially from the South. Since the late nineteenth century these merchants (also called factors) had dominated the southern cotton industry, supplying credit, purchasing the crop from farmers, and selling it to mills. But by the early 1900s technological improvements, higher volume, and lower marketing costs gave large merchants a competitive advantage, and they increased their control over the cotton market by a series of mergers. By 1921 twenty-four firms marketed 60 percent of the nation's crop. Anderson Clayton alone, the largest cotton merchant in the world, handled 14 percent of the United States crop.[19] These merchants followed cotton west and established offices in Los Angeles, which, because of its proximity to the cotton areas in Mexico, Arizona, and California, became the state's cotton financial center. In the early 1920s, the principal merchants were the national firms of Anderson Clayton Company; McFadden and Sons; Bulley and Son; A. Kempner; and the Japan Cotton Trading Company of Forth Worth. But by 1925, the national merchants' hegemony was being challenged by the California-based merchants Major J. G. Boswell; Mauldin and Company; L. S. Atkinson and Company; and Claggett and Gooch.[20]

California cotton gins were at first small independents, operated by local growers or oil mill companies. But within a few years ginning was taken over by larger companies, many dominated by the merchants.[21] By 1925, California boasted thirty-four cotton gins, two cotton compresses, and four cottonseed-oil mills.[22] Ginning companies were crucial to the financial operation of the industry. As in the South, ginner-owned finance companies made annual loans to farmers. In return, farmers accepted a lien on their first crop of cotton and their horses, mules, tractors, and tools, and they agreed to sell their cotton to ginners at a preestablished price and give the ginner an option to purchase the cotton seed.[23]

Cotton-growing followed the pattern already established in California agriculture of large, highly capitalized and mechanized farms worked by low-paid migrant laborers. The pattern of large landholding dated back to the seizure of huge Mexican land grants following the war with Mexico in 1848. By the 1920s some of these landholdings were still intact, such as the Miller and Lux lands, which dominated the west side of the Valley and totaled over 200,000 acres. Yet under pressure from expanding agricultural markets, land prices soared to as much as $100 per acre, and investors began to sell off the vast tracts of land on the west side, including large portions of the Miller and Lux holdings. This part of the Valley became the home of mammoth cotton farms.[24] By 1924 the Kings County Development Company, for example, owned 34,000 acres here and the Boston Land Company had acquired 40,000 acres.[25] The companies in turn either leased the land to tenants or sold it outright to buyers. These cotton ranches were among the largest in the nation: by 1929 over 30 percent of the nation's large-scale cotton farms (farms valued at over $30,000) were in California.[26] One was owned by Corcoran speculator, banker, and rancher J. W. Guiberson, who expanded his ranch to 750 acres in 1923.[27] United States president Herbert Hoover owned a 1,260-acre ranch. By 1926 the largest was the Tagus ranch, owned by financier-industrialist Hulett C. Merritt, which grew over 1,000 acres of cotton, along with peaches and other crops.[28]

These ranches represented high capital investment per acre. Herbert Hoover sunk a quarter of a million dollars into his ranch, and it cost an additional $100,000 per year to operate. Cotton demanded extensive irrigation and, by the late 1920s, cost, on the average, $49.50 an acre to install, plus monthly payments. The Hoover ranch alone had nine electric wells.[29] Tractors and other implements represented an added cost. By 1920, more California farmers had invested in tractors than had farmers in any other state.[30] These investments raised productivity, enabling growers to expand cultivation while reducing labor costs. By 1927 the *Pacific Rural Press* reported that with efficient use of "tractor power and tools, one outfit with a two man daylight shift [could plant] 100 acres per day, 6 rows at a time, and [cultivate] 70 acres 4 rows at a time."[31]

Cotton production was dominated by large merchants, ginners, and landholders such as J. G. "Colonel" Boswell. By 1924 the Georgia-born retired officer was one of the largest growers in the Valley. Boswell came from a cotton-growing family whose business predated the Civil War. He had left the family plantation to become a cotton merchant in the expanding southwestern cotton market of the 1920s, setting up offices first in Arizona and, by 1923, in Los Angeles. Whether lured by profits or the desire to reestablish some tie to the land, in 1924 Boswell and his younger brother bought 400 acres in the new cotton area of Corcoran and leased a gin to process cotton. From the beginning the Boswell operation depended on migratory wage labor. He hired 150 migratory workers and their families and erected a labor camp to house them. Boswell rapidly expanded his holdings. By 1925 the brothers farmed 5,000 acres, operated three cotton gins, and ran a seed oil mill, an enterprise one local newspaper lauded as the "largest industrial unit of the cotton industry of the San Joaquin Valley." By 1938 Boswell owned thirteen farms, and by 1940 the Boswell enterprise was the largest business in Kings County.[32]

The size of landholdings, land speculation, and the costs of production, which favored big investors such as Boswell, made it difficult for small farmers to survive. Land prices and taxes were based on potential profits, and farmers unable to afford machines or wage labor remained at a competitive disadvantage and were often driven out of business. Cotton reinforced the established pattern of a rural society that was marked by an absence of family farms and a class hierarchy that was reinforced by racial differences. While over half of the cotton farms were under fifty acres, they never became the dominant form of cotton farming in California.

Many of these small farmers had come to California in an earlier migration of people out of depressed cotton areas. In the late teens and early twenties it was, briefly, possible for ex-tenants or small farmers to get land in the San Joaquin Valley. Workers who signed on for the Kern County Land Company, for example, were sometimes given land or were able to buy land cheaply. As a cotton-gin worker later remarked, by the mid-1920s "the gin company wanted me to farm—almost anyone could get a farm at that time."[33] The development of the Valley's west side was dominated by large growers and had few small farmers. But on the east side, along the belt now linked by Route 90, there were clusters of white small farmers near towns such as Tulare, Visalia, Madera, Delano, Earlimart, Pixley, Bakersfield, and Arvin (see map 2). By the late 1920s, however, agricultural prosperity had raised land values to a point that made it impossible for those without substantial capital to buy land. Innovations in technology and the monopolization of seed, production, and processing further limited the number who could enter, and remain in, cotton farming.

The history of Jim White's family reflects the precarious economic position of the small farmer. White's family came from Mississippi, where they had been antebellum cotton growers until they freed their slaves during the Civil War and moved out to Oklahoma Territory to sharecrop cotton. In 1918 they moved to California, settling in Kern County "to get away from cotton." For White's father, "cotton was a poverty crop. When you said you were a cotton raiser people would look askance at you. It had a stigma attached to it." But within a few years the Whites were again raising cotton. Like many other small farmers, though, White's father was never able to establish a solid foothold in the business. He moved back and forth on the lower rungs of the economic structure, between sharecropping, tenant farming, and wage work. Unable to finance themselves, the Whites borrowed money from a local ginner to rent sixty acres. High costs and low returns kept them in debt to the ginner. It was not until 1929 that they could afford a tractor, and they were usually too poor to hire wage labor. Most of the fieldwork they did themselves, and, when on occasion they hired workers, they worked side by side with them in the fields, thus establishing a very different relationship with their workers than had their better-off neighbors.[34]

These tenants and small farmers provided an important social base in cotton communities. Over the years they established families, social networks, and organizations. The lines between labor, tenants, and small farmers were fuzzy: tenants and small farmers often supplemented their income by stints as agricultural workers and, when they hired labor, tended to hire local white workers. The largest group of small farmers

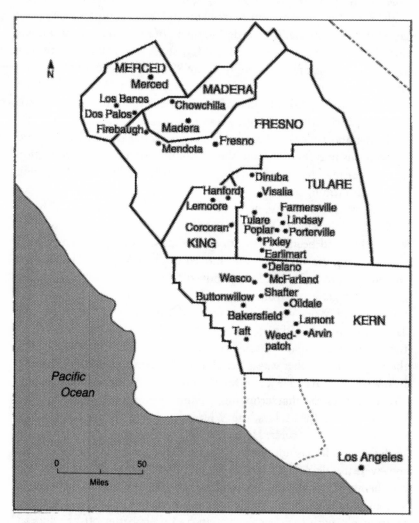

Map 2. Towns of the San Joaquin Valley

was in Kern County. Among poorer whites, some worked as farmers, tenants, or field hands. A substantial number worked part-time in the oil fields of Kern County. Unlike other cotton areas, in Kern the presence of mines and oil fields had created an industrial work force. Fluidity between occupations reduced the barriers that normally separated industrial, field, semi-skilled, and skilled workers. As a result, their interests overlapped. Merchants and other small businessmen relied on these workers for patronage. Most of them shared a similar southern or southwestern cultural background and socialized as friends and neighbors. The structure of work and residence thus facilitated community and labor support for cotton field workers.

COTTON AND CALIFORNIA POLITICS

Larger growers and ginners dominated the economic, political, and social life of Valley towns. They exerted subtle and not so subtle pressure upon poorer whites, which ranged from influencing where they bought a car to how they voted. Virulent and long-festering conflicts between larger growers and processors and the smaller farmers undermined industry unity. In areas where cotton interests dominated the economy, they also dominated the politics. On the west side the political control was more complete. Even on the east side, where there were more small farmers, large growers had an inordinate political power that corresponded to their economic position. They were able to use the power vested in Boards of Supervisors and in law enforcement and other agencies to support their positions and insure their control over local political life.

The strength of large cotton interests was enhanced by their relations with other industries, financial institutions, and state politics. Agriculture was the biggest industry in the state and had given rise to auxiliary industries, such as packing and canning. Railroads, trucking, and finance also depended upon agriculture. Banks' central role in financing cotton-growing, ginning, and marketing made them actively concerned in production and labor questions. As a result, other industrial and economic interests were involved in agriculture to a degree unmatched in other states. This explains the presence of seemingly nonagricultural businesses on the boards of agricultural organizations and the active participation of Chambers of Commerce in agricultural matters.

In part because of their interlocking interests with industry and finance, agricultural concerns enjoyed political power rare in other states. This was enhanced by the peculiar nature of California politics, with its weak political parties. The Democratic and Republican parties had both been divided by factionalism which, since the 1920s, had led to a proliferation of cross-party alliances and blocs. These blocs often represented particular interest groups that were more responsive to lobbying than political parties tended to be. California politics thus became marked by an abundance of lobbyists who wielded inordinate political power. By 1937, 286 registered lobbyists worked on only 130 legislators.[35]

The Progressive movement's attack on bossism had helped discredit party government, while also introducing a series of reforms that weakened the influence of political parties. Structural features of the political system that stemmed from the Progressive era, such as registration laws, cross-filing, and rigid constitutional requirements enacted to counter entrenched parties, contributed to the nonpartisan orientation of California politics. Legislators tended to be independent of governors from their own party, and the executive had a hard time exerting leadership over his party. By the 1930s this heritage was firmly entrenched in California politics, party organization was weak, and most party leaders had no strong ties to official party machinery.[36]

The power vacuum left by weakened political parties enhanced the power of the farm lobby, and agricultural organizations usually presented legislation directly, rather than submitting it as part of a party platform. In 1921 the Agricultural Legislative Committee, composed of 112 farm organizations, was organized to formulate agricultural legislation.[37] Several organizations served as interrelated lobbying or pressure groups: the Agricultural Department of the powerful State Chamber of Commerce coordinated activities of farm organizations, such as initiating conferences on policy issues; the Agricultural Council of California, formed in 1919, helped draft and pass legislation favorable to agriculture; and the California Farm Bureau Federation, formed in 1919, was a major instigator in agriculture's political involvement within the state.

State and local Chambers of Commerce became crucial to the development of agriculture. As organizations of businesses interested in the economic development of the state, they approached agriculture with an eye toward modernizing the industry. In the early 1920s, the Chambers of Commerce helped spearhead the introduction of cotton, pushing cotton as an anecdote to community-wide economic problems. Parker Friselle of the State Chamber of Commerce advocated diversification and slow but solid economic growth, and he formulated plans to supply the industry with labor. George Clements of the Agricultural Bureau of the Los Angeles Chamber of Commerce, a Red-baiting and flamboyant character, was instrumental in obtaining labor and regulating its supply; he played an important role in suppressing labor unions in the 1930s. Branches of the state-wide Chamber of Commerce helped coordinate local efforts and initiated and supported cotton's efforts to regulate and standardize production. They also provided the mechanism by which to deal with the federal government over issues of labor and immigration that concerned agriculture. These organizations became critical in addressing problems in production and labor brought on by specialization.

THE CRISIS OF SPECIALIZATION

Agricultural production, dependent on nature, was inherently more unstable and risky than production based on machines. That such production could not be rushed, slowed, or stopped, short of destroying the crop, prevented growers from responding to changing market conditions. Moreover, a year's investment was tied up in each harvest and could be lost through the caprices of nature—bad weather, pests—or a lack of workers. Harvests were times of crisis, anticipation, and short tempers.

Specialization helped offset these risks by creating crops of superior quality and yields that commanded higher prices. Yet specialization magnified the inherent instability of production. Specialization increased the need for labor, accentuated fluctuations in labor demand, and increased growers' vulnerability to labor shortages. Their dependence on a few cash crops made growers more vulnerable to market fluctuations and natural disasters.[38] From the vantage point of the industry as a whole, specialization led to conflicts between short-range interests in increasing individual profits and the rigid requirements for specialization to succeed. Ginners, to process more cotton, set gin saws at high speeds that tore the long staple and made the lint uneven for milling. Some growers planted non-Acala cotton, with which they were more familiar.[39] Foreign seeds became mixed with the Acala, creating uneven quality in the cotton lint, which made it poor for milling. By 1925, the *Pacific Rural Press* warned that "mills that have tried our cotton are laying off. It does not spin well. It gums up the machines. It does not spin out . . . because of poor ginning." Another paper reported that cotton "has been discredited" on the European market.[40] Mills bought less cotton, and the price for Acala fell from a 1923 high of 31 cents a pound to only 13.9 cents.

Falling profits encouraged ginners and merchants, backed by investors and civic boosters, to establish industry-wide regulations to insure the standardization that was crucial to cotton's long-term viability in the state. The careers of the men who backed the One Variety Act, designed to standardize cotton growing, represented the extensive interlocking interests of federal agents, banks, utility companies, media, large land-holders, and processing companies which were brought to bear in efforts to regulate the industry. W. B. Camp, the primary proponent of the act and the federal extension agent from 1919 to 1929, who had developed the Acala seed, also established the Camp-West-Lowe cotton ranch with his brother. By the late 1930s the brothers expanded into ginning and financing to become one of the largest cotton interests in California. In 1929 he left the USDA to become an appraiser of cotton lands for the Bank of America. He later took charge of the bank's California Lands, Inc., a Depression-born arm of the bank responsible for foreclosed crop land which by 1939 produced 1.2 percent of the state's total crop. From 1933 to 1936 Camp was the head agricultural economist for the regional section of the Agricultural Adjustment Administration.

Parker Friselle was the influential manager of the experimental 500-acre Kearney Farm near Fresno that was operated by the University of California. A leading member of the state Chamber of Commerce and a long-time proponent of slow growth and diversification, he was active in attempts to organize the agricultural industry. In 1917 he had formed the Valley Fruit Growers. By 1926 he chaired the agricultural committee of the Fresno County Chamber of Commerce, and in 1932 he became chairman of the state-wide Agricultural Committee of the state Chamber of Commerce. He was a founding member of the Agricultural Labor Bureau.

Large business, processing, and landholding interests also had a hand in the passage of the bill. Stanley Pratt founded the San Joaquin Cotton Oil Company in 1922. In 1927 the company was bought out to become a subsidiary of Anderson Clayton, and in 1939 Pratt left Anderson Clayton to form a rival company, the Producers' Cotton Oil Company. Hugh Jewett and Hal Woodworth, both cotton ginners from Kern County, were active supporters of the bill. Woodworth later became director of the California Cotton Growers' Association and chairman of the Citizens and Growers Committee of the San Joaquin Valley, formed in the wake of the 1933 cotton strike. Woodworth, along with Camp and Friselle, was instrumental in forming the Associated Farmers. Emory Wilson, of the San Joaquin Light and Power Company, who encouraged cotton production in order to increase electrical use, campaigned for the bill along with Arthur Crites, a Bakersfield banker and secretary of the Kern County Mutual Building and Loan Association. Jastro, the manager of the mammoth Kern County Land Company, actively supported the bill, as did H. H. Clarke. Clarke managed the Chandler farms of the Imperial Valley and Mexico, which belonged to the owners of the *Los Angeles Times* and was one of the largest cotton landholdings in Mexico.

The One Variety Act, supported by these men and the utility companies, bankers, land companies, large growers and ginners, the Farm Bureau, the USDA, local Chambers of Commerce, and the California Development Association, passed the state legislature in 1925, making Acala the only strain that could legally be planted in the San Joaquin and Sacramento valleys.[41] Camp supported the One Variety Law because:

> If you have . . . just one variety of cotton, Acala, we can standardize. It's all one length and easier to process. The gins can be organized in such a way . . . to handle one type of cotton properly. The cotton mills want a standardized product . . . [because] they want to be sure that they're buying . . . cotton that is all the same thing. . . . Acala cotton was chosen because it was the most productive and the best quality. . . . But . . . *standardized quality* . . . is the reason why we need just the *one variety.*[42]

The act was an important step in protecting and promoting the cotton industry: by insuring buyers a uniform product, it increased prices for California cotton.[43]

This crisis simultaneously encouraged merchants to extend control over other aspects of production. Merchants bought gins to process the cotton they purchased; for instance, in 1926 Anderson Clayton bought three gins and began constructing a fourth because "they cannot get the cotton ginned to their satisfaction, and . . . have decided to erect or buy gins throughout the valley and gin the cotton to suit themselves."[44] In 1927 they bought the San Joaquin Cotton Oil Company. By 1928 Anderson Clayton owned five gins; by 1930 they owned eleven; and by 1939 they owned or leased forty-six gins and ginned 35 percent of the crop for Arizona and California.[45] The expansion of interests such as Anderson Clayton increased the concentration of control within the industry, which in turn facilitated imposing regulations on growers. This was done through control over loans. Gin companies financed the majority of farmers, and Anderson Clayton controlled an increasing share of that financing. In 1939 alone, the company advanced $6,500,000 to growers in Arizona and California.[46] While Anderson Clayton preferred to leave the riskier business of growing to farmers, other merchants bought land and financed sharecroppers. The merchant firm of Claggett, Gooch, and Harris bought or leased over 3,300 acres in the Porterville area, operated local gins, and financed growers who sold their crops back to the company.[47]

Local growers expanded into processing. The huge Tagus ranch planted 1,800 acres of cotton and ginned and shipped it on the premises.[48] Wylie Giffen built a cotton gin on his 6,000-acre "plantation" to process cotton grown on his and neighboring ranches, and he built three more gins in Fresno, Madera, and Chowchilla.[49] Banker and cotton grower J. W. Guiberson built a gin to process the cotton around Corcoran, as did his neighbor, J. G. Boswell.

In 1927 financially independent cotton growers began to invest in cooperative marketing to extend their control over financing and marketing and to undercut gin companies and merchants. Cooperatives offered a locally based alternative to merchants' control and marked increasing participation by local growers and California-based capital in marketing, finance, and political influence. California growers, long familiar with the idea of commodity cooperatives, had already developed them in local areas to finance gin construction and to market cotton.[50] In 1927 growers organized the Valley-wide San Joaquin Cotton Growers' Association, reorganized the following year as the California Cotton Cooperative Association. The composition of the California Cotton Cooperative reflected the preponderance of large growers, financiers, and ginners in these associations: J. M. Hansen, Corcoran cotton grower; Clarence C. Selden, former head buyer for a Los Angeles cotton merchant; L. W. Frick, cotton grower from Kern County; H. E. Woodworth, ginner and supporter of the One Variety Act; J. W. Guiberson, Corcoran grower, banker, and ginner; Wylie Giffen, Mendotta grower and ginner; and H. V. Eastman, influential in the American Cooperative Association, the national body with which the cooperative was associated.[51]

Mergers and consolidations continued the trend toward monopolization. Within twenty years, the forty merchants who had begun to trade in 1910 had dwindled to four: Anderson Clayton; Pacific Cottonseed Products; J. G. Boswell; and Globe Grain and Milling Company.[52] Despite intra-industry competition and conflicts between cooperatives and merchants, the trend was toward a greater concentration of power in the hands of fewer organizations and corporations.

This represented an increasing concentration of financial control which, while funded by banks, came to rest with the central financier-

ginners. California's state-based banking system enabled local capital, more sensitive to local needs, to finance agricultural expansion. In 1924 the Bank of Italy's vice president recommended financing cotton as "one of the best lines of business in California," and the bank began to lend money to large ginners, merchants, and cooperatives.[53] By 1929 the Bank of Italy, reorganized in 1930 as Bank of America, financed over half the cotton crop. The bank's cotton investments, over $10 million in that year, gave them a direct interest in the long-term success of the industry and in its labor costs.[54]

The bank, by refusing loans to farmers and lending only to large ginners and merchants, increased the financial control of ginner-owned finance companies, which became the only source of funding for many farmers. The financial leverage helped enforce specialization, by requiring borrowing farmers to grow Acala cotton and refusing loans in areas considered unsuitable.[55] Financial pressure was applied to force farmers to agree with other stipulations as well, such as setting labor costs. The lending contracts regulated cotton prices by requiring farmers to sell the company their entire crop at prearranged prices, an arrangement which prompted an inspector for the California Division of Labor Statistics and Law Enforcement to remark that the system insured that "there is no competition either in ginning charges or prices on seed."[56] By 1933 over 45 percent of the cotton crop was financed by the ginners.[57] The control over finances and thus over production created an undercurrent of resentment which exacerbated intra-industry conflicts.[58]

Ginners and merchants established voluntary organizations to set production regulations. As early as 1923, merchants had organized the California-Arizona Cotton Association to establish industry-wide standards and had founded the *California Cotton Journal*.[59] In 1925 the California Development Association (CDA) hosted regional conferences to discuss "stabilizing the industry." The CDA subsequently passed the following recommendations to insure standardization and the market reputation of California cotton: exclusive planting of Acala; state and federal government classification; and standardization of weights.[60] The CDA helped create an organization called the California-Arizona Ginners and Crushers, which worked to standardize oil and ginning practices, facilitated cooperation among ginners, and pushed for standardized receipts, tags, and storage and baggage ties, and for accurate weighing policies and ginning specifically geared for Acala.

Conflicts over seed, ginning practices, and marketing would continue. Yet the One Variety Law and the growth of organizational responses to enforce specialization laid the basis for cooperation within the cotton industry. Vertical integration and increasing financial control contributed to, and was enhanced by, the success of specialization. By the late 1920s, cotton-growing had become geographically concentrated in the lower Valley in the counties of Kern, Kings, Fresno, Tulare, and parts of Merced and Madera. Cotton production was specialized by settling on one variety and by instituting uniformity in processing, classifying, and marketing. The process of specialization and the resultant increase in the concentration of industry control and organization were to have a profound impact on workers. The methods and the series of organizations that had been formed to standardize production were later applied to obtaining workers and depressing wages.

THE WORK FORCE

Since the 1870s, California agriculture had depended on a large force of low-paid migratory workers. The introduction of cotton, however, had far-reaching consequences for agricultural labor: it altered labor demands, changed migrant patterns, and contributed to the development of new methods of labor recruitment and labor relations.

Growers had several requirements for cotton pickers. They wanted skilled workers who could pick quickly and adeptly, choosing the right bolls without injuring the plants. They also needed a high volume of workers for cotton, the most labor-intensive of California crops. Labor demands had grown incrementally with the rapid expansion of agriculture in the 1920s. By 1925, 35,000 pickers were needed for the fall harvest: 12,000 were migrants from outside the Valley.[61] And as the demand for labor increased 400 percent in cotton, demands increased in other crops as well, adding to the overall drain on the available labor pool.[62] Cotton growers needed these workers for long periods of time: the fall cotton harvest lasted from September through January. But the increased demand for cotton workers, coinciding with a general expansion of agriculture, upset what growers regarded as an already precarious balance between the availability of labor and their need for workers.

The source of this labor had long been a subject of passionate debate. Some growers and communities opposed the expansion of cotton, claiming it would attract sharecroppers and reproduce the grinding poverty of southern cotton areas, placing a blight on California agriculture as a whole. Certainly cotton attracted workers from the cotton-depressed Southeast, and when cotton had moved into the Imperial Valley in 1910, those workers had followed to pick cotton. Yet cotton, like other California crops, quickly came to rely on migratory Mexican workers. Mexican laborers, many of them skilled pickers from the Mexican cotton districts of Durango and Baja, began to cross the border to pick cotton in Texas and the Imperial Valley. After World War I, they replaced Anglos. In the 1920s they migrated north into the San Joaquin Valley, attracted by higher wages.[63] Cotton's expansion in the 1920s depended on Mexicans.

By 1926, Mexicans made up 80 percent of the work force in cotton. On ranches of 300 acres or more, they were often 95 percent of the work force. Varden Fuller has argued not only that the industry came to depend on Mexican labor but that it was the availability of Mexican workers that enabled the cotton industry to expand in the first place.[64] Discussions in the press and private correspondence emphasize that Mexican labor was crucial to the cotton industry. As George Clements, manager of the Agricultural Committee of the Los Angeles Chamber of Commerce, wrote Governor C. C. Young, "We are totally dependent . . . upon Mexico for agricultural and industrial common or casual labor. It is our only source of supply."[65]

Statistics on the origins of migratory workers are scarce and inexact. Growers' inaccurate claim that workers migrated annually from Mexico and thus placed no burden on social services went uncontested by a public who preferred the fiction it offered. As the *Pacific Rural Press* breezily argued in disputing the allegation that cotton was importing peon Mexican labor: "Peon? isn't the word peon a little out of character when applied to a Mexican family which buzzes around in its own battered flivver, going from crop to crop, seeing Beautiful California, breathing its air, eating its food, and finally doing the homing pigeon stunt back to Mexico with more money than their neighbors dreamed existed?"[66] Yet personal accounts, oral histories, newspaper accounts, and reports agree that most Mexican workers came from Southern California and the Mexican communities around Los Angeles and the Imperial Valley. Another group migrated from Texas. Only a small group came directly from Mexico.

Yet despite the availability of these particular groups of Mexican workers, growers depended on a large labor pool fed by a steady flow of Mexican immigrants to depress wages and keep workers disorganized. Constriction of the labor pool increased competition and drove wages up. The expansion of labor needs and the large number of undocumented Mexicans residing in California and subject to deportation threw growers into direct confrontation with the federal government over the issue of Mexican immigration. In 1925 federal enforcement of the 1924 immigration law led to a sharp decline in Mexican workers. The quota acts of 1921 and 1924, designed to limit European immigration, excluded the Western Hemisphere and hence Mexico. Yet visa stipulations, the establishment of a border patrol, and an $18 head tax (a prohibitive sum for workers) hampered legal immigration. This, along with the already easy migration across a hardly discernible border, increased the number of undocumented Mexican immigrants to an estimated 85 percent of the Mexican population in California.[67] The government began to enforce immigration laws, threatening undocumented workers with deportation. Immigration officials raided cotton and lettuce ranches in the Imperial Valley, deporting workers without documents. Their activities gave rise to reports that picking in the Imperial Valley "is practically at a standstill . . . due to stringent immigration laws." Growers blamed the raids and border checks for "jeopardizing crops valued at millions."[68]

Worried by these reports, San Joaquin Valley growers began to compete for workers, attempted to entice workers away from other ranches, and placed advertisements in local and southern newspapers to attract pickers, spurring rumors that southern blacks would be imported, much to the consternation of those who had feared all along that cotton would eventually "saddle [California] with a negro problem."[69] The merchant-sponsored *California Cotton Journal* urged the industry to "demand" the cooperation of local schools in releasing schoolchildren to pick cotton for the harvest, a suggestion that exacerbated public fears that cotton would lead to severe social problems.[70] The *San Francisco Daily News* editorialized, "If [child labor] is what the 'cotton atmosphere' means we want none of it. If for a $20,000,000 industry we must pay the lives of little children—bent little backs, premature age and stunted minds—let the industry go hang. The price is too great."[71]

The smaller labor pool and increased demand drove wages up by one and a half cents a pound to an average wage of $1.65 per 100 pounds of cotton, for workers still an abysmally low wage but for growers an ominous precedent. By February the shortage of workers willing to work at the still-low wages, compounded by an early fog, left hundreds of acres unpicked. There were concerns that unless the labor problem was solved, cotton would not be able to expand. Parker Friselle warned, "We are having difficulty in handling 95,000 acres. . . . I don't believe two thirds of the first picking was completed. . . . That cotton is being damaged by showers . . . the grade of cotton is going down."[72] By the end of the season, the financial loss due to labor problems was placed at between

$375,000 and $1,000,000.[73] Economic losses and increasing public concern about the social effects of cotton-growing gave the industry three choices: to seek solutions to the labor problem, to change the labor system entirely, or to reduce the amount of cotton grown.

THE AGRICULTURAL LABOR SYSTEM
AND THE AGRICULTURAL LABOR BUREAU

Since the late nineteenth century, California agriculture had been based on cheap labor. In the 1880s wage rates stabilized at a point so low it was impossible for laborers to make a living wage. Yet, simultaneously, the value of land increased. Thus the agricultural industry became capitalized upon the profits anticipated from the exploitation of cheap labor. Cheap labor became so deeply embedded in the industry that its elimination would have necessitated a readjustment of the entire capital structure of California agriculture.[74] Specialized agriculture had also led to a seasonally uneven demand for workers. Large numbers of workers were needed for short periods of intensive work during harvests; when the season ended, those workers were expected to disappear. The problem plagued California agriculturalists: the labor system demanded that labor had to be cheap and workers readily available, but these often conflicting requirements led to an inherent instability in the system.

To meet the demands of agriculture, the labor pool had to be large, made up of workers who were unorganized and composed of the almost chronically unemployed who would be willing to migrate in the manner required by the system. Since the nineteenth century, California growers had depended mostly upon nonwhite immigrant workers: Native American, Chinese, Japanese, Hindustanis, Filipinos, and Mexicans. Poverty, lack of political rights, vulnerability to deportation, and low social status made them vulnerable to exploitation. Since each group was stigmatized as foreign and despised as inferior, their low wages and barely tenable working conditions were rationalized: they were foreigners who, growers claimed, neither wanted nor deserved an "American" standard of living.

Yet the migratory labor system was hardly satisfactory to growers. The fluidity and disorganization required to keep wages low exacerbated the growers' problem of getting enough workers on time to harvest perishable crops. Factors outside the growers' control, such as competition from other sectors or federal pressure on Mexican immigrants, could reduce the number of workers available. There was no guarantee that enough workers would show up, and the disorganization of the labor pool contributed to a maldistribution of labor, in which some areas had large numbers of workers, while others had far fewer. This uncertainty fueled growers' hysteria over "labor shortages." "Labor shortages," with few exceptions, meant not a lack of labor but a labor pool too small to insure both abysmally low wages and a readily available supply of workers.[75]

Each introduction of a new crop increased labor demands and elicited outcries from Valley communities and segments of the agricultural industry that the labor system had to be changed. The decade of the 1920s was no exception, and following the 1925 debacle several options were proffered. The Chambers of Commerce, for example, favored limiting cotton-growing to small areas where it could be picked by family or local labor. They urged farmers to diversify production in order to stagger labor demands while reducing dependence on any one crop. But these proposals, while lauded, were ignored in practice. The cotton industry found it more profitable to expand production, increase control over workers' movements, and depress wages. Control over the labor force became as important as the cost of labor.

This required cooperation within the industry. The idea of a centralized labor bureau had been introduced during World War I when fruit interests, led by Parker Friselle, organized the Valley Fruit Growers Association and attempted to establish industry-wide wage rates. Yet participation was voluntary, and without unanimous cooperation the attempt failed. The financial losses of 1925 and the trend toward centralization helped intensify efforts to establish an industry-wide organization. When Parker Friselle proposed an industry-wide labor bureau in 1926, it was supported by cotton ginners, who contributed $2,400. In April the Agricultural Labor Bureau (ALB) of the San Joaquin Valley was established.

From the beginning the bureau represented larger agricultural interests, as was reflected in the fifteen board members drawn from the fruit and cotton industries, the county Chambers of Commerce, Farm Bureaus, and civic organizations. Among cotton interests the representatives were: Stanley R. Pratt, representative of the San Joaquin Cotton Oil Company, a subsidiary of Anderson Clayton; J. W. Guiberson, cotton grower, president of the First National Bank of Corcoran, and co-author of the One Variety Act; J. A. Pauley, cotton grower and representative of the Kern County Farm Bureau; and Parker Friselle.[76]

The cotton industry became the major contributor to the bureau. Cotton interests had good reason to be interested in a labor bureau: they hired more workers and paid more for labor than any other crop in the state. By 1929 wages comprised 30.6 percent of the crop's total value.[77]

Led by the gins, cotton interests provided an increasing proportion of the bureau's financial support. In 1926 cotton interests paid one-third of the ALB budget; by 1930 their support had increased to two-thirds of the budget, over six times more than any other contributor. These subscriptions came from the largest interests: in 1929 the merchants' association, the California-Arizona Cotton Association, pledged $1,125; the San Joaquin Cotton Oil Company paid almost half of the cotton subscriptions between 1932 and 1938. From the beginning, the bureau was dominated by and the agent of large cotton investors.[78]

The bureau had four major tasks: to recruit workers and distribute workers evenly across the Valley to avoid a maldistribution of labor; to set and enforce a standardized wage rate to prevent competitive bidding and depress wages; to prevent or settle strikes or walkouts; and to represent agricultural interests in dealings with the federal government over immigration policy.

In 1926 cotton interests, through the bureau and other organizations such as the state Chamber of Commerce, joined agricultural groups to press for a more favorable immigration policy. J. W. Guiberson, the cotton grower from Corcoran, represented the bureau at congressional hearings in Washington, D.C. Emphasizing their dependence on Mexican labor, Guiberson said growers sought "labor only at certain times of the year, at the peak of our harvest, and the class of labor we want is the kind we can send home when we get through with them. . . . We must go into Mexico for the picking of our cotton."[79] They wanted a federal-contract labor program, similar to that of World War I, that would import workers who would be "administered and cared for under government supervision" and then sent home at the end of the harvest.[80] These efforts to revive a contract labor program were unsuccessful until World War II, when the bracero program was enacted, although the control over workers such a program offered remained, to growers, an attractive solution. Growers' immediate problems with immigration were handled, not through legislation, but through an informal arrangement with the Department of Labor, which agreed to stop detaining and deporting Mexican workers in the Imperial Valley. Both in the short and the long run, the bureau acted as the mediator in what would be a long interaction between growers' representatives and the federal government over immigration issues.

The ALB recruited workers, acting as a more centralized and streamlined version of the older contracting system. They hired Francisco Palomares as manager, to recruit Mexican workers, distribute them to ranches, and deal with labor conflicts. Palomares had worked as a Taylorized version of a labor contractor for California agriculture since World War I, when he had been the special agent in dealing with "Latin races" for the California Commission of Immigration and Housing. He later contracted Mexican workers for the Spreckels Sugar Company for over seven years and, following this, had recruited Mexican workers for the Arizona Cotton Growers Association.[81] Palomares personally directed the efforts to contract workers to the fields, traveling to Los Angeles, where "our greatest supply of field labor originates,"[82] and to the San Francisco Bay Area. His visits were preceded by advertisements in Spanish-language newspapers and, by 1930, printed circulars, such as the following:

AVISO IMPORTANTE! CHANZA LIBRE

El representante del Agricultural Labor Bureau of the San Joaquin Valey [sic] Inc. está aquí para informar a las familias y hombres solos que quieran salir a la

PIZCA DE ALGODÓN

en los Condados de Kern, Tulare, Kings, Fresno, Madera, Merced, California. Este año las cosechas son muy buenas y

EL PAGO ES MAGNIFICO

Las condiciones inmejorables y la temporada de trabajo es muy larga. Ocuran a la Oficina de Empleos Del Gobierno situada en _____ y pregunten por Don FRANCISCO PALOMARES, Representante del AGRICULTURAL LABOR BUREAU of the SAN JOAQUIN VALLEY; tendrá gusto en darles todos informes.[83]

(IMPORTANT NOTICE! FREE OPPORTUNITY

The representative of the Agricultural Labor Bureau of the San Joaquin Valley Inc. is here to inform families and single men who want to go to

PICK COTTON

in Kern, Tulare, Kings, Fresno, Madera, and Merced counties, California. This year the crop is very good and

THE PAY IS MAGNIFICENT

The conditions are superb and the work season is very long. Come to the Government Employment Office at _____

and ask for Don FRANCISCO PALOMARES, representative
of the AGRICULTURAL LABOR BUREAU of the SAN
JOAQUIN VALLEY; he will be pleased to give you all the
information.)

Palomares sought workers in pool rooms, mutual aid societies, and
neighborhood halls in Mexican colonias to explain the work and conditions. These trips by Palomares and his agents became regular events
until the Depression. Evidently he was effective. By early September
1926, Palomares counted 101 automobiles filled with workers headed
for the cotton fields as they crossed from Los Angeles into the San
Joaquin Valley. Palomares claimed that the bureau placed 12,500 workers, or nearly half of the cotton work force.[84] Once they arrived, workers
were sent to bureau offices in Valley towns, and from there they were
directed to local ranches.

The bureau undermined what little ability workers had to renegoti-
ate wages. Heretofore workers had often resorted to brief, often spontaneous and localized strikes over wages and conditions: taking advantage of the press of the harvest, as picking was about to begin they
would refuse to work. It was a haphazard, unorganized, and not always
effective form of collective bargaining. Yet so long as growers were
worried that no new workers would quickly be available, it often
worked. But the bureau established universal wage rates for the industry, which eliminated competitive bidding and depressed wages. At the
beginning of each season, ginners, large growers, and investors met to
establish wage rates for the upcoming season. While the rates had some
variation—more was paid if the worker supplied his own housing, and
the rates were higher for the later and less productive pickings—the
wage rates applied across the board. The industry's lending system
made it difficult for growers to pay above the rate, for loan contracts
were granted only upon assessment of an itemized budget that included
a set wage for labor. As one small but self-financing farmer testified, her
neighbors were "never able to pay the wages we do because they said
that the gins would not let them pay higher wages."[85] The combination
of economic pressure and self-interest worked. A 1928 survey reported
that only two out of a hundred ranches studied paid higher than the
bureau rate.[86] In the first year of its existence, the bureau dramatically
increased industry control over wages: it was able to depress wages
from $1.65 per 100 pounds picked in 1925 to $1.25 in 1926. From
1926 to 1949, the bureau was able to maintain a standard wage rate
throughout the cotton industry except for the strike years of 1933 and
1937.[87]

Moreover, the bureau enabled the cotton industry to expend its collective resources on individual labor disputes and thus to reduce, if not eliminate, any form of collective bargaining, to stave off labor conflicts, and to quell disputes. A strike in 1926 indicates how effective the bureau could be. Mexican workers on four ranches coordinated a strike, probably in response to the lowered wage rate. Palomares reported that he moved in quickly, personally visiting striking workers, and that he was able to settle the strike by "quick action . . . [in] explaining matters to the pickers in their own language." But it is unclear that striking Mexicans were convinced, and, perhaps more to the point, Palomares in fact imported Mexicans from Los Angeles to replace recalcitrant workers.[88] From now on workers had to deal with the bureau and, through it, with the entire cotton industry. As worker Luis Lima remembers, before the bureau was established, labor crews had used local and spontaneous strikes as a form of collective bargaining. These strikes were possible "cuando las asociaciones no asociaban todavía a todos los rancheros. Pero eso fue . . . cuando ya se asociaron . . . nos pusieron control a nosotros. Un solo *trust* [fijo el] precio de algodón." ("when the associations hadn't yet organized all the growers. But when they got organized, they were able to exert control over us. One trust set the price of cotton.")[89] After 1926, a successful strike had to stop production throughout the Valley in order to force the entire industry, not only a few growers, to agree to a rate change. By altering the terrain of labor negotiations, the bureau laid the basis for the industry-wide strikes that erupted in the 1930s.[90]

The bureau marked a milestone in the commodification of agricultural labor and embodied the impersonal labor relations which had come to dominate California agriculture. As George P. Clements, manager of the Agricultural Department of the Los Angeles Chamber of Commerce, said in 1925:

> The old fashioned hired man is a thing of the past. There is no place for him, and the farmer who does not wake up to the realization that there is a caste in labor on the farm is sharing too much of his dollar with labor. . . . California requires fluid labor. We are not husbandmen. We are not farmers. We are producing a product to sell.[91]

COTTON LABOR CAMPS

Industry-wide measures, such as those imposed by the bureau, were not the only steps taken. Large growers found that the isolation of west-side ranches, the size of the harvest, and the length of the harvesting period made it even harder to get and keep workers. The car, introduced into migrant life in the 1920s, aggravated growers' problems by making Mexican workers more mobile than any other group that had picked crops. Workers who had migrated on foot, ridden the rails, or driven horse-drawn wagons had faced a trip of several weeks from southern California and, once in the Valley, their lack of freedom of movement isolated them on cotton ranches and subjected them to the whims of growers and contractors.[92] In 1921 the road linking the San Joaquin Valley to Los Angeles and southern California was completed, and within a few years caravans of automobiles, laden with children, household goods, and supplies lumbered over the Tehachapi mountains. Although the car enabled workers to arrive more quickly, it also enabled them to leave with relative ease.[93] By 1925 growers were complaining of Mexicans' shiftlessness and rootlessness brought on by their auto-induced "lack of stability."[94]

Growers took steps on their ranches to reduce their vulnerability to labor shortages and to increase control over workers. Some growers developed cotton labor camps where they exerted control through the living arrangements, a hierarchy of supervision, and a form of ambivalent paternalism. Laid out in the middle of fields, miles from the nearest town, and accessible only on dusty roads which rain often turned to impassable mud, the large, isolated camps began to grow, as Carey McWilliams noted, into something akin to rural company towns.[95] Growers initially erected tents to house workers. Wood floors were later put under tents on some ranches and, by 1928, a few growers, such as J. G. Boswell, began to build small wooden cabins. Small towns took shape as camps swelled to accommodate the growing number of workers. Dirt spaces separating tents and cabins became worn by use into rough streets. Makeshift schools were set up in tents. Merchants came by wagon to sell goods and, on some ranches, growers or contractors set up company stores.

The size of the labor camps varied from year to year, but by the late 1920s they housed a larger concentration of workers for longer periods than ever before. By 1925, 150 pickers were counted on one Corcoran ranch, and by 1927 a surveyor counted 150 workers and their families living in J. W. Guiberson's camp, 180 families at J. G. Boswell's, and 280 families on the San Joaquin Cotton Syndicate's ranch. The east-side ranches tended to be smaller; a Madera ranch, for instance, housed 82 people.[96] For growers, the improved conditions of these camps attracted more workers, helping insure a steadier labor supply. But the camps also enabled growers to increase their control over workers' lives.

Schools on the labor camps attracted migrants. Mexicans traveled as families, yet wanted to send their children to school, at least part-time, and "often refuse[d] employment in a cotton camp unless there [was] a school open for their children."[97] Local communities balked at allowing migrant children in their schools: townspeople feared "the contamination of their children, morally or culturally, by the children of Mexican greasers."[98] To pacify Mexican demands for education while maintaining racial segregation, growers on large ranches cooperated in establishing camp schools.[99] Ranch schools expanded with cotton production. Kings County, for example, established its first "cotton school," as they were called, in 1923 and by 1927 had six migrant schools.[100] The schools perpetuated child labor, for children attended school in the morning and were released at noon to work in the fields.

The lending practices of the convenient company stores where workers could purchase food, clothes, gas, and supplies kept workers in perpetual debt. These stores sold inferior merchandise at inflated prices. The Giffen ranch store, for instance, sold milk diluted with water.[101] Yet workers had little choice. The distance from towns and the practice of doling out pay in chits redeemable at the company store made it difficult for them to shop elsewhere. For example, on the Tagus ranch workers could, theoretically, convert chits into money for a charge of $0.10 per $1.00, but for workers making an average of $0.15 an hour, that differential was prohibitive. Manipulated bookkeeping combined with the inflated prices to keep workers in perpetual debt. As one union organizer remarked, "There were many people in there, been there for five years and never seen any money at all." Even Carlos Torres, a loyal foreman, complained bitterly that "in two years I got a check for four cents. In *two years* I got four cents! Four cents. Everything else was paid to the store."[102] Keeping workers indebted helped to block easy migration from the ranch. Luis Sálazar had to work for a year at the Tagus ranch before he could afford to leave.[103] While the Tagus store was larger than those on other ranches, the same practice occurred elsewhere.

In the camps, workers found their work and lives watched by a grower or foreman. The stores, schools, and slightly better housing which attracted them to the ranches and contributed a stabilizing influence also became instruments of coercion. They were more than workers; they were also store customers and tenants. As tenants, they could be evicted without notice. Moreover, the ranch's isolation reduced their contact with the outside world and limited the news they received about new or better-paying jobs elsewhere. Ultimately, the control growers were able to impose intensified the relative powerlessness of workers.

As ranches expanded, owners specialized management tasks and created a hierarchical chain of command to supervise workers. Workers labored in large crews supervised by Spanish-speaking Mexican labor contractors, foremen, and a series of assistants called *mayordomos*. In season, the contractor had five or six assistants: the *sucaro*, to check for cotton that workers might have left behind; the *pesador*, who weighed the cotton; the *campero*, who cleaned the camp; and the assistants who hauled water and chopped down trees from nearby areas to provide wood for fuel.[104] Hired managers oversaw all or part of the ranch operations.[105] The expansion of operations and the hierarchy of management reduced or obliterated the possibility of personal contact between workers and employers. Although personal relations still played some role, especially between labor contractors and workers, the division between management and workers polarized the groups. The extent of this effect varied, however. On smaller ranches, where farmers often labored next to the workers they hired, relations remained fairly personal. But it was the large, impersonal ranches that hired the majority of workers, produced the most cotton, and set the pattern of social relations that dominated the industry.

The stratified hierarchy on the labor ranches of the 1920s reemphasized the crucial role played by race in enforcing class divisions. The owners and top managers were white: foremen, contractors, and workers were Mexican. Southern-born growers superimposed attitudes toward blacks onto Mexicans. As one Southern-born grower remarked, "Southerners didn't think [of] a Mexican or Nigger as anything but a slave."[106] Growers and managers used racism to justify treating Mexicans as inferior, paying them low wages, and offering only barely tolerable working and living conditions. The racism of local townspeople, law enforcement, and society as a whole reinforced growers' power over workers.[107]

Yet in the 1920s a form of paternalism or, on some ranches, a quasi–welfare capitalism underlay some growers' attempts to control workers. Both seem oxymoronic to capitalist agriculture: paternalism or welfare capitalism is usually associated with industries that are dependent on a stable work force and ignored by those who can easily replace workers. Yet the *relative* shortage of experienced, low-paid, and available workers in the 1920s led some growers to begin to see the benefits of employing a limited paternalism to obtain and keep good workers.[108]

This produced an ambivalent paternalism; it fluctuated in response to the supply of labor and was used to tie employees, especially full-time and long-term workers, more tightly to the ranches. Certainly this paternalistic relationship was coercive, unequal, and ultimately based on the threat of economic or physical violence. Yet it also suggests that there was a negotiated interplay between workers' needs and those of growers that played a role in their relations.[109]

Growers who ignored workers' desire for respect could find themselves with a shortage of workers or with labor problems. Francisco Palomares, manager of the Agricultural Labor Bureau, consciously urged growers to be more paternalistic, and used the method in his own dealings with workers. The Spanish-speaking Palomares listened to the Mexicans' problems and attempted to convince growers to act with some sympathy toward workers, if only in the growers' own interests. He attempted to explain to growers that "little acts of kindness" and respect would go a long way toward stabilizing their work force and preventing strikes. Workers flocked to the ranch of a grower who, after having been reprimanded by Palomares, paid the funeral expenses for the child of a destitute worker.

The quality of the housing offered was sometimes influenced by these ideas, for, as Parker Friselle told a gathering of cotton growers, "[Those of you] who have decent quarters are getting the laborers."[110] The most pronounced example was the Tagus ranch, where the owner, Hulett Merritt—a crony of Rockefeller's from the iron mines of Minnesota's Mesabi range—believed that improving conditions would increase productivity.[111] R. N. McLean, a missionary, complimented the Tagus ranch for recognizing that "contentment and well-being are factors in production."[112] In 1938, Ralph Merritt pointed out that better housing was not "philanthropy . . . [but] to me it is just good business. . . . Our costs charts . . . show that our permanent residents, happily housed and cared for, give us a much lower unit cost, almost by one-third, than those of the indefinite itinerant labor."[113]

By means of the camps, then, growers could increase their control over workers' lives both in the fields and in the camps and, potentially, reduce or limit labor disputes and interfere with the worker's ability to leave at will. All in all, the camps helped insure a steadier labor supply for the grower and were one answer to the chronic problems growers had faced with the labor system.

The measures growers took to obtain and then control workers, while heavily weighted in the growers' favor, laid the basis for structural changes among the work force as well. Centralization of control over labor, through the Agricultural Labor Bureau, contributed to the commodification of labor and the relative powerlessness of workers. Working relations became more impersonal, hierarchical, and controlled. The change in working relations exemplified by the bureau and the ranch structure laid the groundwork for more widespread labor conflicts. Yet in the 1920s, the camps gave cotton workers a greater degree of stability than they had hitherto known. Growers' attempts to insure a steady labor supply led them to recruit the same contractors and workers year after year. As jobs became steadier, workers migrated in more definite and stable patterns. The longer harvest period encouraged workers to stay longer on ranches. Some began to live on the ranches year-round, venturing out for work on other ranches during lulls in cotton work. Growers encouraged the practice because it insured them a supply of labor; workers liked it because it gave them a home base. Steadier employment patterns contributed to broader and more continuous contact among workers. Those who came yearly or lived on the ranches developed social relations that nurtured friendship, trust, and kinship ties. Workers would later use the stability cotton work gave them in the 1920s to their own advantage.

Thus, in the 1920s, market demands to make cotton-growing a profitable enterprise within the business world of California agriculture profoundly influenced the industry's relations with workers. The organizations used to enforce specialization of the crop were applied to labor and helped contribute to the commodification of labor and shaped the world which farm workers encountered.

CHAPTER TWO

Sin Fronteras

Mexican Workers

Migrant agricultural workers are often pictured as historical nonentities, helpless victims of a rapacious system, who lack the roots that nurture families and friendships and form the basis for social and political organization. Certainly the agricultural labor system shaped the parameters within which Mexicans lived and worked, and it was indeed an unstable and chaotic labor system, which paid them little, disrupted their lives, and demanded endurance and imagination to survive. As part of the cheap labor pool, Mexican migrant workers were easily exploited. Most were not United States citizens, and the estimated 80 percent who crossed the border without documents not only lacked political rights but lived under the threat of deportation. The lack of formal power increased their vulnerability to exploitation. And racism helped isolate them from other workers and both intensified and justified their exploitation in the fields and the surrounding society.

In the 1920s and early 1930s, 75 percent of the cotton work force were Mexicans: on larger ranches they constituted 90 percent or more. Mexicans adapted to some of the seemingly insatiable demands of the agricultural industry and resisted others. In the process, they helped form patterns of work and shaped their own lives within the chaos of the labor system. They attempted to stabilize their lives, to form and maintain families, to raise their children, and to establish the social networks and communities that would permit them to help themselves and each other. To understand the interplay of class relations in cotton requires an examination of the lives of Mexican workers. In part the stability they sought came from the structural changes in the cotton industry, as the nature of cotton work in the temporal window of the 1920s laid the basis for more regular patterns of migration, living, and work. Yet other sources of stability that outlasted the relatively prosperous years of the 1920s had deeper roots, in their own experiences as workers and Mexicans and in their families, networks, and communities.

THE MEXICAN BACKGROUND

The Mexicans who picked cotton in the 1920s and 1930s were veterans of the upheavals of the expansion of industrial and agricultural capitalism both on haciendas and in the mines, fields, and industries of Mexico and the United States. The majority came originally from Mexico, but it was a Mexico in which the transformation to a capitalist economy, well underway by the 1890s, had become an essential part of their work, lives, and migrations. It had shaped them as individuals, workers, and members of their families and communities. It had, in essence, shaped the culture of Mexican cotton workers. Although they were stereotyped as passive and malleable agrarians, many had in fact joined rural uprisings, labor unions, and armed forces of the Mexican Revolution. In the process they had adapted, through their families and social networks, to cushion the blows of dislocation and migration. The memories of these experiences, and the means they had adopted to survive, they brought with them to the cotton fields of California.

To understand this, we need to trace the origins of Mexican workers in cotton. Although the lack of records makes this difficult, we can make some generalizations. According to Francisco Palomares of the Agricultural Labor Bureau, by 1926 the majority of Mexicans who picked cotton had come originally from Mexico. This coincides with national figures which reported that, by 1930, 64 percent of the Mexicans then in the United States had immigrated since 1915, and 94 percent of the population had been here only since 1900. California had received what a 1930 report called a "disproportionate share" of this immigration.[1]

The Mexicans who worked in the San Joaquin Valley cotton fields in the 1920s and 1930s were not peasants but seasoned workers who had labored on both sides of the border. They had come originally from areas of Mexico that had been disrupted by capitalist expansion since the 1880s: the central plateau area of Mexico, which encompassed Michoacán, Jalisco, Aguascalientes, Zacatecas, and Guanajuato; and the western Mexican states, near the U.S. border, of Baja California, Sinaloa, Sonora, and, to the north, Chihuahua, Coahuila, Durango, and Nuevo León.[2]

The expansion of capitalist investment and the spread of massive production for the market that was to propel the Mexican economy changed the lives of rural workers, peasants, artisans, and crafts workers. By the late nineteenth century, foreign investment in Mexico had accelerated capitalist development in agriculture, mining, railroads, and industries. In rural areas large haciendas that produced for the market were expanding to swallow up small, independent villages. The independent Mexican peasantry disappeared, and by 1910 over nine and a half million people, 96 percent of Mexican families, were landless.[3]

As the growing network of railroads began to import relatively cheap manufactured goods into local areas, thousands of weavers, shoemakers, tobacco workers, leather workers, silversmiths, and other artisans and craftsmen were displaced. The growth of large textile mills alone, which replaced the traditional hand looms, forced 29,000 textile shops out of business. The pattern was repeated in other crafts; between 1895 and 1910, an estimated 30,000 artisans joined the industrial proletariat.[4]

Displaced rural workers found employment in Mexico's industries as the demand for labor grew. By 1910 almost 700,000 worked in mines, railroads, and textiles. No record exists of the number who scraped by as beggars, shoeshine boys, or prostitutes in the growing urban areas.[5] As in other industrializing areas, women became part of the wage-labor force. They made up one-third of the factory workers and were the mainstay of the budding textile industry.[6] Juana Padilla, for example, daughter of a rebozo maker, migrated to Mexico City with her brothers: "Fuimos a México para trabajar en las fábricas donde podíamos trabajar. Yo tenía catorce años. . . . Estaba muy joven. . . . Trabajé en la fábrica de colchas, y de rebozos, y las tarjetas postales y los zapatos, y una fábrica de cigarros." ("We went to Mexico [City] to work in the factories, where we could get work. I was fourteen years old. I was very young. I worked in a factory that made bedspreads, a rebozo factory, a postcard shop, a shoe factory, and a factory where they made cigarettes.")[7]

Workers who entered Mexican industry faced conditions reminiscent of those in England or the United States in the early days of industrialization. They labored for long hours in airless rooms under dangerous conditions. Women in textile factories sweated over the looms for fourteen to sixteen hours a day. Miners hacked away in dark mine shafts for twelve-hour shifts. Employers docked paychecks for misconduct (euphemistically called "dancing without music") and for defects in the final product. In isolated areas of the mills and mines, the company store sold workers poor goods at inflated prices, gouging them still further.[8]

Whether they worked in rural areas or industry, Mexicans had become well acquainted with capitalist relations by the end of the nineteenth century. Roughly 80 percent remained in rural areas, yet these were hardly peasants tied to the land. Capitalist relations encouraged the development of a mobile, impermanent labor force which (as in California) would appear as needed and then disappear. An increasing number of dispossessed workers began to work as day laborers, hired hands, or sharecroppers, such as José Bañales, part Purépecha Indian, who left his village to work as a sharecropper and paid laborer on a hacienda near San Francisco Angumacutiro in Michoacán.[9] One Michoacán hacienda with forty-two resident workers hired an additional twenty-five workers from nearby towns for the harvest.[10] Work and patterns of hiring borrowed from traditional social relations were carried over into the wage economy and would resemble in many ways the work they would later do in California's cotton fields. Migrant workers were often assembled into crews of fifteen to twenty people under the direction of a leader from their village who acted as both labor contractor and supervisor. This leader reported to a mayordomo in charge of a section of the hacienda who reported, in turn, to the general manager. Workers were paid wages, often in the form of chits redeemable at the company store.[11] A worker on the La Guaracha hacienda remembers: "We worked from sunup to sundown. . . . The salary they gave us was three pesos a week and of this the company store took one and a half pesos, leaving us with only one and a half."[12]

Displacement, whether from the land or from traditional occupations, set in motion a process of migration in search of work. People first migrated within local areas, but by the late nineteenth century the expansion of roads and a railroad system that linked previously isolated areas of Mexico broadened the scope of migration.[13] By the early 1900s Mexicans were migrating into northern Mexico and across the border into the United States, drawn out of the rural areas by wages in newly opening industrial areas and the availability of work on the railroads and in mines and textile factories.[14] Many worked their way gradually, migrating north within Mexico to find work, usually on the railroads, in the mines, or in agriculture, staying awhile in one place, moving north again, finding more work, and, finally, crossing the border into the United States.

Mexican workers were lured north by higher wages in the expanding industries and agriculture of the southwestern and midwestern United States. In Mexico, people found themselves caught in the pincers of rising inflation and low wages. In 1908 workers received, on the average, 7 centavos less than in 1895. By the 1910s workers on one hacienda received only 31 to 37 centavos a day.[15] Industrial wages, although higher, were below the 5 pesos considered necessary for daily subsistence: in 1927 they still ranged from 1.29 to 1.786 pesos a day.[16] The low wages failed to match inflation rates. Between 1877 and 1903 the price of corn increased 50 percent, and the cost of beans and corn doubled between 1910 and 1920.[17] Beans sold for over $1 a pound.[18] By 1910 workers found their purchasing power to be four times higher in the United States, where they could earn 50 cents a day picking crops and $2.00 a day working on the railroads or in the mines.[19] By 1907 the stream of departing migrants from northwestern Michoacán became so pronounced that employers had begun to complain of losing their work force.[20]

Although estimates of Mexican migration are notoriously inexact, they agree on the rapid increase in immigration during and after the revolution. According to one source, the number of Mexicans migrating to the United States jumped from 17,760 in 1910 to 51,042 in 1920 and to 87,648 by 1924. By that year two million Mexicans had migrated to the United States.[21]

Because of the chaotic conditions of rapid capitalist expansion, the seasonal nature of some occupations, and uneven wage rates, Mexicans sought work in a variety of industries. As they had in Mexico, Mexican men in the United States worked on the railroads, in factories, in agriculture, in mines, and as day laborers. By the time they reached the San Joaquin Valley, cotton workers had had a broad variety of work experiences on both sides of the border. Braulio López, for example, had worked on the Mexican railroad. In the United States he worked as a miner in Arizona and New Mexico before settling in Los Angeles, where he laid tracks for the electric streetcar line and worked in an Anaheim cement plant. But these jobs were part of a work mosaic, held together by migration: he mined gold in the mines of Randsburg, California, labored on construction of the road that linked San Diego and Los Angeles, and migrated with his family to pick crops in the Imperial and San Joaquin valleys.[22]

Roberto Castro, of the Boswell ranch, had worked on the railroads in Louisiana, Oklahoma, and Texas before settling in Corcoran, California. Juan Magaña, originally a laborer on the Tierra Negras hacienda in Guanajuato, labored on Mexican railroads and in 1918 began working for the Southern Pacific Railroad before picking cotton in the San Joaquin Valley in 1927. Elias Garza, a displaced sharecropper from Michoacán, worked in a Mexican sugar mill, labored on Kansas railroads, and finally settled in Los Angeles, where he worked at a packing plant, in a stone quarry, and at a railroad station before he picked cotton. Gilbert Hernández made the transition from artisan to industrial worker to entrepreneur. Trained as a printer in Mexico, he worked on the Mexican railroads and in the mines of Cananea; in California, he worked as a day laborer, picked cotton, and, eventually, ran a pool hall. José Bañales from San Francisco Angumacutiro left the hacienda in 1907, laboring on the railroad in Kansas and cultivating plants for an Anaheim nursery before he began to pick cotton.[23]

The extent of migration and the variety of jobs meant that these workers were experienced proletariats, had acquired skills in many areas even though they were considered "unskilled labor," and had endured and adapted to a variety of working conditions. Thus when we speak of migrant cotton workers of the 1920s and 1930s, we are speaking as well of workers who were miners, day laborers, engineers, artisans, and steel hands. The experience and consciousness forged in the mines, on the railroads, in factories, and in the fields of Mexico and the United States were transferred to the cotton fields of the San Joaquin Valley.

COMMUNITY, CLASS, AND NATIONAL IDENTITY

By 1926 the majority of cotton workers came from or through the towns around Los Angeles and the Imperial Valley. A small number worked in these areas temporarily or moved through them on the way to other jobs. Yet the majority of cotton workers made their homes in Los Angeles or the Imperial Valley, and it was in these communities that they formed the social networks they carried with them to the cotton fields. Los Angeles, which had remained a predominantly Mexican city until the 1880s, was still a magnet for Mexican migrants. The city's railroads linked the city to Mexico and other parts of the United States. *Enganchadores*, labor recruiters, set up makeshift offices and signed migrants up to work in the Colorado beet fields, midwestern mines and railroads, and the agricultural fields of California and other states. The expanding economy had sought Mexican workers to build Los Angeles and its infrastructure. Mexicans were the laboring backbone of the city by the late 1910s and they could find work there as unskilled laborers, industrial workers, and agricultural workers in the fields which still composed most of what would become cosmopolitan Los Angeles. By 1920, 33,644 Mexicans lived in Los Angeles, making it the largest concentration of Mexicans anywhere outside of Mexico City. By 1930 the population would reach 97,116 in the city and 167,000 in the county.[24] New arrivals gravitated to the Mexican center of town, called Sonoratown, which was nestled close to the old placita and the railroad tracks and was readily accessible to labor recruiters, contractors, and ranchers seeking day laborers. Economic expansion and the growth of the interurban railway system in the 1920s encouraged Mexicans to move into the eastside communities of Belvedere, Boyle Heights, and Maravilla, to other barrios in communities such as Anaheim, El Monte, Pacoima, Azusa, Pomona, San Fernando, Watts (known as Tejuata by Mexicans), and to the barrio of Sotelo in West Los Angeles.[25]

Another concentration of pickers came from Mexican barrios in Brawley and Calexico and adjacent ranches in the reclaimed desert of the Imperial Valley. Mexicans had lived in the Valley since the late nineteenth century, when they had been hired to construct the railroad and pick crops. Adjacent to the Mexican border—the Mexican town of Mexicali adjoined Calexico on the California side of the border—the Imperial Valley was an easy destination for workers going north. By 1927 over 20,000 Mexicans lived and worked in Brawley; most were residents. By the 1920s Mexicans in the Valley worked primarily in agriculture, the main industry in the area. Some had learned to pick cotton here, when the Imperial Valley was the center of cotton production. When wages dropped in the 1920s, many of these migrated north to the San Joaquin Valley, where wages for picking were higher.[26]

Mexican barrios in both areas were overwhelmingly working class. Social and family ties cut across the faint class lines that began to emerge in the 1920s and 1930s. Even if store owners, barbers, small merchants, and contractors did not work as laborers, they had children, parents, uncles, cousins, or compadres who did. These ties generated in them concerns and obligations that slowed or prevented the development of an entrepreneurial class divorced from its working-class roots and community. Many of these entrepreneurs came from a working class background. Juan Magaña, for example, had worked on the railroads and picked cotton before opening a small grocery store in Hanford.[27] Others depended on working-class clients. Members of this merchant group saw themselves, and were viewed by many, as representatives of the community. They tended to be literate, some were bilingual, and most had experience dealing with the Anglo world, its bureaucracy and its growers, business, and government.

Institutionalized racism, segregation, and economic hardship reinforced the bonds among Mexican workers and underscored their alienation from Anglo society. By the 1920s Mexican barrios had become segregated enclaves within Anglo society and, as Governor C. C. Young's report concluded, Mexicans tended to live together because of shared language and class and in response to prejudice and segregation.[28] Housing segregation was reflected in the uneven distribution in school populations: over 90 percent of the students in ten Los Angeles schools were Mexicans. Eight schools in the Imperial Valley were attended mostly by Mexican children.[29] Institutionalized racism contributed to low incomes and poor housing conditions and took a painful toll in death and disease. The infant mortality rate in Los Angeles from 1916 to 1929 was more than twice as high among Mexicans as among Anglos.[30] In 1927, Mexicans comprised 11 percent of the total population, yet accounted for 14 percent of all deaths in the city.[31]

Segregation, working-class status, and the geographic mobility of Mexican men and women reinforced their identity as Mexicans, lessened and narrowed their contact with Anglo-Americans, and reaffirmed the need to rely on each other in an Anglo-dominated society. Their lack of institutionalized social and political power underscored their reliance on each other. Thus as Mexican barrios grew they became less, rather than more, attached to Anglo-American institutions. The expansion of the Mexican population, rather than leading toward assimilation, helped create a stronger identification as working-class Mexicans.

Thus it was not surprising that in the 1920s and 1930s people called themselves Mexicans or Mexicanos, for what else were they? The majority were Mexicans by birth and citizenship: in Los Angeles, for example, 84 percent were Mexican-born. More significantly, few wanted to become naturalized citizens because, as C. C. Young's report claimed, they were "proud of their country of birth and slow to assimilate."[32] While in 1910 almost 46 percent of other immigrant groups became naturalized citizens, only 3 percent of the Mexican population wanted to be "Americans." By 1920 the number had risen only to a paltry 5.5 percent.[33] The proximity to the Mexican border, the constant influx of new immigrants with its consequent sense of contact with home villages, and the ease of returning to Mexico contributed to a sense of *Mexicanidad* (the quality of being Mexican) even after long periods of residence in the United States. In 1927 Manuel Gamio noted a proliferation of Mexican flags and pictures of Mexican heroes prominently displayed in most Mexican houses, "giving patriotism thus an almost religious quality."[34]

National identity remained strong among the second generation. In 1932 an elementary school teacher reported that second-generation Mexican children singing the popular "We will be true to the red, white, and blue" instead sang "red, white, and green," substituting the colors of the Mexican flag for those of the U.S. flag. While other youngsters painted their sneakers the colors of the American flag to honor the 1933 Olympics taking place in Los Angeles, Mexican children painted their shoes red, white, and green.[35]

Their identity was also expressed in the tenacity with which Mexicans held on to the Spanish language. Some children among the often bilingual merchant class did learn English, but the working-class language was still overwhelmingly, and usually exclusively, Spanish. In a 1921 study of over a thousand Mexican families in Los Angeles, over half the men and almost three-quarters of the women did not speak English.[36] To an extent this reflected the lives of working people with little time to learn a second language, but for some it was a conscious choice. Guillermo Martínez, a field worker and young teenager in the 1930s, remembered his "complete animosity" toward learning English because it was the language of the conquerors. He felt his refusal to learn English was an expression of his dignity as a Mexican, although a teacher finally lured him to the language with the sonnets of Shakespeare.[37]

This identity as Mexicans did not preclude deep divisions and conflicts within these communities. Mexicans differentiated among themselves on the basis of ethnicity, regional identities, class, and length of residence in the United States. Recent immigrants derisively called United States–born Mexicans *pochos*. U.S.–born Mexicans in turn called new immigrants *cholos* or *chimacos*.[38] While understanding why their compatriots migrated north to earn more money, Mexicans were scornful of those who began to assimilate. One merchant who considered becoming a United States citizen rejected the idea, afraid his clients would consider him a traitor and boycott his business.[39] Popular ballads, such as "The Renegade," reflected the tension between the envy of economic advantages to be had in the United States and distaste for changing class, cultural, and ethnic identity:

> You go along showing off
> . . .
> This happens to many
> That I know here

When they learn a little American
And dress up like dudes. . . .
But he who denies his race
Is the most miserable creature.
There is nothing in the world
So vile as . . .
The mean figure of the renegade.
And although far from you,
Dear Fatherland,
. . . A good Mexican
Never disowns the dear Fatherland
Of his affections.[40]

Economic changes had only begun to affect the community in ways later reflected in conflicts between Mexicans and the second generation who by the 1940s were called Mexican Americans. Strains had surfaced that caused friction within the community.[41] But, overall, Mexicans operated within a community bound together by common identities and by notions of cooperation, sharing, and reciprocity.

FAMILY, SOCIAL NETWORKS, AND MUTUALITY

The family was the basic social and economic unit of society, and it included not only immediate kin but an extended linkage of grandparents, cousins, and other relatives. *Compadrazgo*, a system of godparentage, linked families and individuals together through fictive kinship. In compadrazgo, people entered a relationship that entailed family-like mutual obligations and responsibilities. Traditionally, godparents or compadres were chosen at the birth of each child and for significant rites of passage. At its most extensive, compadrazgo could link one hundred people together. Genetic and fictive kinship formalized mutual need for support, and the interlocking networks became the basic economic and social glue of the society.

Economic upheaval and migration and the Mexican Revolution of 1910 to 1920 had disrupted family and social networks in Mexico. Migrating for work had separated parents and children, husbands and wives, siblings, relatives and friends, often for long periods of time. An increasing number of Mexicans met and married people from other areas or states of Mexico and bore children in yet other parts.[42] The Mexican Revolution accelerated social and economic disruption. Two million were killed as armies fought over the countryside, burning fields and driving off inhabitants. Agricultural production halted in some areas.

Factories closed and unemployment increased. The thin lines of communication workers had developed were severed within Mexico and
between Mexico and the United States. The revolution loosened, weakened, and sundered families. People lost track of each other, sometimes
temporarily, sometimes permanently. Mateo Castro left the ranch in
Tlazazalca, Michoacán, in 1909 to work in the United States. His son,
Roberto, followed in 1912 but did not see his father until 1922. Roberto's mother, who remained in Mexico, never saw Mateo again.[43]

Male migration had shifted the gender balance in many parts of rural
Mexico. Vicente Trinidad Navarro remembers that, in his Michoacán
rancho, "había como siete mujeres por cada hombre" ("there were about
seven women for each man").[44] Changing social conditions involved
women in new ways. Women participated in industrial strikes, such as
those that erupted at the Río Blanco and other textile factories in the
Orizaba-Puebla textile industrial area. Women joined the revolution as
camp followers who cooked, nursed, and provided sexual and emotional
comfort, and fought and were even executed in the course of battle.
Women of the Partido Liberal Mexicano, such as Josefina Arancibia,
concealed guns in baby carriages and ran them across the border into
Mexico.[45] Rural women facing advancing armies seized their children
and joined the stream of refugees. Juan Gutiérrez-García fled rancho
Ancihuácuaro with his mother to avoid the encroaching revolutionary
armies of Chavez Inez García: his father had left in 1910 to work on the
railroad. Fourteen-year-old Juana Padilla, her father killed by Zapatis-
tas, worked in Mexico City factories and later sold bread with her
mother and grandmother to soldiers at the train station in León as the
Padillas worked their way north.[46]

To an unparalleled extent, the upheavals of migration and revolution
led working-class Mexican women to shoulder increased responsibilities,
migrate, and survive on their own. Women migrated without men, making their way to the United States, where they faced an uncertain future.
Inés Amescua fled the revolution in 1915 along with a sister and younger
brother, working her way through New Mexico and Arizona before
settling in Brawley, California.[47] In the United States, Mexican women
worked. Single women were more likely to work outside the home in
garment factories, canneries, laundries, packing houses, and fields.[48]
Although married women tended to work in the home, where they
cooked, cleaned, and cared for the family, they often took in boarders
and cooked, sewed, and washed clothes for money.[49] Yet desertion,
widowhood, and economic hardship forced women into the labor mar-

ket; for instance, Guillermo Martínez's widowed mother cleaned houses and washed laundry to support her children. Women often worked in agriculture, for, as Rosaura Sánchez points out, although cultural traditions frowned on Mexican women working "outside the home as waitresses, maids or laundry help, [they] were nevertheless needed to work in the fields during the seasonal harvest to which families migrated. Here women were culturally 'protected' and simultaneously exploited by the growers, as much as the male members of their families."[50] While some stayed in the camps to care for young children (often augmenting their income by selling food or doing laundry for single men), many others joined their families in the fields, taking their children with them.[51]

The changes in women's lives affected gender relations. Women living without men and women who entered the labor market had more physical freedom and increased access to economic resources. Although incomes were usually pooled, wage work outside the home increased women's contacts, lessened their reliance on the family unit, and increased their direct economic value within the family. Women found they had legal recourse against abusive mates, which gave them some support, even though they hesitated to use it. The changing images of women in the United States influenced Mexican women who, like their Anglo counterparts, saw tantalizing images of flappers on the screen in darkened movie theaters. One study reported that young Mexican women in Los Angeles went to the movies at least once a week, and even a small mutualista in rural Imperial Valley boasted among its few assets a "motion picture machine."[52] The increasingly consumer-oriented society of the 1920s offered clothes and makeup that copied cinematic images, and many younger women adopted short skirts, bobbed hair, and makeup. One confused man lamented in a corrido that "Even my old woman has changed on me. / She wears a bobtailed dress of silk, / goes about painted like a piñata / and goes at night to the dancing hall."[53]

The frustrations and upheavals of industrialization, migration, and revolution were most sharply felt within the families. The nuclear family unit was sometimes unsteady, and both men and women deserted their mates.[54] The compadrazgo networks were disrupted and shrunken, and there were signs that the extended family was breaking down. Richard Griswold del Castillo, for example, points to the increasing number of Mexican orphans who were institutionalized in Los Angeles instead of being cared for by kin or compadres. Yet these disruptions, while unsettling, simultaneously increased the individual's reliance on family and friends to adapt and survive, thus strengthening the bonds of family and kinship networks. Capitalism enforced the individual's dependence on the family unit and in turn itself depended on the family to nurture, educate, recruit, and train workers. As a result, the family remained the primary institution that mediated between the needs of the workplace, society, and the desires of the individual.[55] Compadrazgo networks did not become as extensive in Los Angeles as they had been in Mexico, but they remained a crucial social link within Mexican barrios.[56]

These families and networks operated with mutually understood concepts of mutual need, reciprocity, and obligation that had been nourished in isolated communities in Mexico. These concepts persisted in California, not out of an abstracted idealism, but out of the gritty need to cooperate if the individual and the group were to survive.[57] Family needs usually superseded those of the individual.[58] Children often worked to help the family, and they usually turned their wages over to their family, even among the second generation. In a 1929 study thirty-eight out of forty young Mexican women in Los Angeles, born on both sides of the border, gave all or most of their pay to their parents.[59] Children took over household responsibilities, the burden falling heavily on females. Girls were often pulled out of school to take charge of the household on the death or absence of the mother or other older women. Lillie Gasca-Cuéllar went to school only a few years, "pero no fui muchos años porque mi mamá murió y yo tenía que cuidar a mis hermanitos." ("But I didn't go for many years because my mother died, and I had to take care of my little brothers.")[60] Sabina Cortez left school at fourteen to help her family during the Depression while her brothers, one younger and one twin, remained in school.[61]

The need for mutual assistance encouraged social mores which dictated that help be extended to a broad network of people linked by blood, fictive kinship, or friendship. Families supplied housing for new migrants, and people could rely on family members or compadres for food or even money. While aspects of mutual aid underlie any society, the importance of reciprocity was more powerful among immigrants. This (and the implied contrast with the non-Mexican community) was reflected in a 1932 study of Belvedere, near Los Angeles, which noted that "The communistic spirit of the Mexican is found in his attitude toward other Mexicans. During the present stringent times, when hunger and poverty are common, social workers report that as long as there is a little food in the Mexican district they know that it will be shared by all as far as it will go."[62]

Mexican communities were formed out of and based on these family and social ties. Mexican immigrants had already begun to form communities by the 1910s, some from remnants of Mexican communities dating back to the nineteenth century, when the Southwest had been the northern outreaches of Mexico. Mexicans, like other immigrants, migrated to where they had family and friends. Soledad Regelado explained that in her family's moves to Los Angeles, Hanford, and then Corcoran they were simply moving where "tenemos familiares" ("where we had relatives"). Family members moved in with one another, then found housing down the street or around the block, lacing whole areas with extended kin. Families separated by long periods of migration or the revolution were reunited. José Padilla moved to Brawley in 1927 to join his uncle and, the following year, sent for his wife, Juana, and their two children. Barrios began to reproduce communities of Mexico. The effect (and intention) of these settlement patterns was to recreate, at least partially, Mexican social networks. Barrios became known by the origins of their members. Migrants from San Francisco Angumacutiro in Michoacán migrated to Anaheim and San Fernando in California (although many kept a close enough tie with their origins to have their children's baptism registered in the Angumacutiro church). These communities thus became enclaves of social groups from Mexico.

Much like other immigrants, Mexicans extended the idea of mutual aid into community organizations, called mutual aid societies or *mutualistas*. The swell of immigrants in the 1910s expanded the membership of older mutualistas in California such as La Sociedad Progresista and La Sociedad Hispano Americano, and immigrants established new mutualistas, such as the Sociedad Mutualista Benito Juarez in El Centro, founded in 1919, or the Sociedad Mutualista Miguel Hidalgo, established in Brawley in 1922.[63] Mutualista membership was broad, and while most of its members were workers, contractors and small merchants also joined in pooling their meager resources to provide insurance, financial aid, and burial assistance. Mutualistas charged dues—members of the Miguel Hidalgo Society of Brawley paid $2 a month, for example—yet they operated more on concepts of mutual aid and reciprocity than on a legalistic interpretation of membership. Mutualistas often accepted memberships of people sick or dying or gave help to indigents. In Brawley the Miguel Hidalgo Society helped indigents pay for medical care, transportation to Mexico, and funeral costs. Perhaps indicating shifting values within the community, some Brawley Mexican businessmen complained of the practice; significantly, however, the mutualista continued to extend aid to nonmembers.[64]

Thus although conditions strained family and social ties and affected gender relations, the hardships simultaneously increased individuals' dependence on kinship and social networks. Ideas of mutuality and reciprocity that had been part of Mexican society were transferred to and institutionalized in workers' communities in the United States to form an important material bond among working-class Mexicans. These social networks, fired by the concept of mutual aid, were crucial in the development of Mexican communities, work crews, and organizations, and they became the vehicle and the expression of Mexicans' ability both to adapt to and to resist the conditions of life and work in California cotton fields.

COTTON WORK

In size and in the concentration of workers, cotton ranches and work resembled Carey McWilliams's descriptive term "factories in the field." Yet because the cycles of harvests, not machines, dictated work, the expansion of cotton-growing did not significantly change the work process or increase specialization. The work was hard. Workers began to pick as soon as the night dew had evaporated from the cotton bolls. On fields irrigated the day before, workers slogged through brown mud that oozed into their shoes and slowed their movements. For ten hours a day, they worked mile-long rows of cotton plants, each plant four to five feet high. Picking required skill, strength, and endurance. Workers stooped, stood, and often crawled to reach the bolls. Jessie de la Cruz remembers that she would cut the straps of her twelve-foot sack lengthwise and "tie the sack around my waist, and the sack would go between my legs, and I'd go down the cotton rows, picking cotton and just putting it in."[65]

Experienced pickers used both hands, quickly picking cotton bolls clean from their casings, avoiding branches and twigs, and working up and down the plant, stuffing cotton in a sack with one hand while picking bolls with the other and pausing only to pack the white bolls tightly into the bag. When the bag was full, the worker hoisted the hundred-pound sack, slid one end back, hefted the other end over one shoulder, and walked to the end of the row, where a contractor's assistant weighed the bag and, deducting for twigs and debris, tallied the weight on a list beside each worker's name. The worker then picked up the bag, climbed a rickety ladder, and emptied the sack of cotton into the wagon. The work took its toll. Workers' fingers were cut by sharp thorns, and their backs ached from bending over all day and carrying the heavy bag.[66]

While the pay was always low, the income varied depending on skill and experience. They worked under a piece-rate system; from 1925 to

1930, the pay varied between $1.00 and $1.65 per hundred pounds. Piece rates acted as a built-in speed-up mechanism: income depended not only on how much cotton was in the field but on how fast and clean workers could pick. Adults could, on the average, pick between 180 and 300 pounds a day. Only under optimal conditions could the most experienced workers pick over 500 pounds. Strength enabled most men to pick more than women, but an experienced woman could easily outpick a novice.

Cotton picking was often labeled unskilled, yet experience, speed, strength, and dexterity made the difference between picking 80 or 400 pounds. It took experience to judge which bolls were ripe and to pick efficiently. Over the years, Mexicans developed skills in cotton picking. Child labor was a form of apprenticeship. Children began picking by the age of seven or eight, and younger children collected cotton dropped by adults or picked bolls and piled them in heaps for adults to bag. They learned to spot cotton ready for picking and acquired picking techniques. Jessie de la Cruz remembered, "I was just a little kid, but I remember going out there and picking cotton and just making little piles on the ground, and then my brothers or my uncles, who were older than me, would come up and put it in their sacks. That's when I was just a child, and that kept on for many years."[67] As a result, many were experienced pickers by their early teens, because "we just grew into it."[68] Some Mexicans were already experienced cotton workers when they began to work in the San Joaquin Valley: growers sought out Mexicans who had worked in the cotton-growing area of Laguna in Mexico, in Baja California, or in the Imperial Valley.[69]

COMMUNITY OF WORKERS IN COTTON

Migration is disrupting to workers' lives, and migrant workers are often viewed, from the outside, as lacking the homes, communities, and social ties needed to nurture families, friendships, and formal organizations. Yet workers were able to draw upon several sources that brought stability into their lives.

Within the migrant stream there were, and still are, different degrees of migrancy, intensified or alleviated by variations in economic conditions. In the 1920s most cotton workers lived in communities, often migrated to the same job or within the same circumference of jobs year after year, and each year returned to their home communities. Few were homeless or migrated without destination, though the number of such migrants was to increase in the Depression. Migrants should thus be seen as attached, as if by a cord, to their home community, and they migrated without severing that cord. Workers usually migrated with family members and friends and thus recreated pockets of their home communities on the ranches on which they worked. These communities and their interlocking ties helped them both adapt to and resist the conditions in cotton.

Cotton picking offered workers a degree of stability for several reasons. Pay for experienced cotton pickers was as good as or better than other jobs. While California brick makers earned $3.00 to $4.00 a day, railroad workers made $2.48 to $4.75, and cannery workers earned from $2.80 to $4.00 a day, cotton pickers could earn from $3.00 to as much as $5.00.[70] The schedule of cotton picking, which began in September, peaked in October or November, and continued into January and February, provided work during the slack months in other crops, closing the gap between the grape harvest—the last work in summer—and the first work in spring. Encouraged by growers, workers who had hitherto returned to southern California in winter instead extended their stay at the labor camps or returned year after year to the same ranches. This steadier supply of jobs reduced the uncertainty of migration and stabilized migration patterns. The affordable car increased their ability to migrate considerable distances to find jobs, thus encouraging seasonal migration and, by making it easier to take a group of people, encouraged family labor.[71] By 1925, 90 percent of Mexican cotton pickers labored along with their families.[72] Migrating and working together reinforced the family as a social and economic unit and, as Douglas Foley remarked about Mexicans in South Texas, "drew the family together in a struggle for survival."[73]

Workers still faced uncertainty. As Arnold de la Cruz said,

> I know sometimes you didn't have a job. For instance, if you moved from one place to another, well, somebody would drive out there first and check around and then we'd move out there and try to find work. And if there was no work we'd stay there for two or three days while they went around further to look around until they found another place for us to move to. We'd go up there without finding any work and sometimes we used up what little money we had saved, we knew we'd need it for gas, so we'd save enough until we could keep on going, until we could find a job.[74]

Within the work force, workers found different levels of stability. Few workers in the 1920s simply wandered from job to job, a pattern of the bindlestiffs of the 1910s that would reemerge with the Depression. At the other end of the spectrum, only a few got full-time jobs on ranches. Most established a pattern of working with some regularity on the same ranch or ranches for increasingly long periods of time.

The example of José Bañales's family illustrates how cotton work became part of a stable combination of jobs. José Bañales worked in an Anaheim nursery four to five months of the year. But agriculture paid more, and when the crops ripened the Bañales family migrated to work in the fields of Los Angeles and of the Imperial and Coachella valleys further south. They first went to the San Joaquin Valley in 1922, when agricultural expansion lured them north for the grape harvest. In 1924 they invested in a Model T Ford. Joining a contractor, the Bañales migrated north that spring to chop cotton near Corcoran. They picked other crops over the summer, then returned to Corcoran for the harvest. They preferred to work on the large Corcoran cotton ranches because "it was more of a steady job. A lot of the other areas . . . we'd eat, we lived, but when we moved out of there we had just enough to get home. Corcoran was the place we used to make more money picking cotton."[75] The first year they stayed four months on the Harry Glen ranch. From 1924 to 1938 they returned to the Glen ranch to pick cotton. In 1938 they moved to Corcoran permanently.

This increased stability of jobs facilitated both the re-creation of older social networks and the formation of new subcommunities within the work force. Labor contractors often recruited a number of workers from the same community: Francisco Gallardo, for example, contracted Juana Padilla, her family, and a hundred other people from Brawley to work on a San Joaquin cotton ranch. Often these networks had a common origin in Mexico.[76] Mexicans from the area around San Francisco Angumacutiro in northeastern Michoacán, for example, migrated to and settled in Anaheim and San Fernando, California. When they began to work in cotton, families and friends formed labor crews, signed on with contractors they knew from Mexico, and migrated to Corcoran ranches. This type of migration, dictated by the needs of social networks, developed clusters of workers in work crews, on ranches, and in areas of the Valley with multi-layered interpersonal relations which stretched back to home communities in California and ranches and towns in Mexico. These dense social relations helped define relations among workers, and among workers and contractors.[77]

Work on the same ranches, labor in the same crew, and meetings on the road contributed to the formation of new ties among workers. Men fraternized in pool halls and bars such as the Monte Carlo in Delano, where they met, talked, told stories, joked, and exchanged information.[78] Others met at the ranches.[79] These conditions contributed to a sense of communities in cotton which deepened as workers formed and solidified friendships, lived together, married and had children. Over the years these new bonds, interlaced with older communal and familial bonds, reinforced their ties to one another to create overlapping sub-communities within the work force.

Women's informal networks became an important part of Mexican communities, both in barrios and on ranches. Because women migrated with their families, they usually had sisters, mothers, aunts, cousins and friends from their home areas to rely on. Women who migrated without other women and were not yet part of ranch life often felt isolated; for instance, Lillie Gasca-Cuéllar, migrating for the first time to Corcoran in 1933 with her husband and a new baby, remembered how lonely she felt on the Boswell ranch.[80] Yet women also began to form new networks, bound together by common concerns, that incorporated women they met on the ranches. People frequently shared food, and women, who prepared it, were in charge of this exchange. They cooked for each other when they were sick, helped take care of each other's children, provided support among themselves, and facilitated mutual aid within the community.

The re-creation of subcommunities from Mexico or southern California on the ranches helped reinforce values and social relations. These networks were crucial to growers and contractors in hiring labor and, as we shall see later, in the relation between contractors and workers. Recreating these networks also meant that family ties, friendships, animosities, loyalties, and other associations were partially re-formed on the ranches. They provided emotional and material support, buffered their members against poverty and migration, eased the pains of displacement, insured survival, and enabled the pooling of limited resources. Workers found jobs for each other. They helped each other cope with illness, birth, death, and family crisis. Families and friends received from each other a level of support out of the reach of workers who migrated alone.

Unlike the Imperial Valley and Los Angeles, where agricultural workers were an integral part of resident Mexican communities by the 1920s, in the San Joaquin Valley the pattern of agricultural work had helped create two separate, although sometimes touching, Mexican communities. One was the community of workers on the west side, where Mexicans lived in labor camps, with a small resident population in towns such as Corcoran that came from, worked with, and catered to the camp communities. The other was the Mexican community that lived in towns, mostly on the east side, where barrios developed in the towns of Visalia, Fresno, Bakersfield, Tulare, and Delano.

These Mexican communities had developed as agriculture expanded. Its residents had initially come as agricultural workers in the late nineteenth and the early twentieth century to work for the Kern County Land Company or Miller and Lux, or as construction workers on the Southern Pacific Railroad. As agriculture expanded in the 1920s, those who had saved enough to become part of a small entrepreneurial class began to invest in stores, pool halls, cantinas, and other services that catered to the growing Mexican working population. Tómas Salas Arambula, for example, who migrated to Delano in 1922, started a pool hall and restaurant in a rented wooden building with a tent top; by 1929 he ran the Monte Carlo dance hall and a movie theater.[81] Juan Magaña worked for the Southern Pacific Railroad until 1927, picked cotton, and by mid-1934 was able to establish a small grocery store that catered to Mexicans.[82] By the 1930s, a parallel, and often illegal, economy had developed, much to the consternation of the Mexican consul, who complained that a mutual aid society in Fresno ran a pool hall, sold bootlegged liquor, and operated two prostitution houses, "trayendo mujeres mexicanas de Los Angeles" ("bringing Mexican women from Los Angeles").[83]

In the 1920s, cotton workers on the ranches existed relatively independently of Mexican communities in the towns and surrounding areas. Cotton ranches were often isolated, and there were few independent Mexican communities, at least on the west side. Relations with established Mexican communities began to expand in the 1930s as some who had lived on labor camps began to move into town. Jessie de la Cruz's family worked on the Giffen ranch on the west side near Mendotta and, although they went into Fresno for Mexican holidays or to see a movie, they did not get to know any local Mexicans, not even those who sometimes picked on the ranch, "unless somebody happened to move out of the labor camp and into the town."[84] By 1931 an increasing number were indeed moving to town. Fresno boasted the largest Mexican community in the Valley, with a resident population of 6,000, which grew to 10,000 each harvest.[85]

Mexicans who lived full time in the Valley had formed mutual aid societies, working organizations, and patriotic associations, and these had grown in number as the labor force expanded and more Mexicans settled in Valley towns.[86] No Mexican organizations in the San Joaquin Valley were recorded from 1917 to 1923, but between 1927 and 1933

La Opinion reported thirty-five.[87] Workers in camps sometimes formed
separate mutualistas, an indication both of the permanence of these camp
communities and of their isolation. The DiGiorgio farms had a branch
of the Sociedad Juarez Mutualista Mexicana, and the Tagus farms had
a company-run mutualista. But it was not until the late 1930s, as workers
began to settle in Valley towns, that branches of the Progresistas and
other mutual aid societies would be established in the cotton towns of
Arvin, Hanford, Tulare, Selma, and Corcoran.[88]

Mexicans on the cotton ranches, especially those isolated on the west
side, came into contact with other populations of the Valley only tan-
gentially. Mexican workers had limited contact with the Mexican consul
at Fresno in his role as representative of the Mexican government. By
1931 this post was held by Enrique Bravo. Bravo, typical of Mexican
consuls, was from the Mexican upper class, schooled in Mexico and the
United States; he came to Fresno in 1931. Bravo assisted Mexicans with
discrimination problems and complaints about exploitation and withholding of wages, and established local Comisiones Honorificas as the
consulate arm in various parts of the Valley.[89] Workers met the Anglo
community through such officials as police, growers, nurses, and teachers who came to the ranches, representatives of bureaucratic offices they
had to deal with (and tried to avoid), and, occasionally, priests and
ministers. Mexicans also had contact with non-Mexican merchants who
catered to the migrant trade. These merchants were often non-Anglos
themselves. In Tulare, Portuguese and Japanese as well as Mexican store
owners sold goods primarily to Mexican workers. Some extended credit
to long-time customers. And, because they were often treated as racial
minorities and excluded from Anglo-dominated towns, some shared the
grievances of Mexican workers, sympathized with their situation, and
eventually supported them during strikes.

The interlapping of these communities and subcommunities reinforced the ethos of mutuality (even if it came under strain at times) and
supported preexisting social networks and families, which were in turn
reinforced by the nature of cotton work itself. By the 1920s and 1930s,
then, Mexicans workers had formed various overlapping communities
that intersected and reinforced each other: the communities of the cotton
ranches; the community of cotton workers on the road; and the community formed through the ties that bound them to their home communities in southern California and Texas, and, ultimately, to their
original communities in Mexico. The Mexican community of workers
was rich, formed by interlocking ties which could both support and

control workers and would be invaluable in their adaption and resistance to cotton work.

LABOR CREWS, CONTRACTORS, ACCOMMODATION, AND CONFLICT

The relationship between contractors and workers was pivotal to the functioning of the camps. Contracting was an old institution in California agriculture. As middleman between workers and growers, the contractor recruited and transported workers, negotiated payments, and sometimes provided shelter. While small labor contractors, called crew leaders, were workers who negotiated for their own small crew, the increasing demand for labor created the need for a labor entrepreneur who provided labor and supervised work on the large ranches, at first seasonally, then full-time. Growers began to hire contractors who doubled as labor recruiters, foremen, and supervisors.[90] They recruited workers, employed a series of mayordomos to help run the operation, organized labor crews numbering in the hundreds, supervised field work, ran the labor camps, and often operated the camp store. On some camps the contractor organized gambling, prostitution, and the sale of bootleg liquor.[91] By 1933, for example, contractor Librado Vidaurri of a Corcoran ranch oversaw two crews of about five hundred people and supervised a mayordomo and two assistants.[92]

Contractors initially primarily recruited family members and friends. But as the crews expanded in the 1920s, contractors began to hire workers from outside their own social and familial networks. Mateo Castro hired migrants from Texas and Arizona, and Felix Ybarra recruited single men from Los Angeles. Relations between contractors and workers, now based more on market demands, became increasingly impersonal. Steady work on the ranches removed contractors from the insecurities faced by pickers. With money gleaned from these concessions, contractors began to buy property and earned enough money to push them just over the edge of the working class. By the mid 1930s Narciso Vidaurri, who had started his working life as a miner, bought three acres of land and built a little house. By 1938 he had saved enough money to build a small store. As contractors came to rely more on their position within the camp for their income, they became increasingly dependent on the growers' goodwill.[93]

Yet the contractors' success depended on getting good pickers and successfully managing camps and supervising the work.[94] While ultimately such success depended on the size of the labor pool, it also depended on obtaining experienced workers and minimizing dissatisfaction with the wages and conditions set by growers.[95] As a result, the relationship between worker and contractor was mediated by mutual obligations and expectations. Workers obeyed the contractor and paid him a cut of their earnings: in return, he obtained jobs for them and acted as intermediary with the English-speaking world. Some contractors lent money to workers or helped them in personal ways. José Bañales, for example, stayed with his contractor, with whom he had a twenty-year relation, because, according to his son, "my dad knew he had a job with [Ybarra] . . . [and] Ybarra used to help him out with financial problems."[96] Lending money was not necessarily altruistic, for the practice could keep workers in debt while simultaneously seeming to comply with an ethos of mutuality.

The worker-contractor relationship was governed not only by working relations on the cotton ranches but also by the interplay between the pull of familial and social obligations and ties that had their genesis in Mexico. The contractors' relation to workers was often a partial replication of the Mexican class structure. Contractors who had been village leaders or hacienda mayordomos in Mexico frequently hired workers who had labored under them in Mexico. Mateo Castro, for example, had been the mayordomo on the Jamanducuaro ranch in Tlazazalca, Michoacán, before becoming Boswell's contractor. Felix Ybarra, for example, traveled to the United States with the crew he had worked with in Michoacán. Other contractors had begun as workers but, over the years, had developed the networks (and often a facility with English) that enabled them to work as contractors. Narciso Vidaurri had mined copper and worked in the Del Monte canneries before joining his brother, Librado, in 1924, to contract labor. By 1933 the Vidaurri brothers contracted and supervised over 150 families on the Peterson ranch.[97] For workers who had labored on haciendas, cotton work resonated with those earlier experiences, and aspects of hacienda relations became transposed onto the labor camps of the San Joaquin Valley. While the analogy should not be exaggerated, it was understandable that Edward Bañales remembered that some contractors would "rule their people just like living on a hacienda."[98]

Contractors who had been officers in one of the revolutionary armies often supervised their ex-soldiers and transferred military discipline, loyalty, and obedience into work in the cotton fields. Ex-officers often marched workers into the fields in military formation.[99] Ramón Peña wore the baggy pants and boots of a military officer into the fields, conveying a visual military presence and authority in relation to workers. One worker recalled that another former officer of the revolution ran the crew "as if he were running an army."[100] The military experience of both contractors and workers often helped enforce discipline and hierarchical relations. As Edward Bañales recalled, "Their men still obeyed them as if they were still generals."[101] Yet this relationship could have its drawbacks: those disenchanted or inexperienced with the military could well resent the harsh discipline.[102]

Contractors used a mixture of encouragement, paternalism, threats, and violence. In the fields, contractors, riding horses or walking along the rows, pushed workers to work faster, admonished some to pick the bolls cleaner, ordered others to repick a row. According to Edward Bañales, contractor Ramón Peña "chased workers on his horse. . . . and if he'd see a row where there was too much pluma [cotton bolls] left behind," he would order workers back to repick it.[103] Contractors pitted crews against each other to speed up work. The threat of violence underlay their control. Many contractors carried guns, both to protect the wages they often carried in cash or coin and also for self-protection against disgruntled workers.[104]

The labor crews were the basic unit of production, formed of family and friends, and they were the basis of society in the camps. Crews were bound together by notions of workers' solidarity and mutual aid, albeit at a very pragmatic level that sprang from their common interests. Familial obligations frequently overlapped with those incurred by workers. They worked together, pooled incomes, and dealt with problems as a unit.

Workers usually traveled as families or groups of families, helping each other on the long trip, whether crossing the Imperial Valley desert or the Tehachapi mountains, or fixing flat tires. People found jobs through friends and family members. On the job, neophytes took lessons from family members and friends on efficient techniques of cotton picking and learned from them the expectations and peculiarities of the particular contract and employer.[105] Workers who ran into trouble, be it from the contractor, the law, or personal conflicts with other workers, could usually rely on family members.

Within work crews, commitments to family, compadres, and friends could enforce work discipline. Crew members established standards of performance and pressured each other to meet those standards in order to insure their collective income. The piece-rate system acted as a speed-up mechanism and forced group members to pressure each other to pick cleanly, meet quotas, and keep up the work pace, for the earnings of the individual depended on the rate of the group. Contractors skillfully used family relations to pressure workers and held family members, in particular the head of the family, responsible for the others' performance. José Bañales was called on the carpet and expected to discipline his son, Edward, for talking back to the contractor. If contractors could rule with what one worker called "an iron fist" they did so in part because family members helped keep each other in line. Pragmatic calculations and a sense of mutuality often restrained individual impulses and prevented walkouts that threatened group interests. Edward Bañales remembers that although "every year there were disputes in labor . . . there were always people with families and they'd break the strike."[106]

Yet this sense of mutual reliance and communal obligation was double-edged and at times contributed to the transformation of work crews and social groups into labor organizations and laid the basis for collective action. Within the crew, workers shared experiences, work, and grievances and forged a group identification and history that reinforced family and social ties. Running a crew provided the crew leader—usually the father or oldest male—with leadership skills and experience in bargaining with contractors over conditions and wages. This sense of a group economy could thus escalate the problem of one worker into a concern of the entire crew. When a friend of María Hernández's was fired, for instance, her father and the rest of the crew walked out. Growers were surprised when Mexicans, whom growers often lauded as being malleable workers, joined walkouts. But both acquiescence and rebellion were based on pragmatic calculations. Workers endured what was necessary to ensure food for their families. Because a crew acted—as did a family—as a social and economic unit, it could control its members and yet, at times, also act as a labor organization. The same group interest that kept workers in line could transform an individual's problem into a group concern and motivate the group to organized resistance. Edward Bañales, who had decried the "people with families" who would break a strike, also remembered that when he fought with his contractor and walked out in disgust it was "families and friends [who] would quit with you." As we shall later see, the breadth and the relative success of the 1933 strike were rooted in the social groups and work crews and their ability to act as a unit.[107]

The brutal conditions on the camps, from deplorable housing to the lack of beds, ovens, toilets, showers, and running water, were a constant source of sickness, accidents, strain, and conflict. Shoddy tents offered little protection from blistering summer heat or the rain, wind, fog, and freezing temperatures of fall and winter. Jessie de la Cruz, who in the late 1920s lived in a tent, remembered that "our bedding was always damp, [and we had] no beds. . . . There was no floor, just the ground. And when it rained, water would just come under and everything was muddy in the tent. We had to walk stooped over because if our heads touched the canvas it would spring a leak. . . . Staying warm was hard." As Lillie Gasca-Cuéllar saw it, life for women "es muy dura. Sufrió uno mucho. Mucho trabajo. No teníamos estufa. No teníamos camas. Dormíamos nomás con cartónes, no teníamos casa—y a veces en las calles durmiéndonos." ("[Life] is very hard. We suffered a lot. Lots of work. We didn't have a stove. We didn't have beds. We had only cartons to sleep in, we didn't have a house—and at times we slept in the streets.") Belén Flores remembered, "Nosotros sufrimos mucho, en los campos trabajando. Que tengo un sobrino que dice que 'en México también hay mucha gente pobre.' . . . Pero muchos tienen su lugar a vivir . . . Pero nosotros aquí en los campos . . . nos daban carpas y agua corría abajo. . . . No tenía estufa para tener caliente. Todos mojados . . . Nosotros sufrimos mucho." ("We suffered a lot, working in the camps. I have a nephew who said that 'There are lots of poor people in Mexico also.' . . . But at least people have a place to live . . . But us here in the camps . . . they gave us tents and the water ran underneath them. . . . I didn't have a stove to stay warm. Everything was wet . . . We suffered a lot.") Wooden floors under tents or uninsulated wooden cabins offered only slightly more protection.[108]

There were few basic utensils for living. Few workers had the luxury of wood stoves; workers used oil drums or other makeshift stoves to cook, heat water, and keep warm. Water for drinking and bathing came from a single outside faucet shared by the whole camp. Most camps provided only cold water, and that was frequently contaminated. Indoor plumbing was nonexistent, and sanitation facilities were inconvenient at best and frequently unhealthy. On the Peterson ranch in the late 1930s, seepage from open toilets leaked into the drinking water, creating a foul stench. One worker remarked that "you couldn't even smell [the water], let alone drink it."[109] Luis Lima resented that "uno tenía que comprar los sacos" ("you had to buy your [own picking] sacks") and that the grower Peterson did not give his workers houses in which to live. Drinking water was notoriously bad. Luis Lima remembered, "No podía tomar la agua usted. Ya traía aroma como petróleo." ("You couldn't drink the water. It smelled like petroleum.")[110]

Poor sanitation, contaminated water, inadequate housing, and hard work led to sickness: colds, flu, chronic lung problems, bronchitis, pneumonia, diarrhea, tuberculosis, and typhoid. Jessie de la Cruz remembered that as a result of the conditions people "had a lot of illnesses." She herself came down with typhoid in 1930. "We were . . . at the Barlow ranch. . . . [Housing] was just a tin shack . . . used for storing boxes. It was very cold, and we had no running water, and there was something wrong with the water, and I caught typhoid."[111] Pickers caught and often died of valley fever, a disease peculiar to the San Joaquin Valley.[112] Children often suffered the most from the poor conditions and malnutrition. Sometimes small children fell asleep in the cotton racks and were smothered to death by loads of cotton or died when the flammable cotton caught fire.[113]

Workers resented the racism, low wages, and abominable conditions. Jessie de la Cruz, a young cotton worker, remembers that they were treated "like slaves. . . . The farmers treated the horses and the cows . . . better than the farmworkers. At least they had shelter . . . but the farmworkers, we lived under the trees."[114] Luis Lima argued that the cause of this treatment was, in part, the capitalist structure of the cotton industry, and he compared their work with those of slaves:

> Toda las compañías que había aquí en el Valle de San Joaquin eran compañías fuertes que estaban protegidas por la Wall Street. Nomás que los encargados que había aquí y rancheros que tenían terrenos—esos eran los que militaban. Y de esa manera a nosotros nunca. . . . Cuando pagamos en el rancho nunca nos daban: el transporte para ir a pizcar o volver o nada, ni agua tampoco. Nada. Era un escalavo voluntario, uno. Por aquí nosotros no les importábamos a los rancheros.

> (All of the companies that were here in the San Joaquin Valley were strong companies that were protected by Wall Street. Only the people in charge here and the landowners—these were the ones who had the power. And so we got nothing. . . . When we were paid at the ranch, they gave us nothing: no transportation to get to the fields or return or anything, no water, either. Nothing. You were a voluntary slave. Around here, we meant nothing to the ranchers.[115]

In the 1920s, agricultural worker Juan Berzunzolo said, "I have left the best of my life and strength here, sprinkling with the sweat of my brow the fields and factories of these gringos, who only know how to make one sweat and don't even pay any attention to one when they see one is old."[116] These long-term resentments became a major factor in labor disputes. By the 1920s a report noted that when the Mexican worker "realizes the employer assumes no responsibility except during working hours, he falls an easy prey to agitators. He comes to believe that his employer regards him merely as a thing, a tool for agriculture."[117]

Their experiences in Mexico and the United States contributed to an intersection of class and national identity that helped shape migrants' response on the ranches. Mexicans were aware that foreign investors dominated Mexican mining, timber, industry, and transportation and held over 120 million acres of choice agricultural land.[118] In both countries they labored in mines, factories, and fields owned and supervised by North Americans or Europeans, were relegated to the lower-paying jobs, and were paid less than North Americans doing the same work.[119] North of Mexico, Mexicans worked on land that had belonged to Mexico until the United States forcibly annexed the area in 1848.

Grievances on cotton ranches were underscored by the historical animosity between Mexicans and Anglo-Americans, which surfaced in oral histories and in Mexican popular culture. Workers remembered the postwar, 1848 acquisition of half of Mexico by the United States, and they were well aware that they were working on land which had once belonged to Mexico. Guillermo Martínez's mother taught him that "this was our land," stolen by "the barbarians."[120] Cotton workers told legends and sang corridos about the nineteenth-century California Mexicans Joaquín Murietta and Tiburcio Vásquez, who fought Anglo invaders. Those who came via Texas had heard of Gregorio Cortéz and the continued skirmishes between Anglos and Mexicans that plagued southern Texas well into the 1910s.[121]

Conflicts on ranches resonated with the class and national discrimination and struggles the workers had experienced on both sides of the border. Edward Bañales remembers "all kinds of fights [because] . . . the workers didn't like to be bossed by a white man." Gilbert Hernández, describing a heated argument with an American manager who was trying to cheat him, said, "All the Mexicans came to my help and we came mighty near lynching him. After that . . . I went around with my pistol." In that case, workers' complaints forced the grower to replace the manager with a Mexican, which evidently settled the dispute. Although

Gilbert Hernández was picking cotton in Texas at the time, similar incidents no doubt occurred in California.[122]

The workers also had conflicts with Mexican contractors and supervisors. They had experienced the erosion of personal relations in the new haciendas, factories, mines, and industries of Mexico. In 1909 a working-class poet penned an unsigned circular lamenting, "Today in Mexico the patron is not the father / that yesterday he was. / He is the terrible predator / without feelings / who increases our work / in exchange for a trifle."[123] Women at the Santa Rosa textile mills complained more bluntly that they were treated like "slaves from Africa."[124] Frequent conflicts with Mexican contractors on cotton ranches surfaced in spontaneous local disputes. Workers retaliated against low wages or contractors who rigged the scales by wetting the cotton or loading picking sacks with rocks, pebbles, or mud to make them heavier. In some cases, such conflicts led to tighter supervision; in others, workers were able to change some conditions. In the early days, for example, contractors were in charge of measuring the length of cotton rows, crucial to workers paid by the acre to chop cotton. Contractors often gave workers rows longer than an acre. The ensuing conflict between workers and contractors over the measurements became so intense that growers, under pressure from workers, started to plant uniform rows evenly divided and marked off into quarter-mile sections.[125] There were overt conflicts as well. Edward Bañales threw rocks at labor contractor Leandro Madero and then quit, "tired of being pushed around." Workers called foremen who cheated them "traitors," implying they were betraying class and national loyalties. Some foremen were threatened or attacked.[126] Sometimes workers simply refused to work for particular contractors until a dispute was settled.

The Depression of the 1930s worsened conditions. Workers lost other jobs, forcing them to rely increasingly on a diminishing amount of work in agriculture. In 1932 Sabina Cortez's father lost his job in the clay mines, which precipitated their move into agriculture. Eduardo López, who had combined cotton picking with work in a cement factory, lost his factory job and took the entire family to the Valley: "It got so hard during the Depression. There was no work. When my dad died we had to leave. We never had any money. I never had any work."[127] In 1934 Refugio Hernández's father gave up sharecropping in the Imperial Valley to become a migrant worker. Children dropped out of school to help their parents. Wages dropped, and workers had a harder time collecting what was owed them: by 1929 the California state government reported that because of the Depression growers had failed to pay $10,000 in wages owed to workers.[128] Unable to buy food, families resorted to hunting rabbits, catching fish, and scrounging for mustard greens and mushrooms amid the Valley scrub.[129] When Jessie de la Cruz's family "didn't have anything to eat" in the winters after picking was over, "we would go out with the ditch bags and pick mustard greens and mushrooms. My oldest brother would go out and look for fish. I don't know how we survived. And that was not only my family but all the families around us."[130]

To add to their misery, hundreds of Mexican families living in San Joaquin Valley camps were marooned when the Fresno River overflowed and flooded 25,000 acres in 1932. Approximately 2,500 workers were evacuated, but 600 remained, facing starvation and surrounded by water that had "assumed the proportions of a vast inland sea."[131] Lack of money for gas reduced migration. Some stopped migrating north, and others simply remained in the Valley year-round. The López family, for example, stopped migrating north between 1930 and 1932 because "there was no money" to go until in 1933 they pooled their money with another family for the trip. Once there, the truck broke down, and "we never made it back to Anaheim. I've been here in Fresno since 1933."[132]

Deportations and repatriations further brutalized the community. Between 1929 and 1939 over one million Mexicans returned from the United States.[133] Raids on homes and popular public areas and the fact that Mexican consuls helped in the repatriation efforts made workers even more anxious about dealing with government agents. By 1931 an estimated 50,000 Mexicans had already been deported from California.[134] Some Mexicans returned willingly, as Jessie de la Cruz remembers, "because of the hardships they were going through here. They said if they were going to suffer hunger and everything, then 'we want to suffer it back with our families in Mexico.'"[135] Many were bitter. Mr. Piña of the San Joaquin Valley said, "My father left his best years of his life in this country because he worked hard in the mines and in the fields and when hard times came around, we were expendable, to be thrown like cattle out of this country."[136]

But Mexicans had carved out a world for themselves within the system of agricultural labor in California's cotton fields. Through their friends and families they had shaped networks in the cotton communities and labor camps. They faced the tenuous life of migrant workers, but some were able to take advantage of the structural changes introduced by cotton: they were able to stay longer in one place, work more steadily, and maintain close familial and social ties. Even the most transient worker, through tenacity and effort, had been able to forge bonds and make connections. These bonds helped train all workers and protected them against the harshest aspects of work. Workers used them to seek jobs and to modify aspects of their work. And Mexican cotton workers used them to wage the largest agricultural strike to that date in the history of the United States: the cotton strike of 1933.

CHAPTER THREE

"As the faulting of the earth . . ."

The Strike of 1933

In September 1933, over 18,000 cotton pickers associated with the Cannery and Agricultural Workers Industrial Union (CAWIU) went on strike. The strike marked the culmination of a wave of thirty-seven agricultural strikes in California that affected 65 percent of the state's crops. Seventy-five to ninety-five percent of the cotton strikers were Mexicans.[1] The strike extended over one hundred miles and four counties, spreading from the southern tip of the San Joaquin Valley in Kern County north into Tulare, Fresno, and Kings counties. It lasted twenty-seven days and was the largest, longest, and most bitter agricultural conflict to that date. Three Mexicans were killed. And ultimately the Mexican, the Californian, and the federal governments became involved in its resolution.

The cotton strike was a social earthquake. Paul Taylor wrote of this strike, "As the faulting of the earth exposes its strata and reveals its structure, so a social disturbance throws into bold relief the structure of society, the attitudes, reactions and interests of its groups."[2] The strike revealed the strata of workers' networks and internal organization, as well as the knowledge, born of their experiences, which they brought to bear in the conflict. The strike exposed the class structure of the cotton industry and exacerbated its internal tensions. And the strike became the unintentional testing ground for the New Deal's policy toward agricultural labor. In short, the cotton strike throws into bold relief the elements that were to mold class relations in cotton into the late 1930s.

PRELUDE TO A STRIKE

By 1932 the Depression had thrown the cotton industry into crisis. Cotton prices fell precipitously, from 20 cents a pound in 1927 to only 6 cents a pound in 1932.[3] Small farmers, unable to pay expenses, lost their land, and an unrecorded number of tenants and sharecroppers were forced into the migrant stream or joined the growing ranks of the unemployed. Wages for cotton chopping were cut by more than half, and cotton pickers, paid $1 per hundred pounds in 1928, received only 40 cents in 1932. Because wages were cut in other crops as well, workers in 1932 found their annual income further reduced from a level already substandard. The National Industrial Recovery Act (NIRA) was signed into law in the late spring of 1933. Agricultural workers were excluded from the legislation under the president's Re-Employment Agreement, yet their belief that the federal government supported their organization under section 7a of the NIRA encouraged them to strike against formidable odds.

Between April and December 1933, thirty-seven strikes, involving 50,000 workers, erupted in California's agricultural fields. Under the leadership of the CAWIU, twenty-nine resulted in gains for workers and an increase in wages.[4] The strikes began in the Santa Clara Valley fruit orchards and, following the path of migration, extended through the berry, sugar beet, apricot, pear, peach, lettuce, and grape harvests. As the strikes continued it became clear that workers would carry the strike wave to cotton. By May some of the 6,000 Mexicans on strike in the El Monte berry fields vowed openly to carry the strike north into the San Joaquin cotton fields. One promised, "If there is no strike in the San Joaquin Valley now, there will be one when we get there."[5]

Cotton growers and the Agricultural Labor Bureau were tensely aware that Mexicans were threatening to carry the strike to the Valley. Francisco Palomares made futile efforts to prevent strikers from migrating north, and he canceled a contract to import a thousand workers from strike-ridden El Monte because he "did not desire to import any Mexican labor agitation."[6] George Clements of the Los Angeles Chamber of Commerce admitted that 400 still carried union cards and despite the strike's end were "strike minded," "union mad," "just as on strike as they ever were"; he warned that "anyplace they might go there would be strikes and trouble."[7] Attempting to stop further strikes, the Los Angeles Chamber of Commerce tried to deport strikers or convince them to repatriate voluntarily to Mexico.[8]

Despite those efforts, the first of the series of strikes that were to convulse the San Joaquin Valley from August through October began at the Tagus ranch at the beginning of the August peach harvest. This strike against the Valley's largest peach grower and a major cotton producer foreshadowed the cotton strike that followed. A small group of Mexican workers at Tagus had organized a union and were preparing to strike. When Pat Chambers, an organizer for the CAWIU, arrived to encourage a strike, they joined the CAWIU. Led by these workers, 700 Tagus workers went on strike, protesting forced buying from the company store and demanding a pay increase, a forty-hour week, and union recognition.[9] The strike spread, and by mid-August 4,000 workers were on strike. In an effort to prevent a general strike, other growers began to raise wages, eventually pressuring Tagus to raise wages as well. The capitulation of the largest grower forced the ALB to raise the wages for the entire industry.[10] The Tagus strike laid the groundwork for the cotton strike. The basic union structure was organized: the CAWIU established union headquarters in Tulare and, following the strike, formed nineteen small CAWIU locals covering the length of the Valley.[11] The Tagus strike's success encouraged workers to strike in the grape and cotton harvests. And Tagus workers provided a core of leadership for the cotton strike. Veterans of the Tagus strike and CAWIU organizers remained for the beginning of the cotton harvest in late September.

A two-week delay in the cotton harvest provided time to prepare for the strike and for news to spread to isolated ranches, company towns, pool halls, and Mexican barrios in the small agricultural towns. As workers waited for the crop to ripen, Pat Chambers and other organizers traveled to outlying communities and ranches, meeting with workers to discuss wages and living and working conditions.[12] An organizer recalled:

> Word spread that there was something the agricultural workers could do to better their condition. . . . Individual workers came in to the union head-quarters . . . and delegations came in from the ranches and we advised them to organize which they did. . . . They organized locals in tents, in hovels, and in holes in the wall. Strikers who had taken part at the Tagus ranch had experience and formed the nucleus. The strike cry began to ring through the valley . . . "not a pound for less than $1 a hundred."[13]

The workers faced formidable obstacles. Because cotton wages were set for the entire industry, a strike would be effective only if it could stop production throughout the entire cotton belt, which encompassed over a hundred miles and two thousand ranches. Despite the difficulty, workers were pressing for a strike. According to strike organizers Pat Chambers and Caroline Decker, even without the CAWIU "there would have been strikes anyway. Strikes were breaking out all over. . . . Those strikes would have happened union or no union. It was the historical moment for them."[14]

On September 19, seventy-eight delegates met at CAWIU headquarters to formulate their demands: $1 for picking a hundred pounds of cotton; abolition of the contract-labor system; no recrimination against strikers; and hiring only through the union.[15] They sent their demands to the ALB and threatened to call a strike if the demands were not met.

Two days later, growers and ginners met at the ALB to set the wage rate for cotton picking. The cotton industry was stronger financially than it had been for several years: following the announcement of the Agricultural Adjustment Act, cotton prices had risen by as much as 75 percent over the previous year. Growers and ginners were primarily concerned with holding wages down, stopping the spread of the strike wave, and undermining the CAWIU.[16] The bureau rejected a minority suggestion that wages be increased to 75 cents, but, faced with a shrinking labor pool and a series of strikes, they raised the rate to 60 cents, which was 20 cents more than in 1932 but was still disproportionately low in light of the increased cotton prices.[17]

When the wage rate was announced, 2,000 to 3,000 workers spontaneously walked out of the fields.[18] Pressured by walkouts, the union declared a strike on October 4. Thousands joined; a thousand walked off the Tagus ranch, and another thousand went on strike around Wasco in Kern County.[19]

Workers struck in response to immediate conditions, but also in response to long-term dissatisfaction, hope, and the momentum from earlier strikes. Their expectation of government support contributed to the strike's duration, intensity, and extent.[20] The National Recovery Administration (NRA) had already mediated agricultural strikes in El Monte and, in the San Joaquin Valley, the Tagus peach strike and the grape strike. Section 7a became as important a factor in precipitating strikes in agriculture as it was in other industries. One hundred and thirty Mexicans near Fresno struck, demanding a minimum wage under the NRA.[21] Organizer W. D. Hammett, arguing for the strike under NRA provisions, said, "We . . . are entitled to enough to live on and we're behind the president when he says that concerns unable to pay a living wage to their employees should go out of business."[22] Mexican strikers prominently displayed a handwritten sign in Spanish declaring their support of the NRA.[23]

The strikes of 1933 revitalized the CAWIU.[24] In 1931 the Communist party had revived the CAWIU from the ashes of the Trade Union Unity League.[25] The CAWIU was small, understaffed, and, as one organizer pointed out, "a union in name, a rallying point, but as far as organization . . . it was actually non-existent."[26] Mexicans became the backbone of the union. As spontaneous strikes erupted, the union allied with Mexican leftists and strikers. In some strikes the CAWIU contended with a Mexican union for leadership; in others it was the sole union. But it was only as strikes spread that the CAWIU took shape. As Caroline Decker, CAWIU organizer, pointed out, the union "was in our heads and in our ideals but it . . . was when strikes began popping all over the state that the union began to take on an identity."[27] Networks of workers that had developed over years linked seemingly disparate areas and became the organizational underpinning for the small, if ambitious, union. Although Anglos went on strike, especially in Kern County, this chapter will focus on the Mexican workers who composed 75–95 percent of the cotton strikers.

THE BACKGROUND OF THE MEXICAN WORKERS

How Mexicans responded depended on their past experience in social conflicts as well as the immediate situation they faced. In the areas of heaviest out-migration to the United States, Mexicans had engaged in widespread, if sporadic, peasant rebellions against the expropriation of land by the expanding haciendas.[28] Workers in Mexican industry formed unions and workers' organizations. By 1906, 80,000 belonged to mutual aid societies, the only legal arena for labor organization. The ideology of these groups was an amalgam of anarchism, mutualism, and cooperativism. Mexicans organized strikes in the 1890s and 1900s which became larger and more militant as the work force expanded, inflation deepened, and repression increased.[29] In the 1906 strike against Cananea Consolidated Copper Company, a subsidiary of the American-owned Anaconda, 2,000 Mexican miners struck under the leadership of the anarcho-syndicalist Partido Liberal Mexicano (PLM).[30] Cotton workers had been involved in these conflicts.[31]

Cotton workers who had come from the Laguna area of Durango, the center of Mexican cotton production and a major source of migrants to Los Angeles, may have been part of social conflicts there. A closer look at the Laguna suggests the response by Mexicans to the dislocations of industrialization. The rich Laguna area of Durango paralleled the San Joaquin Valley in several important ways. The region's economic mainstay was cotton. Between 1880 and 1920 cotton production in Mexico doubled and became the country's principal agricultural product. Mexican entrepreneurs and the Anderson Clayton company, which was based in the United States and entered Laguna shortly before entering the San Joaquin Valley, invested heavily in Laguna's cotton. Laguna's economic development depended on the migration of laborers, and by 1910 its population had swelled to over 200,000. Free towns developed on its fringes and became centers of unrest among workers. The forms of protest, based on a tradition of revolt by small landowners in the area, ranged from banditry to organized revolt and support for the PLM.[32] Mexicans who had labored in the United States were influential: in 1908 reports surfaced that an armed group, the "Mexican Cotton Pickers," was planning to attack the town of Coahuila from their stronghold on the Texas border.[33] Migrants to the San Joaquin Valley from the Laguna brought this tradition and experience with them to the United States.[34]

The Mexican Revolution of 1910 to 1920 was also a formative experience for Mexicans. Increasing dissatisfaction with enclosure, the upheavals of industrialization, and a growing rupture over Porfiriato politics increased the social conflict that led to the revolution, the first mass social revolution of the twentieth century. Rural workers demanded the return of communally owned land. Industrial workers went on strike for higher wages and improved conditions, and, not infrequently, included a demand for Mexican ownership.[35] As the Mexican historian Adolfo Gilly argues, "The Mexican people . . . burst onto the historical stage and lived for a time as its main protagonists. Feeling themselves to be the subject, and no longer the mere object, of history, they stored up a wealth of experience and consciousness which altered the whole country as it is lived by its inhabitants. It was impossible to ignore or depreciate this change in the decades that followed."[36]

While Gilly refers to those who remained in Mexico, the revolution also decisively influenced Mexicans who labored in the California cotton fields in the 1920s and 1930s. This is clear in oral histories taken from strikers, in which the legacy of the revolution overshadowed haunting personal memories, and remained a bond among Mexicans in the United States, despite whatever conflicts persisted between factions. Belén Flores, for example, remembered bodies stacked in grisly heaps along the streets following the battle of Torreón, and soldiers who threatened to cut off the soles of anyone who withheld grain from incoming troops.[37] Yet these memories coexisted with a sense of participation and connection to the revolution. Refugio Hernández proudly told a multigenerational story of struggle: his grandfather fought the French for Mexican independence, and his father fought with both Villa and Zapata. Both Guillermo Martínez's father and Luis Lima's had joined the army of Francisco Villa. Manuel García had fought with Carranza. Juana Padilla's father had served in the federalist army. The revolutionary national heroes, such as Zapata and Villa, personified the Mexican nation, sanctioning armed struggle and social change. The image of the *soldaderas*, women who followed men into battle, became symbolic of women fighting for their people. However ambivalently the actual soldiers and soldaderas were received, over the years their image became a focus for Mexican nationalism and a rallying cry in conflicts.[38]

In the United States, Mexicans organized in the mines, agricultural fields and urban industries. Ignored by the American Federation of Labor (AFL), they organized ethnic unions or allied with the sympathetic Industrial Workers of the World (IWW) and, later, affiliates of the Communist party.[39] Mexican workers drew on this experience when they worked in agriculture. Since the turn of the century, Mexican agricultural workers in California had organized in ad hoc organizations that usually disappeared at the end of the season. By the mid-1920s Mexicans in California had established more permanent agricultural workers' organizations. In 1927 the Federation of Mexican Societies, meeting in Los Angeles, formed the Confederación de Uniónes Obreros Mexicanos (CUOM). By 1928 the union claimed 3,000 members in twenty-two locals in southern California and helped spur the formation of the first permanent Mexican agricultural union. CUOM faltered within a few years, but was revived in 1933 as the Confederación de Uniónes de Campesinos y Obreros Mexicanos (CUCOM), becoming one of the major agricultural unions of the state.[40] CUCOM reflected the structure and ideology of the Mexican union, the Confederación Regional Obrera Mexicana (CROM), which sent CUCOM funds and by at least 1933 (and probably earlier) listed CUCOM as an affiliate.[41]

In these social conflicts Mexicans were exposed to diverse, class-conscious organizations and ideologies. One of the most important influences was the Partido Liberal Mexicano, formed in Mexico by Ricardo Flores Magón. Under repression by the Díaz regime, Magón fled to the United States in 1904 and recreated the organization in Los

Angeles. Mexican workers in the Southwest were receptive to the PLM, which built upon a tradition of Mexican rural anarchism.[42] By 1914 an estimated 6,000 belonged to the PLM north of the border. In Los Angeles alone the PLM newspaper, *Regeneración*, was the most popular Mexican newspaper in the city, with over 10,500 readers.[43] The PLM worked closely with the IWW on both sides of the border, organizing in the copper mines of Cananea and in the mines of Arizona, and joining in an ill-fated 1911 expedition to retake Baja California for Mexico.[44] In Los Angeles, most of the IWW's 400 members were Mexicans.[45] The PLM-IWW alliance taught a lesson in internationalism that made Mexican leftists more open to working with their Anglo counterparts and laid a basis for alliances between Mexicans and Anglos of the U.S. Communist party in the 1920s and 1930s.

The PLM had declined in size and militancy by 1925, but anarchists remained active in labor organizing.[46] Guillermo Velarde, leader of CUCOM, was the son of a founding member of the IWW and was himself a member of the IWW.[47] Anarcho-syndicalist ideology continued to influence younger radicals well into the 1930s. Porfirio Cervantes, a member of the PLM, gave political literature to at least one member of the Young Communist League in the Imperial Valley. CUCOM's left wing remained largely anarchistic.[48] Their receptivity to later strikes stemmed in part from the influence of the PLM and the Mexican Revolution. As CAWIU organizer Dorothy Ray Healey remembers:

> Mexicans who would come over the border were products of the Mexican Revolution. They were very much influenced by anarchist-syndicalist ideas. . . . I can remember the meetings we'd have when I'd be discussing why they should become communists, and they'd all listen very tolerantly and sweetly. I was telling them something they knew far better than I did as to the evils of capitalism. When I was all done they would all smile benignly and say, "Dorothy, it's all right. When the revolution comes we'll be on the barricades, but don't bother us with organization now."[49]

By the 1920s, progressive Mexicans in the United States began to turn to the Communist party. An intriguing issue is the relation between Mexican communists in the United States and the Partido Comunista Mexicano (PCM). Workers in the PCM maintained sporadic and informal contact with their compatriots in the United States, especially along the border. Perhaps more telling of its influence was the broad circulation of *El Machete*, the PCM newspaper, in the United States, especially in Los Angeles. CAWIU organizer Stanley Hancock remembered that "some party literature came across the Mexican border into the Imperial Valley . . . like *Hoy* and *El Machete*, the daily organ of the Communist party of Mexico."[50] Yet there were few formal ties. The financially strapped PCM, facing increasing repression in Mexico, lacked the resources to maintain broad contacts with Mexicans in the north. The PCM focused more on urban workers than on peasants and more on workers in Mexico than on those who had migrated north. Moreover, by the 1930s the tensions of Comintern politics had badly strained the relationship between the PCM and the Communist party of the United States of America (CPUSA). The PCM, placed by the Comintern under organizational subordination to the CPUSA, chafed at taking orders from the CPUSA and, understandably, viewed this relationship as continuing national subordination to the United States.

As a result, Mexican communists north of the border looked to the CPUSA, not the PCM, for help.[51] Mexican participation in the party has not been adequately researched, yet limited research suggests Mexican participation played a crucial role in organizing and strikes among Mexican workers. Mexican communists formed party cells in at least seven towns in southern California.[52] Younger Mexicans joined the Young Communist League (YCL). Membership was small: by 1933 over two dozen Mexicans belonged to the YCL in Tulare, and by 1934 there were fifteen members of the Mexican Communist party cell in Brawley. Left-wing Mexicans were, with few exceptions, workers: cotton workers, such as José Gómez, Francisco Medina, John Díaz, and Miguel Gutiérrez were members. As agricultural workers who followed the crops, they were crucial in helping to organize and lead strikes, and they acted as conduits between the Mexican workers and the CAWIU. The presence of these organizers helped spread the strikes of the 1930s, and it is open to question how effective the Communist party would have been without them. Just how many communists worked in cotton is open to speculation, but it is clear that they, like their Anglo counterparts, played a disproportionate role in labor organizing.[53]

Among the cotton workers were many who had been exposed to labor conflicts on both sides of the border, had fought in the Mexican Revolution, and had developed experience in organization and tactics. This is borne out in oral histories. Pedro Subia, Braulio López, Luis Lima, and Narciso Vidaurri worked in the mines of Clifton-Morenci, Bisbee, and Cananea, where the IWW and PLM had organized miners.[54] Carlos Torres took part in a 1919 strike by San Bernardino County orange pickers that was perhaps led by the IWW.[55] Luis Lima joined mining strikes in Sonora in 1918 and participated in other agricultural strikes.

Lima's uncle was killed in the large Cananea strike in 1906. Eduardo López helped organize a small melon strike, allied with the CUCOM, in San Juan Capistrano, California, in 1933.[56]

In 1928 the Communist party abandoned its "boring from within" policy and formed the Trade Union Unity League (TUUL) to help create a revolutionary union movement outside the AFL. In 1931 the TUUL spawned the CAWIU. Mexicans, as the backbone of agricultural labor in the state at the time, also became the backbone of the work force organized by the union. In 1928 and 1930, strikes led by Mexicans and the TUUL broke out in the Imperial Valley. Some of the cotton pickers were involved in these strikes and would later join strikes led by the TUUL or the CAWIU.[57]

There were certainly Mexicans who had not been in labor conflicts and had little or no organizational experience. Organizer Leroy Parra felt that "they didn't know they could organize. It was something new to them and it was something new to a lot of us." Yet, even so, Parra argued, "The Mexican people are revolutionary as it is. They had [the Mexican Revolution] behind them that helped them to see the exploitation here in this country."[58] The revolution was a topic of conversation in cotton camps, where revolutionary tales and songs took on new nuances.[59] The revolutionary experience also provided organizational skills, for, as Guillermo Martínez explained, "The experience of pain was carried to the various communities over here . . . [and] the know-how for organizing came, perhaps, through organizing in the army."[60] Even if few joined the CAWIU, Mexicans were receptive to labor organizations and left-wing organizations that "helped the working people."[61] These Mexican workers would be the impetus for the cotton strike of 1933.

STRIKE EXPANSION

Over the hundred-mile cotton belt, workers' responses to the strike varied. On smaller farms, where relations were more personal and farmers often offered 75 cents per hundred pounds picked, many pickers remained at work. And in areas of Kern County where migrant white workers, new to California and desperate for work, agreed to work at whatever rate offered, the strike failed to catch on. One union leader commented, "They work for ten cents a day and think they are getting good money."[62]

But the strike was aimed at the largest growers and the gin and finance companies represented by the ALB. The biggest walkouts occurred on ranches of over 300 acres. Here hundreds of Mexicans refused to pick, pressing their contractors to demand more pay. Contractor and strike supporter Roberto Castro, who took a small group from Corcoran to meet with the CAWIU, remembered:

> Se sufrió . . . Pues, tenía que sufrirse, porque todos sufren cuando es el pobre contra el rico. . . . Nosotros creíamos que si se organizaba la gente, no había que hacer huelga. Nosotros nomás queríamos que pagaran más. Por eso se organizó la gente, se juntó, sin saber si era unión o no era. . . . La gente se unió, porque querían que hubiera más sueldo, un poquito más, para darle de comer a los chamacos. . . . Pues, si no trabajaban, no comían.

> (There was suffering. Well, there had to be suffering, because everybody suffers when it is the poor against the rich. . . . We believed that if we organized the people, there wouldn't have to be a strike. We only wanted them to pay us more. That's why, without knowing if there was a union or not. . . . The people got together because they wanted to get more wages, a little more, to be able to feed their kids. . . . Well, if you didn't work, you didn't eat.)[63]

Growers responded by evicting strikers from the labor camps. The evictions began at the Peterson ranch near Corcoran, where 150 families were on strike. Seventy-five growers, backed up by armed guards, assembled to evict the strikers. Despite the cold weather and growers' threats, the workers still refused to pick. Belén Flores, who lived with her husband and two small children on the ranch, remembered that in a unanimous voice "dijeron ellos, '¡Ninguno trabajamos!' Sí, dijeron '¡Ninguno!'" ("And the people said, 'None of us will work!' Yes, they said 'None of us!'") Growers loaded the strikers' belongings onto trucks and unceremoniously dumped them onto the highway. Within a few days growers evicted over 3,500 from around Corcoran, and other evictions occurred across the Valley.

Before the evictions, organizers, hoping to prevent growers from importing strikebreakers, had urged workers to stay on the camps and refuse to work. Yet the evictions worked to the strikers' advantage. The deportation of 150,000 Mexican workers since 1930 had "caused a marked shortage . . . of agricultural labor" and this, with a slight upturn in urban employment, federal relief, and the strike wave combined to leave growers without workers and their camps empty.[64] The evictions became a crucial tactical victory for the strikers.

Evicted strikers formed camps which became focal points for organization. In the Mexican barrios of McFarland, Tulare, Porterville, and other towns, homes, pools halls, and bars such as the Old Monte Carlo

Saloon in Delano and the Salón Mexicano de Billares in Wasco were converted into strike headquarters.[65] In Arvin 1,500 workers assembled at four main refugee camps. Each resembled, to some degree, the largest strikers' camp, which was in Corcoran, Kings County.[66]

The Corcoran camp began as a refuge for evicted Mexican strikers who, with no place to go, camped on the land of a small farmer on the outskirts of town. Within a week about 3,500 strikers had assembled on the four-acre plot, forming a camp that dwarfed the small town of Corcoran.[67] As Paul Taylor reported, "Each family provided its own habitation—an old tent or burlap bags stretched between two poles and a car. These makeshift tents, in addition to the family car, cooking utensils, bedding and the ever-present dog, represented the total possessions of the evicted strikers—with perhaps a goat or several chickens for the more fortunate."[68]

A barbed-wire fence was built around the camp for protection; its single opening was watched by sentries. A crude waterpipe system, marked by frequent spigots through the camp, provided water, and ten wooden toilets were built for sanitation. The irrigation ditch served as a "collective washtub and bathtub for children." In the center of the camp was a place of assembly, where, under a handwritten sign in Spanish hanging from a canvas tarp hung above wooden benches, the union office stood.[69] Food was donated by strike sympathizers or came from strikers who foraged for weeds and grasses, shot rabbits, fished, or had enough money to buy beans and rice. Women cooked food for single men at a community kitchen. Two temporary schools were set up for the children in large tents and were staffed by local teachers.[70] A Mexican circus, the Circo Azteca, serendipitously camped on the land before the strikers moved in, provided nightly entertainment, as did guitar players and singers among the strikers. The dense concentration of strikers in the camp contributed to keeping morale high, facilitated organization, communication, and planning, and provided a gathering point for forming picket lines.

The size of the cotton belt posed the problem of how to reach the workers. Organizers developed mobile mass picketing. Caravans of cars full of men, women, and children were dispatched from the strikers' camp to intercept possible strike breakers and then headed for the ranches where picking was reported.[71] Strikers intercepted pickers on their way to the fields. Picket lines were set up around ranches and, to entice workers to leave, strikers stood across the road from the fields, blowing bugles and calling on pickers, as fellow Mexicans and fellow workers, to join the strike.

The leadership in the camps and on the picket lines came from Mexican workers. Most accounts attribute the leadership and organization to the CAWIU, but they overlook the relationship among the union, the organizers, and the workers. Sam Darcy, the Communist party's district organizer for California, claims he led the strike and was, as Cletus Daniel suggests, the "dominant secret strategist" who operated from Communist party headquarters in San Francisco.[72] Yet the CAWIU organizers who worked in the San Joaquin Valley, Pat Chambers and Caroline Decker, disagree. While a broad CAWIU strategy had been laid out, tactics for day-to-day organizing took form within the harsh realities of the Valley.

The lack of money and organizers and the breadth of the cotton belt made precise long-distance strategy impossible.[73] Although it sounded impressive in newspaper accounts and party minutes, union organization was skeletal at best. Only six CAWIU organizers were active in the entire Valley. According to Caroline Decker, "You had me and Pat Chambers and three or four people down there. A communist here. A radical worker there. No money. No knowledge of how to do these things. Just all good will and idealism."[74] The union brought together disparate segments, coordinated local strike groups, and maintained a tireless round of daily meetings with local strike committees and camp delegates. Yet, as Pat Chambers pointed out, "Although the directives in some superficial way could come from the outside, the actual organization had to come from the workers themselves."[75]

Until recently, activities outside the formal structures of unions or political parties have often been considered as at best auxiliary and at worst irrelevant to larger social conflicts. Yet an analysis of workers' families, networks, communities, and alliances shows a more complex response that challenges the preeminence of unions and formal organizations as the essential form of working-class organization.

The CAWIU's success was based on workers' families and networks, leaders, organizations, and advice. At their August San Jose conference, CAWIU delegates had voted to "develop leadership" out of the rank and file and work out strategy and tactics by consulting those "working on the jobs, who are familiar with job conditions and [in] daily contact with workers."[76] In many respects, this was a recognition of what already existed. As Pat Chambers pointed out, the "well knit families, [a] clannishness . . . [was] an essential element in getting the people to work together."[77] Personal networks that crisscrossed the Valley facilitated the spread and coordination of the strike. Bars and pool halls along the migrant route were rapidly transformed into strikers' centers.[78] Networks provided support for families, whether through shared food, labor, or emotional support. Women activated their networks in strikers' camps and on picket lines. The organization of the Corcoran camp was based on labor crews, as crew members settled near each other, forming the camp into organizational sections corresponding to the crews. From this, they formed an ad hoc governmental structure which, while following CAWIU directives to develop rank-and-file leadership and strengthen locals, emerged from the structure of Mexican labor crews and concepts of mutual aid that had evolved within the crews.

When the camp grew, union leaders, following the inclination of strikers to settle next to friends and family, laid out the cars, lean-tos, and tents in rows, forming a series of crude streets, each of which corresponded to one of the fourteen labor camps from the ranches. Each had a representative, similar to a crew leader in his responsibility to the rest of the crew, who dealt with relief and with day-to-day problems. Nine camp members were appointed as the central committee, and the relief delegates and the central committee together formed a general council. The council was represented at the CAWIU by organizer Leroy Gordon.[79]

The CAWIU acted as an umbrella group, coordinating the disparate elements of the strike. But it was an ephemeral part of the consciousness of workers. By November the union claimed 7,000 members, although fluctuating membership and migration make it hard to make an accurate tally and suggest that actual membership was much lower. Many of the cotton strikers did not know the union or the Anglo organizers by name. Fewer were dues-paying members; even Roberto Castro, who did know of the union and said, "Yo soy de la unión," did not pay dues. "Cuando se hizo la huelga a todos nos hicieron miembros. Pero nosotros no pagamos nada." ("When the strike started all of us became members. But we paid nothing.") The importance of Mexican leadership and organization helps explain why many strikers did not know the name of the union.

Mexican leaders acted as conduits between CAWIU organizers and workers, organized strikers' camps, directed picketing, and dealt with workers on a day-to-day basis. When the strike began, Mexican leaders emerged from recognized chains of authority within the community; they were contractors, ex-officers of the Mexican army, members of the small merchant class, leftists, and workers experienced in earlier strikes.

Contractors took various positions toward the strike, motivated by personal sympathy, family ties, and the knowledge that they would be looking for workers after the strike. Large contractors sometimes sided with growers, whereas smaller contractors and crew leaders usually sided with strikers. The ambivalence of contractors was reflected in the Castro family: Mateo Castro bitterly opposed the strike, but his son (and assistant) Roberto organized workers and helped form the strike committee in Corcoran.[80] Contractor Librado Vidaurri served on the strike committee, and his brother Narciso stayed in close touch with his crew who now lived on the Corcoran camp, continuing to help with problems and bringing them food.

Men who had been leaders in the Mexican Revolution came forward: Manuel Mireles, a colonel in the Mexican army, took charge of security in the Corcoran camp; Manuel García, a member of the strike committee in Corcoran, had been in the Carranza faction of the army. Other men and women, nameless, who played a part in organizing picket lines and security, had no doubt fought in the revolution. Strikers' experience in the revolution was crucial to the Corcoran camp's survival. Veterans established a disciplined, military-like system of sentries who, armed with guns and crowbars, patrolled the camp and prevented armed growers from invading.[81] There were reports that armed strikers near Corcoran patrolled the highways to keep out potential strikebreakers headed for the cotton fields.[82]

Veterans of earlier strikes formed the core of this leadership. As workers descended on the valley, veterans of the strikes in El Monte, the Imperial Valley, the Tagus ranch and other areas organized among people they knew. Among these worker-organizers were Mexican communists and anarchists. Many had been in the strikes of 1928, 1930, and 1933, where they had allied with the CAWIU and joined with the union to organize the workers. These Mexicans reportedly appeared at union meetings in Shafter and Delano, where they denounced capitalism, church, and state and called for armed conflict with the forces of capitalism.[83] Francisco Medina, for example, was a sixty-eight-year-old communist who had helped organize several strikes and was instrumental in founding the Visalia CAWIU local. He worked closely with Pat Chambers during the cotton strike, coordinated organization around the Valley, and spoke at strikers' meetings.[84] John Díaz was another CAWIU organizer. Several dozen Mexican members of the Young Communist

League were active in the strike. Two reportedly played leadership roles in the Corcoran camp, and their presence was visible enough that the Mexican consul threatened to deport one of them.[85] There may have been a handful of members of the Communist party, such as those mentioned in a 1930 report as "aliens who are members of the Communist Party."[86]

If strike leadership came from leaders within the Mexican migrant community, the organization of the camps, picket lines, and reinforcements for the strike came from families, social networks, and labor crews. A pragmatic sense of solidarity that sprang from common interests and had been reinforced in families and work crews bound them together. Their identity as Mexicans and workers and a shared (if still being created) memory of the past helped to solidify the strike force further. Symbols of a shared cultural memory became crucial to making historic links with the present strike. At the same time, local affiliations as members of Mexican communities, states, and towns still played a role in one-on-one conflicts, and there were divisions among Mexican workers, as members of Mexican communities, states or towns, political factions, or factions within the revolution. Yet the image of the Mexican Revolution became a model of a collective struggle that reinforced their shared work and present experience. This was expressed not only in memory but also in appropriating the tactics, organizing forms, symbols, and names of the earlier conflict.[87] Workers named the camp streets after Mexican towns or Mexican heroes of the revolution. Military officers organized sentries and patrols composed of veterans of the revolution. And, at that distance of space and time, the revolution became didactic myth. People told and retold its stories and sang its corridos. Differences between revolutionary factions were submerged in the context of the strike, and memories stressed commonality, through class, nationality, and a historical antipathy toward the United States.

Newspaper accounts and oral histories emphasize that most strikers walked out. Belén Flores, for example, reported that workers deserted the fields on the Peterson ranch where she worked, a show of solidarity which continued throughout the strike and left growers without pickers: "¿Tu piensas que iba a traer otra gente, si cuando *todos* estabas en huelga? ¿Quién les pizcaba?" ("But you think they could have gotten other people, when *everybody* was on strike? Who would have picked?")[88] Certainly there were strikebreakers. Poverty, fear, and the pressing need to feed their families led some to remain. The strike divided some families, separated friends, and alienated compatriots. Yet, given the harsh conditions of the Depression, surprisingly few left the ranks of the strikers. For an agricultural strike that was spread over a hundred miles, the unity of the strike was impressive.

Social pressures which had helped enforce work discipline in crews now buttressed the ranks of strikers and pressured strikebreakers to join the walkout. Often led by women, strikers yelled to strikebreakers they knew from Mexico, home communities in Southern California, the road, or the labor camps. Public pressure was reflected in children's games, a barometer of social values, as they acted out the conflict, pitting strikers against strikebreakers. Women threatened to poison one lone strikebreaker who had the effrontery to live in the strikers' camp and had eaten meals cooked by the women.[89]

Indeed, women became crucial to the strike. As Temma Kaplan points out, when women's ability to perform traditional roles is impeded, they may organize collectively to secure their rights and, in so doing, politicize everyday social networks.[90] Certainly the precedents for Mexican women taking an aggressive role in community concerns helped propel women's participation in the strike. Women in Mexico had rioted for fair corn prices since the colonial period, had survived or fought in the Mexican Revolution, and had participated in labor strikes. Concerns over caring for their families drew them into the strike, as their ability to feed their children and families was undermined by declining wages and worsening conditions. Some women participated along gender-defined lines: they ran the camp kitchen, helped distribute food and clothes, and cared for children. The union paid little attention to these networks and made few attempts to include them. Some of this neglect was due to general organizational problems, as well as linguistic and cultural barriers, for the only female CAWIU organizer was a monolingual English speaker. Yet despite these impediments, women took part. At least two bilingual "girls" of indefinite age were on the Corcoran camp's central committee and acted as interpreters.[91] One Anglo woman remembers that it was a Mexican woman who taught her how to call out, in Spanish, "¡Huelga, pizcadores!"[92] ("Pickers, go out on strike!") Some attended nightly strike meetings. Except for those workers directly involved with organization, neither men nor women remembered Anglo CAWIU leaders with much clarity. Yet in oral histories, women remembered female leadership, which reflected the strength of their networks and their separation from formal organization. In Corcoran, for example, several women (but no men) mentioned Magdalena Gomez, a financially independent woman who stored and helped distribute food to strikers, suggesting that female perceptions of important networks and leaders, while complementary to those of male strikers and union organizers, often differed from theirs.

From the beginning of the strike, women of all ages—older women with long hair who wore the rebozos of rural Mexico, younger women who had adopted flapper styles, and young girls barely in their teens—went on the picket lines. In Corcoran, they organized and led confrontations with strikebreakers.[93] In a pattern that would reoccur in this and other strikes—suggesting that the idea emerged from women's networks, not from the strike committee—women entered the fields to confront the strikebreakers.

On 24 October a conflict between female strikers and strikebreakers erupted into violence at the Hanson ranch in Corcoran. The *Bakersfield Californian* reported that "tear gas and clubs failed to halt the marching Mexicans, many of them women, who swarmed from their squalid settlement [the Corcoran camp] to ranches where strikebreakers labored in the fields."[94] According to Belén Flores, one of the strikers, the women appealed to strikebreakers as "poor people" and "Mexicanos" and cursed those who remained, comparing them to traitors who had "sold the head of Pancho Villa."[95] They threatened and cajoled strikebreakers, using the female issue of food: confronting scabs at the Hanson ranch, the women warned them, "Come on out, quit work, we'll feed you. If you don't, we'll poison all of you."[96] Exhortations turned to threats and then to violence. Some women, armed with lead pipes and knives—evidently in expectation of using methods more persuasive than verbal appeals—attacked the pickers. The male strikebreakers retaliated. One woman was brutally beaten. When some strikebreakers appealed to the women not to hit them because they were poor and simply wanted to make enough to go home, Belén Flores remembered that the women responded with the voice of communal authority, "Sí, nosotros también tenemos que comer y también tenemos familia. Pero no somos vendidos." ("Yes, we also have to eat and we also have a family. But we are not sell-outs.") It was, as Flores remembered it, a battle between groveling strikebreakers and the strong collective voice of the striking community.[97]

Some women fought aggressively. The men remember that some women "could fight like a man." Roberto Castro remembered women who confronted the California Highway Patrol and stole the keys to the truck, and "una señora le quitó las pistolas a un policía." ("One woman took the guns from a policeman.") Women were arrested throughout the strike.

The interlocking ties of labor crews, families, and social networks formed the fabric from which the strike was cut. Social relations not only from the ranches but also stretching back to home communities in southern California or areas of Mexico were called into play to support the strike. A consciousness of themselves as Mexicans and workers reinforced the sense of a collectivity, emphasized in its breach by the attacks hurled at strikebreakers. Leadership arose from the recognized chains of command. The success of the strike was thus based in large part on incipient organizational structures within the workers' networks that were utilized by the CAWIU. The combination provided an organizational framework that facilitated the development of an effective strategy.

THE GROWERS' RESPONSE

The strike polarized Valley communities, bringing to the surface old conflicts embedded in the structure of California agriculture: conflicts between sharecroppers, tenants, and small farmers and the larger growers, ginners, and finance companies; and conflicts between small merchants and ginners, bankers, and business interests tied to large agriculture and represented in the Chambers of Commerce.

Growers, who had expected a short strike, were alarmed as the strike spread, picking all but stopped, and millions of dollars' worth of rapidly maturing cotton was left in the now-deserted fields. Attempts to import 4,000 unemployed Mexicans from Los Angeles failed. Fewer than 500 arrived, and few of those stayed to pick.[98] Faced with a general strike, large growers and ginners ruled out negotiation, resolving to stand behind the ALB wage rate. But when the ALB failed to stop the strike, growers and ginners organized quasi-vigilante groups called protective leagues in Kern, Kings, and Tulare counties to intimidate workers back into the fields. The purpose, as Forrest Frick, organizer of the Kern County protective league, announced, was to "drive [CAWIU and other] agitators out of the county," stop the strike, and maintain the wage rate.[99] Growers turned to vigilantism with an alacrity motivated by economic interests. The perishability of cotton precluded holding out until strikers were starved into submission: as each day passed, the prospect of losing an entire year's investment loomed larger. The close, if informal, relation between the protective leagues and the ALB was reflected in the leadership. Forrest Frick, a prominent Kern County cotton grower, organizer of the county's protective league, and later director of the Kern County Farm Bureau, and L. D. Ellett, chairman of the Corcoran protective league called the Committee of 14, were both officials of the ALB.[100] The leagues hit a responsive chord in growers faced with fields of unpicked cotton. Over 700 growers and townspeople joined the organizations across the Valley.[101] Armed growers began to intimidate strikers: in Kern County, vigilantes drove out strikers and their families; in the Valley town of Woodville, growers broke up a strikers' meeting.[102]

Small farmers were caught between strikers' demands and pressure from gin and finance companies. For them, the strike was another instance of ginners' ability to dictate wages and labor relations. Already living a precarious existence, and thus highly vulnerable to a strike which could push them out of cotton-growing altogether, they blamed the finance companies for causing "all the trouble by setting the price so low in the first place."[103] As the director of the State Employment Service in Bakersfield noted, "It was the fight between the communist[s] . . . Frick and his bunch and the Palomares bureau [the Agricultural Labor Bureau]."[104] Some small farmers who had worked as pickers themselves and had labored next to workers they hired felt some sympathy for the workers. The CAWIU, recognizing the position of the smaller farmers, did not strike their farms but instead attempted to organize them into a union. While this effort had a limited success, ultimately it floundered.[105]

While small farmers may have felt some sympathy with strikers, their position as small property owners and occasional employers won out. Despite their bitterness toward finance companies, ultimately their marginal position, fear of losing their crop, dependence on finance companies, antipathy toward labor organization, and a staunch anti-communism precluded the possibility of working with the union. Their dependence on loans made them reluctant to oppose the Agricultural Labor Bureau openly. The few who did so met with force, such as the Pixley farmer who refused to hire strikebreakers: the finance company put workers in his field under armed guard.[106] Racial feelings contributed to their reluctance to support strikers. As a result, although smaller farmers tended to pay slightly higher wages, and some would not hire strikebreakers, only a few openly supported the strike. Many were farmers such as Jim White, an ex- sharecropper and cotton picker, whose sense of moral justice, nurtured in a Populist family, led him to offer the higher wage of 75 cents.[107] Morgan, a small farmer and service station operator near Corcoran, allowed strikers to use his land to set up their camp, a move he later regretted.

The strike brought to the surface divisions between the major businesses with direct investments in agriculture and the smaller, more marginal merchant class which had developed to serve farm laborers. In Valley towns where agriculture was the economic lifeblood, major businessmen and the local Chambers of Commerce supported the ALB. In towns such as Tulare, Hanford, and Fresno, which had developed a small merchant class less dependent on agricultural power, smaller merchants and business people donated food and clothes and extended credit to strikers.[108] Of these merchants, most were not Anglos, but Jewish, Portuguese, Japanese, or Mexican, suggesting that class and ethnic interests were mutually reinforcing. In grower-dominated Arvin, Dan Reuben's Davis Quality Market was the only store to give credit or sell groceries to strikers. In Tulare, a Portuguese businessman rented his hall to the CAWIU; in Corcoran, a Portuguese dairyman donated milk and foodstuffs to strikers. Mexican markets and pool halls that catered exclusively to Mexicans supported the strikers and donated food. Mexican restaurants, bars, and pool halls were converted into de facto strike headquarters and food distribution centers.[109]

These merchants undercut growers' efforts to starve workers into submission. Angry that "the merchants gave the farmers more trouble during the strike than any other element in the community," growers physically intimidated merchants and publicly threatened to cut off their trade. Dan Reuben was threatened by armed growers. In Tulare, where a number of merchants and a doctor supported the strikers, the Tulare Farmers Protective Association ran a paid advertisement in the local newspaper addressed to local citizens:

> We the farmers of your Community, whom you depend upon for support, feel that you have nursed too long the Viper that is at our door. These Communist Agitators MUST be driven from town by you, and your harboring them further will prove to us your non-cooperation with us, and make it necessary for us to give our support and trade to another town that will support and cooperate with us.[110]

Growers had more control over local law enforcement officers and elected officials than over the merchants. Growers were influential citizens, and local officials, dependent on patronage or votes, would have found it hard, even supposing they were willing, to take a stand against growers. Charles Wilson, the district attorney of Kings County, admitted,

"The growers were really more trouble and danger than the strikers were. We could control the strikers because they didn't amount to anything and couldn't even vote, but the growers were well known and had lots of influence and we were much more afraid we couldn't control them."[111] As a result, local officials would use quasi-legal means to attempt to stop the strike. One under-sheriff said, "We protect our farmers here in Kern County. They are our best people. They are always with us. They keep the county going. They put us in here and they can put us out again, so we serve them. But the Mexicans are trash. . . . We herd them like pigs."[112] Growers often had more direct control over law enforcement. Sheriff Hill of Kern County said at one point, "I know I can handle the situation but I don't know whether it would meet with the approval of the growers."[113] Until pressured by outside authorities, local law enforcement acted as a slightly more legal branch of the vigilante groups. In Kern County, the sheriff, preparing for the strike, purchased two machine guns and tear gas, and swore in forty-five new deputies. Ranchers, ranch managers, and employees of the gins were deputized during the strike.[114] Sheriffs, worried about growers' support, arrested picketers indiscriminately for picketing, for obstructing traffic on the usually all-but-deserted roads, and on charges of inciting to riot. In Kings County, one Mexican striker was arrested for addressing the audience in Spanish.

Within a week the Valley became an armed camp where, the *San Francisco Examiner* ominously reported, "the atmosphere . . . is that of a smoldering volcano." The *New York Times* said that "the psychology of war" prevailed.[115] Armed growers were infuriated as picketers cajoled and shouted at pickers to desert the fields. One Kern County rancher threatened to "blow to hell every striker who so much as laid a hand on the fences." A grower, later charged with murdering a striker, threatened to shoot pickers who tried to leave his ranch.[116]

Yet the threats and violence, instead of intimidating strikers back to work, served to spread the strike and polarize communities. Although the union denied it, strikers began to arm themselves with sticks (and sometimes guns) to protect themselves and force strikebreakers to stop work.[117] For workers there was now no third alternative other than either joining the strike or scabbing for growers who were threatening people who were at least fellow workers and often friends and family. By 9 October an estimated 12,000 were on strike in Tulare, Kings, and Kern counties.

The smoldering conflict came to a head on 10 October. In the small town of Pixley, Mexican and Anglo strikers gathered in a lot across from the CAWIU hall, listening to Pat Chambers speak. Suddenly a caravan

of ten cars, filled with shotgun-wielding growers, arrived. As the unarmed strikers began to cross the street to seek sanctuary in the union
hall, a grower's gun fired. Fifty-year-old striker Dolores Hernández
pushed aside the gun, was knocked down by one grower, and was shot
to death by another as he lay in the dirt street. Growers, crouched behind
their automobiles, opened fire on the strikers and the union hall. When
the shooting stopped, Delfino D'Avila, who was a Mexican honorary
consular representative from Tulare, and an unnamed Mexican worker
from the Tagus ranch lay dead. Eight were wounded.[118]

At the same time, a confrontation occurred at the Frick ranch in Arvin.
As 250 strikers arrived to picket, thirty armed growers appeared to
protect pickers in the field. For five hours, strikers and growers exchanged insults across the dirt road until finally a fight broke out. One
grower opened fire, instantly killing Pedro Subia, a Mexican striker.
Other growers began shooting, wounding several strikers. And there
were other attacks on strikers around the Valley.[119]

Local law enforcement officials made no attempt to arrest the growers. The Pixley attack was reportedly led by F. G. Kruger, manager of
the Camp West Lowe ranch and a newly appointed sheriff's deputy.[120]
Highway patrolmen and sheriffs on the scene had watched the attack
without interfering; after the shooting they slowly pursued the growers
and, upon reaching them, allowed them to go on. Only when public
pressure mounted did they arrest eleven Tulare growers for the Pixley
killings. To placate growers, they also arrested Pat Chambers and sixteen
strikers for disturbing the peace.[121]

The killings at Pixley and Arvin marked a turning point in the strike.
As long as the violence remained localized and stopped short of murder,
growers had been free to intimidate strikers without interference. The
ambush and killing of unarmed strikers and a Mexican consular representative, photographed and printed in newspapers around the country, was another matter. The outbreak of open warfare brought the strike
to national attention, prompting a flood of protests that brought state
and federal mediators, the Mexican government, relief officials, investigators, news people, and sympathetic delegations into the area.

In response to evictions, arrests, harassment, and the killing of three
Mexican nationals, among them a consular representative, the Mexican
consul, Enrique Bravo, demanded that California governor James Rolph
protect Mexican citizens and disarm growers.[122] Consular representatives had attended several CAWIU meetings, and the consular representative shot to death at Pixley had been there to investigate the situation
of Mexican nationals.[123] Yet Bravo, an anti-communist of the Mexican

upper class, attempted to bring Mexicans to arbitrate without the union. Growers, playing on Bravo's opposition to the CAWIU and hoping he could convince Mexicans to desert the CAWIU, told him they would welcome workers back at the old rate but refused to deal with the union.

Bravo had made little effort to support workers in relation to growers and, like other Mexican consuls, had helped repatriate Mexican nationals. Union organizers, Anglo or not, had been helping them on a day-to-day basis, while Bravo had remained in his Fresno office until the shootings. When Bravo attempted to convince 2,000 assembled workers at the Corcoran camp to settle separately from the union, they met his suggestions with boos. Bravo's attempts to form an all-Mexican union in Pixley to undermine the CAWIU were in vain. He reported to growers that the Mexicans refused to give in, adding that the killings at Pixley had in fact created a "stronger union."

Bravo was correct. Strikebreakers, shocked by the killings, had deserted the fields to join picket lines. Workers at the Tulare headquarters suspended a sign that read: "¡Murieron nuestros compañeros pero prometemos seguir hasta el [sic] morir o vencer!" ("Our compañeros died but we promise to continue until we win or we die!") Picketing was suspended, and ex-strikebreakers joined the 1,200 to 4,000 in Fresno who marched to honor those killed at Pixley and Arvin.[124] Local chapters of the CAWIU, Mexican organizations, Mexicans women's organizations such as La Cruz Azul, and local Mexican mutual aid societies, of which D'Avila had been a member, marched together.

National public opinion focused on the San Joaquin Valley, and the shock of the unprovoked killings began to hit local communities. The violence struck a nerve in those sectors of Valley communities which, while dependent on agriculture, were critical of the social problems it engendered. Ministers in Visalia denounced the violence, which they blamed on the rigid structure of the industry and the influence of the finance companies, and they urged people to contribute food to hungry strikers and their families.[125] Even the Corcoran Journal, a supporter of the growers, wrote editorials denouncing the killings. Public pressure began to build on local, state, federal, and international levels to investigate the killings and bring an end to the strike.

FEDERAL INTERVENTION

In the fall of 1933, the nature of federal intervention under the New Deal was still unclear. California agriculture opposed government control, yet for years they had benefited from government programs and policies. Their long-standing ambivalence toward government intervention was intensified during the New Deal: the issue was not government involvement per se but the control and direction of government intervention. In those fuzzy early days of the New Deal, the cotton strike became the unintentional testing ground for New Deal policy in agriculture.

Before the killings, Governor Rolph had been reluctant to intervene in the strike. The politically powerful agricultural interests and their related organizations resented interference in labor relations. In a struggle between the cotton industry and mostly unenfranchised Mexican workers, it was more politic to leave matters up to local authorities. Yet as the conflict escalated, the federal government had taken an increasing interest in the strike, seeing the unbridled lawlessness as a threat to the newly launched recovery program. Early in the strike, Rabbi Irving Reichart, newly appointed director of mediation and adjustment for the NRA in California, wrote Rolph that violence by vigilantes and law enforcement in the cotton fields were "retarding the recovery program in California" and that "President Roosevelt . . . [was] deeply concerned lest the confidence of the American worker in the New Deal be destroyed by an arbitrary use of power on the part of minor police officials."[126] Reichart suggested that peace officers be restrained from "inciting to violence," and he admonished the governor that state protection be extended "not only to propertied interests but to working men as well."[127]

In an atmosphere of growing tension, Reichart pressured Rolph to intervene, with a veiled threat of federal intervention:

> The high-handed and outrageous methods of the so-called vigilantes, instead of being firmly suppressed by the civil authorities, are aided and abetted by them. Gangsterism has been substituted for law and order in the cotton areas. . . . There is no question but that the present lawlessness in Kern, Tulare, Kings, and Madera counties constitutes a threat against constitutional government in California.[128]

In a statement released to the newspapers the day after the killings, Reichart said, "It is high time for the state to issue an ultimatum. The choice lies between immediate arbitration or martial law. No other course lies open."[129]

To appease Reichart, Rolph lamely admonished local sheriffs and authorities to uphold the law and disarm growers, and he increased the number of California Highway Patrol in the area. This anemic show of state power had little effect. Final decisions remained in the hands of local authorities. Local law enforcement officials, insisting that growers were legally armed because they had rifle permits, refused to disarm them. The California Highway Patrol, allying themselves with local officials, did nothing to stop the violence.

But the killings forced Rolph to intervene. He acquiesced to a meeting with a delegation of strikers who asked for relief, protection for strikers, and the prosecution of the growers responsible for the killings. Rolph then made what became one of the most controversial moves in the conflict. On the advice of federal relief officials, the governor overrode NRA regulations stipulating that strikers not involved in arbitration were ineligible for relief, announcing instead that the state would give food and other necessities to strikers "whenever possible."[130] This put him into direct confrontation with boards of supervisors in Kern and Tulare counties who had refused relief to strikers in the hope of starving them into returning to the fields. The state trucked in milk and food for distribution. Hungry strikers at Corcoran, aware that relief recipients could be deported, accepted food only after assurances that accepting it would not lead to their deportation.

The efforts of California agencies to mediate the strike had been blocked by agriculture's opposition to outside interference with their control over labor. The offer of the California state labor commissioner to mediate was rejected twice by the ALB, which flatly told the governor that they had no faith in the "impartiality" of either the commissioner or the commission.[131]

The shootings, by underlining the state's inability to mediate the strike, intensified federal efforts to settle the conflict. Immediately after the shootings, Reichert and the California NRA appealed to the National Labor Board (NLB) to ask George Creel, western administrator of the NRA and California representative for the board, to step in. The National Labor Board had formally taken over jurisdiction of labor disputes from the NRA, and Creel, a crusading midwestern newspaperman, had been a reporter in the Progressive era and had written exposés on child labor and other concerns. Creel supported Woodrow Wilson and was appointed chairman of the Committee on Public Information (CPI) in World War I, where he cranked out patriotic propaganda.[132] Like other Progressives, Creel favored liberal reforms and envisioned the state working not in the interests of a particular group or class but rather toward what he envisioned as the common good.[133]

In response to Reichart's appeal, Creel brought the power of the federal government to bear on both growers and strikers. Creel's actions reflected not only his personal vision but the uncertainties the federal government was feeling. He was in communication with Robert Wagner, chairman of the National Labor Board, asking advice on positions to take and what role to play in the strike.[134]

Federal de facto, if not de jure, policy on the cotton strike reflected the hopes and tensions of the early New Deal. To appease southern Democrats and the powerful farm bloc, agricultural labor had been excluded from the NRA and section 7a of the NIRA. Yet in many quarters there were expectations that agricultural labor would eventually be included in labor and social legislation. Reichart and Creel both believed that agricultural workers were included under the NRA or would be in the near future. If Creel's assertion that "the industrial disputes in agriculture are under the jurisdiction of the National Labor Board" was not technically true, the power of the NLB and NRA were put at Creel's disposal to establish a mechanism for collective bargaining in agriculture.[135] The two organizations had already established a tenuous precedent for federal involvement in agricultural labor disputes by helping settle the El Monte strike and the Tagus strike, among others. Thus, as the strike wave continued, federal intervention came to be seen as symptomatic of a growing involvement of labor relations in agriculture.[136]

The federal government manipulated growers' and workers' deepening dependence on New Deal programs as a lever to force both sides to negotiate. The Agricultural Adjustment Act, passed in 1933, was paying benefits to growers who agreed to remove part of their crop from production. As Reichart wrote to Lloyd Frick, president of the ALB, the NRA could legally intervene, because the federal government provided economic aid to agriculture. Reichart, warning growers that the government would not countenance vigilante methods, pointed out that their participation in government programs was "conditioned upon a fair deal to those who labor in the industry." He threatened that growers who refused arbitration would lose their benefits, the area would be placed under martial law, and the strikers would continue to receive relief. He warned growers that the government might "deny participation in this relief to certain growers because of their unfairness toward labor or their unwillingness to deal with labor representatives . . . [who are] duly elected spokesmen for their group."[137] As Creel wrote to Friselle, "The answer . . . [to agricultural labor strikes is] some form of collective bargaining between employers and the workers. . . . Collective bargaining is the very essence of the National Industrial Recovery Act."[138]

In areas covered by the NRA, Creel believed that the government could make growers agree to higher wages and cooperate in labor arbitration by stipulating cooperation as a requirement for receiving federal subsidies. The agricultural industry's response was mixed: hoping to be considered for federal subsidies, some areas cooperated with the NRA, while other sectors opposed federal intervention as a threat to growers' control over labor supply and wages.

The result of a federally mediated cotton strike in Arizona in 1933 helped convince growers to agree to federal mediation. Mediated by Edward H. Fitzgerald of the United States Conciliation Service, the settlement stipulated that strikers return to work at 60 cents, with the understanding that wage increases be tied to a rise in market prices.[139] The agreement assuaged growers' fears that federal intervention would work to their disfavor and, when Fitzgerald was appointed conciliator to the California cotton strike, growers finally agreed to a fact-finding commission under his direction.

While the federal government was willing to step in to stop blatant repression by growers, they also expected the union to forego the tactics that had forced employers to negotiate in the first place. The federal government opposed the left-wing CAWIU. Creel blamed growers in part for the CAWIU's success, because "where the men have been dealt with through representatives, and given some understanding of the problems of the producers, the agitators have failed entirely."[140] In a letter to H. C. Merritt Jr. of the Tagus ranch, Reichart assured the grower that the government recognized "the tremendous difficulties and serious handicaps which the cotton growers are facing. . . . Certainly we cannot countenance communistic and radical agitation that strikes at our democratic institutions."[141]

The union, already amenable to arbitration and concerned over Creel's threat to cut off relief, readily accepted federal intervention, despite Creel's refusal to recognize the CAWIU as the legitimate bargaining agent. Under Creel's direction, Rolph appointed a three-member fact-finding commission to investigate the strike and suggest a settlement. The commission was composed of labor historian Ira Cross, San Francisco Catholic archbishop Edward J. Hanna, and Tuly C. Knowles, president of the College of the Pacific. On 16 October an official notification appeared of the impending hearings, ordering strikers back to work and urging growers to reemploy strikers. The notification contained statements approving arbitration from George Creel; Governor Rolph; Enrique Bravo, the Mexican consul; federal conciliator Fitzgerald; F. C. MacDonald, the state labor commissioner; R. C. Branion of the Federal-State Emergency Relief Administration of California; and representatives of the San Joaquin Cotton Committee of the California Farm Bureau Federation.[142]

Although both sides had agreed to arbitration, neither was willing to back down. Growers prepared for open warfare. In Kern County alone, 600 new gun permits were issued.[143] Few strikers returned to work, and strikers, too, armed themselves. At the Corcoran camp, veterans of the Mexican army "looking for a scrap" got some 30-30s and ammunition. The Mexican consul counted over 400 rifles prominently displayed in the five strikers' camps he visited. Sam Darcy, after having asked Mexican and Oklahoman strikers to be prepared for a "business-like demonstration," was met by a caravan of picketers armed with shotguns, rifles, and other weapons—the strikers' understanding of preparation for a "business-like demonstration."[144] Their refusal to comply with the order to return to work forced Creel to informally bring in the CAWIU, hitherto excluded from the negotiations. Creel personally visited organizer Pat Chambers in jail and joined Caroline Decker for a tour of the Corcoran camp. The officials' de facto recognition of Decker and Chambers was a symbolic if relatively empty gesture.[145]

In part to avoid imposition of martial law and discontinuance of subsidies, growers turned to legal means to stop the strike. The state health director and the district attorney, supported by the Corcoran Chamber of Commerce and local groups, ordered the camp closed as a menace to public health. Creel supported this legal maneuver.[146] Two infants and a woman had died in the camp, but conditions were no worse than they had been for years in the growers' cotton camps. Camp conditions, ignored for years, now became brandished as an issue of public health as hysteria was whipped up in an effort to close the strikers' camp. After the strike, the district attorney admitted, "The camp wasn't so bad. It was just about as sanitary as other cotton camps. . . . What difference did the sanitation make to them [the officials]? They just had it in the back of their minds that if they could get the camp broken up the pickers would go back to work at 60 cents."[147]

Local officials threatened to repatriate Mexican strikers. The Tulare sheriff declared that the federal government would throw strikers into a bullpen and deport them.[148] Newspapers began to feature the activities of the immigration service; the *Hanford Journal*, for instance, ominously

reported that immigration inspectors were expected with "more war-
rants for Mexicans arrested in the Corcoran area."[149] The *Corcoran
Journal*, warning that growers would not hire the still-striking Mexicans
back at $1, said, "If the strike continues, it is more than likely that every
last one of you will be gathered into one huge bull pen. . . . Do you want
to face the bull pen? Do you want to be deported to Mexico?"[150] The
threats never materialized. Growers, afraid of losing their work force,
opposed mass deportations and even ignored an offer by a group of
Mexican strikers to repatriate voluntarily. Deportations were used se-
lectively; only after the strike were a handful of organizers, such as
Francisco Medina and John Díaz, deported.[151] With a relative labor
shortage due to repatriations in Los Angeles, a strike, and a crop that
would be ruined if it were not picked quickly, deportations posed a threat
to growers as well as workers.

Although the Communist party formally opposed the NRA, organizers Chambers and Decker felt the New Deal could improve conditions
in agricultural work and facilitate organizing, so they made the NRA
part of their strategy. They had invited an NRA representative to the
August San Jose CAWIU conference.[152] As a result, much of the union's
argument at the hearings focused on the social responsibilities of the
federal government. Saying that "America's hope lies in the National
Recovery Act," union representatives urged the government to cancel
AAA cotton contracts with growers who refused to improve conditions.
Recognizing that low cotton prices made it hard for some farmers to pay
a decent wage, they asked a federal guarantee of a minimum price for
cotton on the stipulation that growers pay not less than $1 per one
hundred pounds. In an effort formally to recognize agricultural labor's
inclusion in the 7a clause, the union called for union recognition and
federal, state, and county support of the right of workers to organize,
strike, and picket. They asked that Pat Chambers and the cotton strikers
be released from jail and that those accused of the Pixley murders be
convicted.[153]

The commission's report represented a victory of sorts for workers.
The commission, finding that strikers' civil rights had been violated,
appealed to authorities to prevent violence against strikers. Accepting the
recommendation of the Federal Intermediate Credit Bank, which held a
government mortgage on the cotton crop, that 75 cents was the most
farmers could afford to pay, the commission recommended that wages
be raised to that level.[154] Creel supported the commission's decision.
Growers, balking at the recommended wage, wired Creel that the terms

were unsatisfactory unless the government guaranteed that picketing would stop.

The findings did little to diffuse the tension. A melee at a ranch near Corcoran and pitched battles between strikers, strikebreakers, and sheriffs toting tear gas had led to renewed cries for deportation, swearing in of new deputies, and appeals to President Roosevelt to send troops. Local growers, seeing no point in wasting time with the federal government, marched on the Corcoran camp to "clean them out," but were stopped by a persuasive sheriff and a closer look at the armed Mexicans surrounding the camp. Picking around the Valley remained at a standstill, and strikers' camps were growing as new workers moved in.[155]

Creel renewed his threat to withhold crop loans and federal payments under the AAA to growers who refused to arbitrate. Under pressure from Creel and from the equally compelling realization that if the cotton was not picked soon it would rot, the ALB agreed to the higher wage. At first strikers rejected the offer, holding out for a minimum wage of 80 cents, union recognition, guarantees of no discrimination in rehiring, and the release of imprisoned strikers. Creel then withdrew relief. He argued that while "relief should be given during the strike, otherwise the Government would be helping to starve protesting workers into submission," after the hearings "it was entirely proper to withdraw relief."[156] At the same time, he mobilized the United States Farm Labor Service in Los Angeles to send 1,000 workers to the Corcoran area to pick crops.[157] The sheriffs never carried out their threat to evacuate the Corcoran camp forcibly, but the discontinuance of relief, the economic and psychological drain of the three-week strike, and threats to import workers led workers in the Corcoran camp to vote to accept the offer of 75 cents. The union called off the strike, and the strikers' camps began to break up.

AFTERMATH OF THE STRIKE

Although the union officially accepted the arbitrated settlement and urged strikers to return to work, an estimated 20 percent of the workers continued to insist that all the union's demands be met and refused to work for growers who had led vigilante groups. Some refused to pick for less than 80 cents and insisted that farmers hire only union members. In Kern County, workers stipulated that farmers sign with the union, and they pressured at least seventy-five cotton growers, mostly small farmers, to sign with the union.[158] Although most workers began to migrate back to southern California, those remaining attempted to keep the union alive, and union meetings were reported well into December. Although the numbers were small, ranging from a turnout of three hundred and fifty in Porterville to thirty in Tulare, their persistence indicates a desire to form a stable organization. Given that the CAWIU was itself small and virtually penniless, and that few workers formally joined the union, the number present at meetings probably represents a larger number still in sympathy with the union. A small farmer, perhaps reflecting this, said of the union in Corcoran, which drew 300 workers to a meeting in December, "All the pickers still belong to the union. They could call a strike in five minutes if they wanted to."[159]

But the story of agricultural organization has not always lain in the development of stable local unions. The temporary nature of agricultural work, migration, the lack of money to support a union, and the power of growers in the small towns undermined efforts to maintain local unions. The migration of workers back to southern California dissipated the potential base for union locals in the San Joaquin Valley. The lack of financial, institutional, and social support further undermined efforts to establish locals. In 1933, workers' efforts to improve wages and conditions rested on maintaining networks and organizations that could support strikes as workers moved to new crops. The momentum that peaked in the cotton strike was carried on to strikes well into 1934. The promise by Corcoran organizers that the strike would move with the workers was borne out in the November strike by 1,500 cotton pickers in Arizona and the January strike of 5,000 Mexicans in the Imperial Valley.

Hope for federal support remained an important impetus for these workers. Cotton strikers reportedly felt "bitter because they were led to understand that the government guaranteed their right to gain a union to bargain collectively . . . [and] the NRA has refused to recognize the CAWIU."[160] Yet by early 1934 federal investigators J. L. Leonard, William French, and Simon Lubin found that Mexicans in the Imperial Valley, home of many cotton workers, "had heard of the NRA [and] they believe that the government is going to protect them and improve their economic status."[161]

The federal government still seemed to offer hope that attempts to organize might result in permanent gains. Creel outlined the same demands later made of industrial workers, "that the right to organize and bargain could be maintained only so far as the state conceived it to serve an overriding goal of industrial peace."[162] Certainly this created a dilemma for unions. Federal support for collective bargaining rested on

their willingness to give up the very tactics that had forced the government to recognize them in the first place. Yet if we recognize that the
CAWIU had yet to develop a stable form or base—Chambers himself
called the notion of a permanent agricultural union at the time a "utopian idea"[163]—the importance attached to federal intervention becomes
clearer. Although the government had not formally recognized the right
of cotton workers to bargain collectively, Creel had de facto recognized
the union, the federal government had supported relief, and the government had intervened more actively in the cotton strike than they had
in any previous agricultural conflict. The real seduction of federal intervention was the promise it held for the future.

In his report on the cotton strike, Creel said, "There is now peace in
the cotton fields and it is my hope to work out some plan of collective
bargaining that will prevent further trouble."[164] By the beginning of
1934, it seemed more than possible that the federal government would
continue to intervene in agricultural conflicts, that it might eventually
recognize an agricultural union (although perhaps not the CAWIU), and
that farm labor might be included in national legislation as well.[165]

Only in hindsight is it clear that the federal government's participation
in the cotton strike marked the farthest extent to which the federal
government would go to alter the relation between farm workers and
growers. Until the end of the 1930s, workers and organizers continued
to expect that New Deal programs would improve conditions for cotton
workers, if not fundamentally alter power relations within the industry.
In relation to the federal government, the fate of cotton workers lay not
so much with union recognition per se as with the class relations already
present in government programs and with the structural changes that
resulted from federal programs and were to exacerbate and then institutionalize the fundamental imbalance of power between workers and
growers. Like other industrial workers, agricultural workers would be
caught in the rationalization of the New Deal framework that rested on
the particular nature of the capitalist state and the specific class relations
that determined their relation to the state. Ultimately, this would force
them into a position different from that of industrial workers.

CHAPTER FOUR

The Mixed Promise of the New Deal

The New Deal marked the beginning of direct federal intervention in California's cotton industry and would decisively change the industry and its relationship with its workers. As federal power expanded to deal with the deepening economic crisis it became crucial in defining class relations. Some New Deal programs, such as the Agricultural Adjustment Administration, shifted power within the cotton industry, affecting relations between growers, ginners, processors, and merchants and, indirectly, workers. Others, such as the Farm Security Administration and federal relief, directly affected the relation between the industry and workers. Government programs, policies, and agencies became the focus for increasingly sharp class conflict, fought out not only in the fields but in relief offices, housing camps, and legislative halls.

Before returning to California, let us look at changes that shaped New Deal legislation that was to affect cotton growers and workers. As the Depression reduced purchasing power, industry began to view a weak agricultural sector as a threat to overall industrial recovery and became more open to state intervention in agriculture. Agricultural interests became more receptive to federally imposed production controls. The agricultural depressions that had plagued farmers since the nineteenth century had also led to calls for federal intervention to regulate prices, but no unanimity had developed over the means to do so. Proposals ranged from protective tariffs to farmer cooperatives and production controls. The Great Depression revitalized demands for federal intervention and made farmers more receptive to production controls. By

1931 farm organizations had rallied behind production controls, and by 1932 the powerful National Farm Bureau was lobbying for federal programs to control overproduction.[1]

Agricultural interests had long played a central role in the nation's economy and politics, and during his campaign presidential candidate Franklin Delano Roosevelt conferred closely with farm interests and the American Farm Bureau Federation. By September 1932, FDR publicly supported production controls to stabilize agriculture and return farmers to parity. He campaigned on the provisions of what would become the Agricultural Adjustment Act and, by assuring major farm interests they would not only write but administer agricultural legislation, he secured their support both for the program and for his candidacy.

The farmers' vote was key to the 1932 election. Supported by the Farm Bureau and other powerful farm organizations, FDR captured not only the conservative Democratic states of the South but also the traditionally Republican farm states of the West. The farm bloc remained crucial not only to the success of FDR's farm bills but also to New Deal legislation and the Democratic majority in Congress and was influential in shaping social legislation. The administration's resultant reluctance to alienate farm interests impeded attempts to pass legislation favoring farm workers.[2]

Agriculture and agricultural workers were vital parts of the New Deal program. Yet workers in agriculture fared differently from industrial workers. Agricultural workers were relatively disorganized, fewer in number, poor, disenfranchised, mostly nonwhite, and concentrated in states where industrialized agriculture held both economic and political power. The agricultural industry they faced was well organized and unanimously opposed to farm workers being included in federal legislation. The agricultural lobby's political power and the importance of the farm bloc vote to the administration made New Dealers hesitant to alienate organized agriculture. As a result, while the cotton industry fared well under New Deal programs, receiving more direct government support than other sectors of the economy, cotton workers were excluded from social and labor legislation which might have supported their right to collective bargaining or given them access to social insurance. Recognized as welfare recipients, farm laborers were ignored as workers and effectively excluded from protective legislation and recognition of their right to collective bargaining. That exclusion and the institutionalization of cotton workers' position as low-paid workers placed them at a great disadvantage in relation to the industry and became integral to the expansion of California's cotton industry.

THE AGRICULTURAL ADJUSTMENT ACT
AND THE COTTON INDUSTRY

Cotton, a major United States commodity, was in desperate condition. By 1933 a carryover of 12.5 million bales of U.S. cotton was helping to depress the world market for cotton, and yet another forty million acres were planted for that season. To brake the crisis, Congress decided to plow under enough cotton to create a scarcity which would, in turn, cause the price of cotton to rise. The Agricultural Adjustment Act, written in part in the Washington office of the American Farm Bureau Federation, was passed by Congress and signed into law in May 1933. The act authorized payments to growers of cotton and other commodities in exchange for reducing production. Voluntary agreements would regulate marketing of crops, and processing taxes would finance government payments. The drop in the amount of cotton on the market would, they argued, raise and stabilize prices and, when coupled with additional subsidies, return farmers to parity, that is, to the purchasing power they had enjoyed from 1910 to 1914.

In California, the act met with a mixed reaction. Large cotton interests that supported production controls joined forces. Across the Valley cotton belt, county farm advisers, extension service agents, representatives of the local Farm Bureau and Chambers of Commerce, and officials of the national American Cotton Cooperative met with small groups of farmers in local halls to explain the act and convince them to plow under a portion of the 1933 crop and sign up for crop reductions for the upcoming year.[3]

Banks, gins, and financial institutions that lent money to farmers supported the Agricultural Adjustment Administration partly because the program stipulated that payments could be made directly to lenders to cover debts incurred by participating growers. As W. B. Camp remarked, the program "was just good banking. The risk was much less where they put it out with the man who was cooperating. Otherwise they didn't know whether they'd get their money back or not."[4] The government guarantee of loan repayment made cotton a more attractive investment and drew the financial community into the New Deal nexus. The staunch support from financial interests also helped secure the program among farmers. By 1934, financiers' refusal to extend crop loans to nonparticipating growers compelled reluctant farmers to sign up with the AAA.

Yet growers were suspicious of government intervention and federal limitations on their acreage. The program was also financially unattractive. The AAA paid growers on a sliding scale based on the estimated yield per acre if the land were cultivated. Yet because the act was written primarily with the southern states in mind, where the yield per acre was often no more than half that of California, the AAA paid only $20 for 275 pounds of cotton per acre. This was a paltry amount for California growers, whose yield often exceeded 500 pounds per acre and whose cotton demanded a higher price than southern cotton. Although the government hoped that California growers would withhold 58,000 acres from production in 1933, by July growers had agreed to cut back only 13,265 acres. As the *Pacific Rural Press* complained, the act was simply "not very attractive to high yield areas."[5]

The Farm Bureau, the state Chamber of Commerce, and W. B. Camp lobbied to have the payment raised for California. By December the government, realizing that the program would flounder at the initial rates, gave California growers a higher premium. The *Bakersfield Californian* crowed that Camp had "secured an additional $1,000,000 for California cotton growers."[6] While southeastern cotton growers continued to get $11 per acre for withholding land from production, San Joaquin growers were to receive $14 an acre. By 1934, growers garnered 3.5 cents per pound for the average yield they would have produced on the acreage in California in the base period from 1928 to 1932. They also received a parity payment at the end of the crop year of not less than one cent per pound. In return, they agreed to remove between 35 and 45 percent of their total acreage from cultivation.[7] These concessions by the government not only gave growers a financial edge but, perhaps more important, provided them with a sense that they might at least influence federal programs.

The New Deal had an immediate effect on the cotton market. By the 1933 harvest, prices for California cotton had increased 70 percent, partly due to the AAA and partly because textile manufacturers, anxious to process as much cotton as possible before the National Recovery Act went into effect, were voraciously buying up cotton.[8] By the end of 1933, cotton had made the largest recovery of any California crop, almost tripling its value to reach $11,124,000.[9]

As it did in parts of the southeastern United States, in California the AAA and federal supports helped increase the economic clout of larger investors while simultaneously undermining the already precarious position of smaller farmers, ginners, and processors. Rising prices and federal support stimulated investment despite the threatened reduction in acreage. Large growers expanded their acreage, and new investors scrambled to buy good cotton land. By June, the *Pacific Rural Press* estimated that cotton acreage would increase 45 percent because growers were "gambling heavily on the influence of inflation and Federal relief schemes." Federal guarantees for lenders enabled growers to borrow money from gin-owned finance companies, which were "throwing thousands into the deal . . . [through] ten to twenty-five dollars per acre advances to planters."[10] The Bank of America's annual report noted that for 1933 "cotton has been the outstanding money making deal of the year."[11]

As the La Follette Committee concluded, the Depression "resulted in a great increase in the concentration of ownership of farm land."[12] Overall cotton acreage declined as falling prices pushed small farmers and ginners out of business. Meanwhile, large investors seized the opportunity to buy land and gins at Depression prices. J. G. Boswell, for example, increased his holdings by 8,000 acres between 1928 and 1933.[13] Federal programs augmented the ability of such large investors to expand. Because AAA payments were based on the size of production, not on the need of the farmer, growers with more acreage received higher government payments. Federal loans, too, were based on the size of production. Gin companies and banks lent money proportionate to the size of landholdings. Higher profits, price supports, and available loans enabled large investors to increase their holdings and invest in capital improvements. Boswell, for example, enlarged his cotton mill and seed storage and, by 1937, owned and operated six gins.[14] Russell Giffen borrowed $25,000 from Anderson Clayton to buy an additional 6,100 acres for planting cotton.[15] Some invested in tractors and gin equipment, and tests began anew on mechanical cotton pickers. Such capital improvements expanded production per acre while still complying with AAA regulations. Overall, these programs helped stabilize the income of large growers and the ginners and merchants who depended on them and gave them some protection from market fluctuations of cotton production.

Government programs helped drive small tenants and farmers out of business both in the South and in California. Small farmers were forced to remove a greater proportion of their land from production and, because they had less acreage than larger growers, they also got lower AAA payments. Others were too small to qualify for the program at all. The increase in the value of cotton inflated land values and taxes. Yet farmers with little collateral and small acreage were ineligible for many loans, both federal and private, and could not afford new investments. They produced relatively less and often lower-quality cotton. Small farmers thus fell further into debt, mortgaged their farms, and then failed to make the payments. Small landowners became tenants. Tenants became farm workers.

The result was an increasing polarization between large and small farmers. Behind the dry numbers lay stories of individual success, hardship, fears, and failure. Overall, the number of California farms over 1,000 acres in all crops increased 4 percent. By 1939, 500 farms accounted for over 50 percent of the cotton acreage, and slightly over 3 percent of the cotton farmers controlled more than one-third of the acreage. Small farms shrank in size but mushroomed in number. Small farmers reduced their acreage, and the number of farms with less than ten acres increased 22 percent. By 1939 the 59 percent of cotton farmers who worked less than twenty-five acres accounted for a meager 18.6 percent of the total land devoted to cotton.[16]

The eightfold increase in AAA payments to California cotton growers from 1933 to 1938 widened the disparity in payments to large and to small growers.[17] By 1938, 43 percent of the cotton payments went to only 2.6 percent of the growers.[18] Whopping payments were made to large growers such as J. G. Boswell ($53,822), Camp West Lowe ($30,263), and Russell Giffen ($43,427).[19] By the end of 1933 it was apparent that the AAA was of dubious value to small farmers. One small cotton farmer wrote that "a good deal of farm legislation passed during the last few years has actively hurt the small farmer, because it was the big growers who pushed it through."[20]

The AAA program ultimately contributed to the disorganization of cotton workers in California by helping propel tenants and sharecroppers of the Southeast and Southwest off the land and into the growing stream of migrant workers heading for California. A destitute sharecropper from Wagner County, Oklahoma, who was camping along the highway near Bakersfield, said of the AAA:

> It throws thousands of fellows like me out of a crop. The ground is laying there, growing up in weeds. The landowner got the benefit and the first tenant says, 'I can't furnish [subsist during the season] any more.' So the share tenant . . . goes on FERA [relief]; he's out. . . . It looks to me like overproduction is better than not having it.[21]

The plight of such tenants was undeniably exacerbated by the Depression and dust storms, yet the AAA, and the mechanization that came in its wake, were the major causes of Anglo-American migration from depressed agricultural areas in the 1930s. By 1935, these displaced migrants would become the primary cotton labor force in California, and by 1937 they glutted the labor market, their numbers undermining efforts to organize and enabling growers to drive down wages.

Large investors were ambivalent about the AAA, favoring price supports but often objecting to limitations on production. Although they remained suspicious of government control, they became increasingly dependent on its programs. Despite sharp divisions and conflicts about how the AAA should be administered and by whom, the chaos of unregulated agricultural production helped win support for the AAA. And agricultural interests continued to support the AAA long after the lack of such unity among industrial interests had effectively invalidated the National Recovery Administration.

Their support was insured when these larger cotton interests became the agencies through which government programs were formulated and administered. Large farm organizations, led by the Farm Bureau, had played a crucial role in formulating AAA policy. Once in place, these organizations became the administrative arm of the policies they had helped create. Federal power was vested in organizations such as local farm bureaus and extension services to administer the AAA. Representatives to the AAA and local farm boards were chosen from the same small group of prominent cotton men who had organized the Agricultural Labor Bureau and who worked with the Chambers of Commerce and the farm bureaus. W. B. Camp, for example, who in 1933 was named regional administrator for the AAA in irrigated southwestern states, had arrived in California as a USDA agent in 1919, worked for the One Variety Act, became a cotton grower, and later acted as agent for the Bank of America.[22] The appointment of men such as Camp to government positions, whether in the federal or the local AAA, made them, in effect, agents of the federal government. As Rexford Tugwell complained, the AAA delivered power over the program to the hierarchy of the Extension Service, the Farm Bureau, and the land-grant colleges, and, indirectly, to the agricultural interests it was meant to regulate.[23]

One of the most infamous organizations to emerge from this period was the Associated Farmers. Founded in October 1933 by large industrial and agricultural interests in the state, it bankrolled and supported anti-union forces, fought the CAWIU, engaged in blacklists and vigilante actions, supported political propositions to limit picketing and speech by unionists, and threw its considerable economic and political clout behind efforts to influence the passage and administration of state and local policies favorable to large agriculturalists.

Ironically, however, the original Associated Farmers grew out of an intra-class battle over the nature of the New Deal and the impact of the AAA on agricultural class relations, and it had had a markedly different agenda from the later, vigilante group. Overlooked in historical accounts, this original Associated Farmers sheds light on the considerable conflict and bitterness among cotton growers over the AAA programs and their administration. In August 1933, three hundred small cotton farmers in Kern County, hoping that more sympathetic sectors of the federal government would correct the imbalanced administration of the AAA and undermine the growing power of larger interests, formed an organization called the Associated Farmers. Similar to trade associations that emerged in the wake of the NRA, the Associated Farmers in its first incarnation described itself as part of the "plain farmers movement."[24] They took on the agricultural elite, castigating the "pseudo-farmer, who merely farms the farmers; . . . who is the friend of traders and financiers, and . . . assumes a leadership to which he is not entitled," and argued that small farmers should "no longer be ground beneath the heel of commodity gamblers, [and] unscrupulous industrialists."[25] Encouraged by New Deal rhetoric, they appealed to the National Recovery Administration to counteract the advancement by the AAA of the power of finance companies, seed monopolies, and large farm interests. They asked for regulation of the industry and proposed a code of fair business practices to be regulated under the NRA. They attacked direct payment of AAA money to lenders and urged that "cotton brokers, cotton processors, oil mill, warehouse and press operators" be eliminated as the ubiquitous "toll-takers between us and government funds."[26]

Run by eager volunteers, the association spread through Kern County, but its life was brief. By September the organization was reportedly trying to protect itself from "any suspicion of external or clique control." By October it had dropped out of sight.[27] It is unclear what happened. Had smaller farmers given up the organization because of the polarization caused by the cotton strike? Had they been pressured economically? In any case, the Associated Farmers disappeared as an organization of small farmers. From this point on, larger interests remained virtually unopposed by smaller farmers in their dealings with the federal government.[28]

The metamorphosed Associated Farmers appeared the following month in the halls of the Los Angeles Chamber of Commerce. This new Associated Farmers, controlled by the very forces the small farmers had organized to combat, emerged from the expanding power of larger interests and their growing desire to fight labor unions and strikes. This organization was formed in response to the 1933 cotton strike. At a meeting in the Los Angeles Chamber of Commerce called by H. E. Woodworth—a cotton ginner, director of the California Cotton Growers Association, and chairman of the Citizens and Growers Committee of the San Joaquin Valley, which had led vigilante efforts against strikers in 1933—cotton and industrial interests joined representatives of Chambers of Commerce to formulate a plan to eliminate the "evils brought to light by the recent cotton strike, and stop unionization."[29]

The new Associated Farmers had a different agenda. The months-long series of strikes made them anxious to stop their spread, eliminate the CAWIU, and change government relief policies. F. A. Stewart of the Anderson Clayton Company attacked federal relief for extending the strike and complained that the state labor commissioner had, by delaying the strike's settlement, caused "great loss to the cotton industry."[30]

The committee passed a resolution containing its programmatic priorities. They called for increased government intervention in several areas: they urged authorities to identify, apprehend, and "eliminate" communist organizers, deport Mexican organizers and strikers, and prosecute labor organizers under the Criminal Syndicalist Act. They asked that relief agencies be "directed" to give relief only to what they called the "involuntarily idle." In a slap at both George Creel and the state labor commissioner, the committee urged that jurisdiction over labor disputes be removed from the National Recovery Administration and turned over to the more amenable Agricultural Adjustment Administration.[31] The meeting touched on two areas cotton interests were to focus on in subsequent years: the elimination of communist labor organizers and suppression of labor organization; and the dismantling or restructuring of federal relief programs. In March 1934, they formally organized the group as the Associated Farmers.

Traveling around the state, the officers of the new organization met with local Chambers of Commerce and businessmen to secure support. These agricultural interests envisioned the new Associated Farmers as the enforcement arm of the Agricultural Labor Bureau. The same men who had formulated the bureau directed the Associated Farmers. The planning meeting was called by H. E. Woodworth. F. A. Stewart of Anderson Clayton and the large Kern County cotton grower L. W. Frick represented cotton interests. S. Parker Friselle, founder of the ALB, became president of the Associated Farmers.[32]

The Associated Farmers claimed to be a group of citizens and farmers. In fact, the membership consisted of industries, railroads, utilities, banks, and major agricultural interests, and the organization was headquartered in Los Angeles and San Francisco.[33] The Associated Farmers was one of several statewide anti-labor organizations formed in the 1930s. Part of a national backlash by employers, California business mounted one of the best-organized attacks on labor in the country. The interlocking interests of industry and agriculture fought unionization through organizations such as the powerful Merchants and Manufacturers Association, the Industrial Association of San Francisco, and the Associated Farmers. Although distinct, these organizations shared financial contributors, cooperated with one another and worked closely through the ubiquitous state and local Chambers of Commerce.

As W. B. Camp said, the Associated Farmers had been "set up for a very firm purpose, to fight communism in agriculture."[34] Parker Friselle reflected similar sentiments in a letter to Francisco Palomares:

> It is my thought that this group can become the nucleus of a state-wide organization and can develop policies for immediately handling whatever situation may develop. . . . Possibly it will be necessary to further organize on a county basis in each county in which strike difficulties are imminent, in order that the citizens may come to understanding in advance many of the experiences of other countries including the type of Communist leaders dealt with, vicious methods which they have used . . . and probably the most important political methods which have been successful in other countries in dealing with acute problems.[35]

Shortly after its formation, the Associated Farmers launched its fight against farm labor organizing. It campaigned to make picketing illegal. As labor conflicts escalated in California, the organization participated in the arrest, trial, and conviction of labor organizers on charges of criminal syndicalism. On 20 July 1934, police raided the CAWIU Sacramento office and arrested major union organizers, among them Pat Chambers and Caroline Decker. Fifteen were indicted. The arrests were related to the San Francisco general strike of 1934 and may have been prompted in part by Chambers's suggestion that the CAWIU stop agricultural products from entering the city. The Associated Farmers contributed to the prosecution by hiring Capt. William Hynes, infamous head of Los Angeles's so-called Red squad, as chief investigator for the prosecution. After a lengthy trial which removed the CAWIU from the fields, eight union organizers were convicted, among them Chambers and Decker. The coup de grace was given to the CAWIU by the Communist party. In 1934 the party line had shifted to the popular front and in March 1935 the Trade Union Unity League and its offspring, the CAWIU, abandoned dual unions, among them the CAWIU, to work within existing institutions.[36]

Until 1937 the Associated Farmers received more financing from industry than from agriculture. It had never been a mass movement. When the cotton strike of 1933 ended, local committees faded. Farmers, concerned with direct problems of production, paid little attention to anti-union groups until strikes directly threatened them. Even large cotton interests ignored appeals for contributions, and the Associated Farmers received little direct support from cotton areas.

Only in 1937, when the ALB was again unable to enforce wage rates and a union was organizing workers in the fields, gins, and cotton compress plants, did ginners actively support the Associated Farmers. By 1937, the major investors in cotton were a small number of financial and processing interests, and 75 percent of new capital investment came from "speculative finance capital, banks and gin finance companies."[37] More important, cotton financing and processing were monopolized by four companies. California Cotton Oil Corporation, Camp West Lowe Ginning Company, J. G. Boswell, and the San Joaquin Cotton Oil Company, the California subsidiary of Anderson Clayton, financed 15 percent of the state's crop and ginned 68.7 percent of its cotton.[38] Anderson Clayton alone owned 35 gins and processed one-third of the state's cotton.[39] When in 1937 the large gin companies decided to support the Associated Farmers, their monopolization over processing enabled them to set a de facto tax of .03 cents on each bale ginned: half of the proceeds went to the ALB and half to the Associated Farmers. Cotton gins, which contributed only $109 to the association in 1936, raised their support to $2,234 in 1937. The cotton counties threw in $1,408. By 1938, cotton contributions had increased to $4,743. The largest donation, $2,866, was from the San Joaquin Cotton Oil Company. By 1937 the cotton industry had become the largest supporter of the Associated Farmers.

Cotton growers also played an important role in the association's 1937 reorganization. It was here that AAA payments provided an indirect, and unintentional, subsidy to the Associated Farmers. The large growers, who employed the majority of workers, received the bulk of AAA payments. In other words, slightly less than half of 1 percent of the Associated Farmers in the cotton counties received over 13 percent of the AAA payments.[40]

Through the 1930s, the Associated Farmers, along with the Chambers of Commerce, the Agricultural Labor Bureau, branches of the Farm Bureau, and other industry interests, led the fight to win greater control over government programs that affected agriculture, and they took a leading role in the fight against federal relief policies in the state. In effect, AAA payments indirectly subsidized the organization that would oppose government programs designed to alleviate the problems of farm workers.

COTTON WORKERS AND NATIONAL LEGISLATION

Congressional liberals such as Robert Wagner supported collective bargaining for industrial workers as being essential to national economic planning. Unduly exploited labor, they argued, would not produce as efficiently as workers included in what was envisioned as a harmonious cooperation among labor, capital, and the federal government. But as Christopher Tomlins points out, this right to organize and bargain collectively was conditional, to be granted only so long as "the state saw it as serving the overriding goal of industrial peace" and the "public good."[41] Industrial labor was able to become part of the New Deal only after considerable pressure and social upheaval and against the reluctance of much of the administration. Section 7a was inserted as an afterthought in the National Industrial Recovery Act and then only because AFL president William Green made it clear that it was necessary in order to obtain labor's political support for the bill.[42] Even so, neither government nor employers believed that section 7a committed them to engage in collective bargaining. Only under pressure of unremitting strikes in 1934 were industrial workers and their unions included under the protective legislation of the National Labor Relations Act of 1935. By the late 1930s labor's concept that the government would support union organizing and collective bargaining had been whittled down to a right only as narrowly defined by the government and subject to governmental regulation and administration.[43]

The ideology and class dynamics that shaped industrial workers' inclusion in New Deal programs led to the virtual exclusion of agricultural workers. The administration carefully worded the president's Reemployment Agreement to exempt farm labor. Although it remained confusing as to whether farm workers were or were not included in section 7a, the administration consistently held that the protection did not apply to farm workers. Instead it argued that these workers would benefit from the Agricultural Adjustment Act as agricultural prices improved and that the income would somehow effectively trickle down to farming communities and farm workers. This analysis was strikingly inappropriate to the reality of industrialized agriculture in California.

Although agricultural strikes of 1933, such as the cotton strike, pressured federal and state governments to intervene and brought the question of including agricultural workers in protective legislation to national attention, the strikes did not last long enough or pose enough of a threat to counterbalance the power of agricultural interests. Yet government agencies, reflecting differing interests, were initially divided over de facto policy. Conflicting aims, programs, agencies, and personnel racked the federal administration, prompting Raymond Moley's whimsical observation that "to look upon these policies as the result of a unified plan was to believe that the accumulation of stuffed snakes, baseball pictures, school flags, old tennis shoes, carpenter's tools, geometry books, and chemistry sets in a boy's bedroom could have been put there by an interior decorator."[44]

Despite the exclusion of agricultural workers from coverage under section 7a, in 1933 it appeared that they would be included in de facto arrangements. Some growers assumed that workers would be covered under the NRA; Kern County grower DiGiorgio, for instance, in July 1933 endorsed the NRA and, anticipating its implementation, increased wages for all workers, even migrants.[45] The California Farm Bureau, inundated by inquiries, felt compelled to announce publicly that "farm labor is not to come under the national readjustment program."[46] When Governor Rolph refused to allow the state recovery board to mediate agricultural strikes, George Creel and NRA officials simply included agricultural workers under the NRA, despite the legislation, and arbitrated strikes under federal jurisdiction. By the end of the October 1933 cotton strike there was a de facto assumption that agricultural workers would be included in national labor legislation. In 1934 Senator Robert Wagner included agricultural workers in the initial drafts of the National Labor Relations Act.

Yet organized opposition was growing against farm workers' inclusion in legislation. Agricultural interests opposed federal intrusion in labor, and government agencies, reflecting these differing interests, were sharply divided over policy. Although the National Labor Relations Board (NLRB) was willing to mediate strikes, other sections of the administration were not. In the 1934 Imperial Valley lettuce strike—in which many cotton workers took part—the Department of Justice refused to intervene even after growers attacked a union meeting and abducted an American Civil Liberties Union investigator.

Wagner, hoping to use the escalating violence and flagrant violation of civil rights to demonstrate the need for agricultural unions and force Congress to include agricultural workers in the NLRA, launched an investigation into the conditions of agricultural workers and the suppression of labor organization in the Imperial Valley. The report, by investigators Simon J. Lubin, J. L. Leonard, and William French, urged the federal government to encourage union organization. But, as Cletus Daniel points out, the report recommended that the "federal government [become] the organizational instrument of agricultural labor in the valley."[47] It ignored the communist-led CAWIU, which had directed the strike.

While the investigators may have intended to undercut the CAWIU, it was more significant that even their limited recommendations were never translated into law. The union was not strong enough to pressure the administration to support labor and alienate the farm bloc. More important than liberal antipathy toward communist unions was the antipathy of the New Deal to *any* unionism in agriculture. In a slap at the NLRB, the federal government ignored the report's recommendations, removed the Imperial Valley strike from the purview of the NLRB, and handed it over to the Department of Labor, which was already antagonistic to agricultural unions. The Department of Labor sent in General Pelham Glassford, infamous for his participation in removing the 1932 Bonus Marchers, as the "labor conciliator." Glassford was openly hostile to the CAWIU. When vigilantes kidnapped Glassford, stuffed him into a bag, and dumped him over a county line they were making it clear—if the point still needed to be made—that any unionism at all was antithetical to agricultural interests and would be met with harsh opposition.

Within Congress, the farm bloc, industrial interests, and consumers opposed agricultural unionization as antithetical to the public good. The higher agricultural prices that resulted from government programs were designed to restore farmers to parity, not to give workers a living wage. As Merle Weiner has pointed out, the interests of business and consumers in holding down the costs of food and basic commodities meant that both had an interest in excluding agricultural workers from social legislation.[48] If agricultural workers were to be covered under federal legislation, both social insurance and labor organization would help to increase their wages. Those increases would, in turn, increase the price of food and other commodities, which would probably lead to demands for higher wages by nonagricultural workers to cover rising prices. Thus, it was argued, low wages in agriculture benefited other industries as well. As a result, when it was suggested that provisions to protect agricultural workers be included under the Agricultural Adjustment Act, both industrial and agricultural interests balked, warning that higher agricultural wages would result in higher consumer prices, which would only aggravate social unrest. As a 1934 AAA report stated, "The mandate of consumer protection contained in the act is second only to that of assistance to producers." Worker protection appeared not at all. Agricultural interests also argued that the organization of agricultural labor would ultimately threaten the food supply of the nation. There was thus, they claimed, no "justification for the enactment of legislation which would permit any labor organization, under the sanction of the government, to dictate terms and conditions under which the nation would be deprived of food."[49]

By the time the bill reached the floor in 1935, the farm bloc and industrial interests were able to prevent farm workers' inclusion in any subsequent drafts. The NLRA specifically excluded agricultural laborers. New York congressman Vito Marcantonio's proposed amendment that farm workers be included was rejected.[50] Efforts to include agricultural labor in federal legislation continued until the outbreak of the war, at each juncture being defeated out of deference to the farm bloc, which threatened to impede legislation if farm workers were included.

By this omission, the government institutionalized farm workers' separation from industrial workers and reinforced their economic and political powerlessness in relation to the agricultural industry. Their excision from the NLRA rationalized their exclusion from social security, unemployment, and other legislation and set the precedent that farm workers would be considered, if at all, within only the most limited parameters. After 1935, pressure to include farm workers focused on expanding the definition of industrial labor to include cannery and packing-shed workers. Field workers remained outside the concern of the government.

COTTON WORKERS AND RELIEF

The New Deal recognized cotton workers only indirectly, as relief recipients. In April 1933 Congress passed the Federal Emergency Relief Administration (FERA) to provide immediate relief to the hungry and destitute lining up on street corners, living in Hoovervilles, or taking to the roads in search of work or simply an undefined betterment of their situation. The program was considered a temporary solution. Harry Hopkins, director of the program, reflected a national sentiment that a person could not accept relief "without affecting his character in some way unfavorably."[51] The government defined eligibility and insisted on the humiliating procedure of documenting the recipient's absolute destitution. Because of the hostility to direct handouts, national policy favored work relief as a way of providing assistance without the stigma of paying money for idleness. In 1933 Congress passed the Civil Works Administration—later revived as the Works Project Administration—to provide jobs.

Migrant agricultural workers were technically excluded from state and local relief, on the accurate if callous grounds that relief to agricultural workers constituted an indirect subsidy to agriculture, and only those who had been residents for over a year were eligible for most FERA payments and other assistance. Yet under the FERA's transient relief program, migrants qualified for federal money regardless of occupation or length of residence; thousands of cotton workers thus qualified for federal funds funneled through state and county offices of the State Emergency Relief Administration (SERA).[52]

Relief became a central area of conflict between workers and the cotton industry. How relief could affect cotton was first felt in the strike of October 1933, when federal relief enabled strikers to hold out. In 1934, the relief issue became prominent in California state politics when Upton Sinclair in his campaign for governor proposed the End Poverty in California (EPIC) plan, which put forward a production-for-use program that would have put the unemployed to work as part of a radical reordering of the economy. Growers joined the rising chorus that Sinclair's election would bring an influx of indigents into the state seeking relief.

Relief substantially changed farm labor relations by 1934. Relief offered little money—unemployed residents with a family of four could receive, on the average, only $40 a month.[53] Yet relief created a de facto minimum wage which challenged the ALB rate and reintroduced a form of collective bargaining. Disgruntled cotton workers making 25 to 30 cents an hour quit their jobs when they could receive 10 cents more per hour on Fresno relief.[54] No longer being forced to rely—at least not completely—on substandard wages improved their bargaining position.

The residency requirements helped to stabilize workers in local areas, to slow migration, and to alter migrancy patterns. By 1935, when growers, expanding their acreage, sought more workers, the availability of relief had helped shrink the pool of skilled Mexican workers. Mexicans who were residents and eligible for relief in southern California were loath to travel north to the San Joaquin Valley, where they complained of low wages and often fraudulent promises made by growers.

Growers blamed relief for the shortage of workers. George Clements of the Los Angeles Chamber of Commerce complained that "with hundreds of thousands of people on the relief rolls . . . California has experienced the most disastrous labor famine in her history."[55] Local relief officials lamented that they were unable to remove clients from the rolls to make them pick cotton.[56] Despite the hyperbole, workers were indeed harder to get, and cotton wages rose to between 90 cents and $1 during 1934–1938.[57] Hardly a substantial wage, it was nonetheless higher than it had been in almost a decade, in large part due to the availability of relief.

The Agricultural Labor Bureau, the state Chamber of Commerce, the Associated Farmers, and large cotton interests joined to eliminate or at least control relief programs. The Associated Farmers attacked federal relief policies, which, they claimed, encouraged unionization. The California farm lobby pressured state agencies to remove workers from relief rolls so that they would pick crops. Frank McLaughlin, director of the SERA, willingly complied and in April signed an agreement with the Reemployment Service of California and the California State Employment Service to supply agricultural workers from the relief rolls.[58] The agreement, by redefining eligibility for relief, narrowed the number of recipients. In May 1935, relief officials removed over 2,396 people from relief rolls, and control over relief policies was turned over to local agencies, which were usually sympathetic to local growers.

Yet some workers avoided being expunged by simply registering as nonagricultural workers. George Clements, while praising McLaughlin for "doing all he can . . . to break down the resistance of the SERA slacker," reproached him that pressure was applied only "to those registered . . . as agricultural labor."[59] Palomares and Clements wanted to "bring every pressure possible upon the administration . . . in the relief of SERA labor for agricultural purposes." This was underscored when, as the 1935 harvest approached, sporadic labor organizing in Kern County reactivated memories of the 1933 strike.[60] In a confidential memo, Palomares confessed that he expected things to be "hot" for the cotton harvest.[61]

Concerned that they still had not solved the problem with relief, agricultural interests met again to develop a long-term strategy. In anticipation of the September 1935 cotton harvest, the Agricultural Labor Bureau, the state Chamber of Commerce, the Associated Farmers, and agricultural interests formed the Agricultural Labor Committee under the auspices of the Los Angeles Chamber of Commerce. They aimed to force workers off Los Angeles relief rolls, extend the attack on relief in the state, marshal public opinion against relief, gain control over its local administration, and, with luck, change federal policy to comply with their interests. The committee launched a publicity campaign. Newspaper editorials warned that the cotton crop could be ruined by a relief-induced labor shortage. The Chamber of Commerce noted that the problem was not the sympathetic SERA, which had a "real understanding" of growers' needs, but the federal government and some recalcitrant local relief officials, who, they charged, were coddling workers. The committee mobilized to remove or at least change the policy of "antagonistic" local relief officials, and they criticized federal policies and the "ignorance and indifference" of federal administrators.[62]

Their attack on federal relief policies coincided with rising national dissatisfaction with direct federal relief. Businessmen opposed work relief, in part because it competed with private industry. Southern planters complained that relief was making it harder for them to find labor. As one farmer wrote to the Georgia governor, "I wouldn't plow nobody's mule from sunrise to sunset for 50 cents a day when I could get $1.30 for pretending to work on a DITCH."[63] In 1934 the work program, the Civil Works Administration (CWA), was demobilized and transferred to FERA, where it continued in a severely diminished form. While the CWA had paid regular wages to all unemployed, the FERA gave jobs only to those on relief and then paid them substandard wages.[64] In 1935, the Roosevelt administration abruptly withdrew direct federal relief for transients and reinstated one-year-residency requirements. On the theory that handouts were demoralizing, direct relief was replaced by a national works project under the newly formed Works Project Administration (WPA). Eligible workers were put on WPA and paid wages which, while higher than direct relief payments, were considerably less than the prevailing rates. In 1935 the federal government turned direct relief over to the California State Relief Administration (SRA), where it became increasingly subject to California political pressures. A union organizer railed that with this move relief policies became "a club in the hands of local administration, who are usually controlled by local agrarian interests."[65]

This shift in federal relief policy, nicknamed Harry Hopkins's "work or starve" edict, had a profound effect in California's cotton fields. By late 1935 the thousands of Anglo migrants pouring into the state, refugees from economic disaster, had no other choice but to accept work at the wages offered or wait until they qualified for relief. As the next chapter will discuss, Mexican migrants who left the Valley after each harvest were replaced by new migrants who put down roots in the hard soil of the San Joaquin Valley. Shack towns, shanty jungles, and auto camps filled to overflowing became a permanent part of the rural landscape of the 1930s.

Only days after Roosevelt approved the $20 million Central Valley Water Project to provide water for crops, thousands were cut from Valley relief rolls. Relief, under SRA control, was used to supply labor to agriculture. McLaughlin assured Palomares that "clients who are able to work in agriculture are removed from our relief rolls."[66] The SRA cut off workers "regardless of the probable lower wage scale [they would receive] on the farm than if they were on relief."[67] With this development, relief no longer acted as a minimum wage for workers but became used as an arm of the ALB.

But even with these cuts, local relief rolls could not supply enough workers. Migrants were needed, but many experienced cotton pickers remained in Los Angeles, where those on WPA were paid an average of $80 to $90 a month for part-time work. The Los Angeles Relief Administration (LARA) had originally agreed to cut WPA workers so that they would pick cotton, but when workers complained of insufficient work and poor conditions and the LARA found itself responsible for returning workers from the Valley to Los Angeles, they balked. When Palomares requested another 500 workers to pick cotton, Harold Pomeroy, director of the LARA, simply refused.[68]

Frustrated, the Agricultural Labor Committee of the Los Angeles Chamber of Commerce met with the national coordinator of the Relief Administration, Colonel Sears, to pressure local SRA offices. The next day, the LARA agreed to release workers from relief rolls.[69] To insure that cotton pickers would be removed, Sears appointed a labor coordinator to oversee the operation. Palomares sent in four men to cull through relief lists and personally remove cotton workers. George Clements of the Los Angeles Chamber of Commerce crowed, "I believe that we have blasted this labor off the relief rolls," and he hinted that uncooperative LARA agents might lose their jobs.[70] The WPA still provided jobs in the state, but McLaughlin was as willing to push workers off the WPA as he had been to push them off direct relief. In March 1936 he promised Palomares that "if needed we will close down WPA projects in agricultural communities to provide farmworkers." Although McLaughlin urged growers to pay workers an amount equivalent to the still substandard WPA salary, he refused to make it a stipulation of employment.[71] The effect of this half-hearted suggestion was pointed out by the *Western Worker*'s comment that "in no single week did a picker make the equivalent of the WPA payments."[72]

To the extent that cotton interests could remove workers from relief rolls at will, the relief program began to act as a mechanism to perpetuate and institutionalize the agricultural labor pool. Residency requirements for relief helped stabilize the work force by keeping workers in the area. SRA administrators removed workers from relief rolls at harvest, creating an unemployed and impoverished work force usually willing to work at prevailing wages. The federal government's failure to provide relief to migrants made the influx of out-of-state workers critical to the labor pool. Denied relief altogether, they formed a separate caste which, out of desperation, accepted wages other workers could afford to refuse.

Yet federal relief programs spurred workers to organize to protect their position as relief clients and contributed to spontaneous strikes to raise wages.[73] In cotton areas, unemployed resident workers formed organizations to defend their status as relief clients. By 1934, CAWIU organizers and other communists formed unemployed councils; by August of 1935 they had modestly expanded to include small meetings of SERA clubs in Shafter, Bakersfield, and other areas.

In 1936, the removal of workers from WPA rolls depressed the cotton chopping rate and set off a small flurry of organizing. In response, Wilson D. Hammett formed the Employees' Security Alliance with approximately 190 agricultural workers. They protested to the SERA about the low rate, demanded that they be represented in discussions to establish wage standards, and pointed to the uneven distribution of federal monies: "The same ranchers who want to hold us to a starvation wage have received from the government, in addition to the prices for their produce, $596,175 in Fresno County alone."[74] The Workers Alliance, successor to the Communist party's unemployed councils, was established to organize relief recipients. The alliance aimed to raise relief rates and lobbied for favorable relief policies. It became, in essence, a union for WPA workers. The alliance made inroads in the San Joaquin Valley, under the leadership of Alex Noral, a former socialist and a charter member of the Communist party who had been active in the 1933 cotton strike and had led a small unemployed organization in the state.[75]

The Workers Alliance succeeded in developing a broad-based rural alliance of small farmers, workers, and their sympathizers in Valley towns. Fresno clergymen protested the low chopping rate, arguing that agricultural work should pay enough to enable a man to support his family on his own wages. "The clergy is of the opinion," they added, "that business here should have progressed far enough in ideals that it would not want to make profits on child labor and that of women in the cotton fields."[76] Small farmers, workers, and local poor joined or supported the alliance. The Weed Patch Grange, an organization of small farmers in a community just south of Bakersfield that was inhabited by white small farmers, oil workers, and cotton workers, passed a resolution supporting WPA workers:

> Whereas, only through raising the level of farm commodity prices and the wage of workers will we overcome the depression, and
> Whereas, the present effort to force unfortunate WPA workers to accept wages even lower than the subsistence allotments set by relief,
> Therefore, be it resolved that the Weed Patch Grange repudiate any self-appointed committee [the Agricultural Labor Bureau] who only represent speculative interest in labor and soil.
> Those starvation wages set by said gentlemen who do no toil will foster self hatred and crime. We protest the use of Kern county public funds and public officials to be used by landowners to intimidate by threat of starvation jobless citizens to work for wages insufficient to provide a decent living.[77]

By 1936 the alliance had become a significant organization in agriculture. Statewide membership jumped from 7,000 in 1937 to 42,000 in 1939, although only 12,000 actually paid dues.[78]

FEDERAL LABOR CAMPS

The other federal program that affected cotton workers was the federally sponsored migrant housing project. The idea of migratory camps for agricultural workers came, not from New Deal progressives in Washington, but from a California liberal, Harry Drobisch, state director of Rural Rehabilitation, a division of SERA. Drobisch, recognizing that the perennial problem of substandard housing was a major cause of strikes, appealed to the California SRA for funding a pilot project. Turned down by the state office, Drobish appealed to the national FERA for a direct loan to develop model camps. Assisted by California senator Culbert Olson, Drobisch received money to build two model camps in the spring of 1935, one at Marysville and the other in the cotton community of Arvin, just south of Bakersfield. Drobisch envisioned the camps as the beginning of a string of migrant camps that could house all of California's agricultural workers and would help to supply workers for agriculture.

In April 1935, Roosevelt formed the Resettlement Administration (RA), headed by Rexford Tugwell, and jurisdiction for the migrant housing project was transferred to the RA. The RA had been formed to alleviate rural poverty by resettling and retraining displaced farmers. It had been developed for the South and for the Dust Bowl states of Oklahoma, Texas, and Arkansas, where it generally worked to reha-bilitate displaced farmers through direct loans and experiments in co-operative farming. In California, where the problem of dispossessed tenants attempting to farm worn-out lands seemed irrelevant to a highly developed agricultural industry worked by wage labor, the RA's focus shifted from rural rehabilitation and cooperatives to housing for mi-gratory agricultural workers.[79]

After the initial loan, the idea of migratory labor camps was ignored by the RA. Washington administrators focused on returning displaced farmers to the land, not on meeting the needs of migratory workers, even if the migrants were in fact displaced farmers. This was partially due to the myopia the government developed whenever it attempted to reform agriculture. Still imbued with the image of the yeoman farmer as the basis for a democratic society, the nation had yet to comprehend the extent of change underway in agriculture. Reformers in the RA who favored collectivized agriculture for displaced farmers ignored the problems of migratory workers, products of increasing capitalization in agriculture and an uncomfortable anomaly to the RA vision of independent farming. The RA budget for October allotted no money to the camps at all.

This federal opposition was partially overcome in October 1935, when Tugwell visited the camps. Touched by what he saw, Tugwell approved the expansion of the project to shelter all of California's 150,000 to 200,000 migrant workers. Yet the tension between those who wanted to resettle farmers and those who favored housing migratory workers persisted. Jonathan Garst, the new RA administrator for region IX, which covered California, Nevada, Utah, and Arizona, was appointed in 1936. Garst opposed the camps. He favored resettling migrants on homesteads and turning already constructed camps over to local authorities. By late 1936, the RA was coming under increasing attacks by conservative legislators who scorned the utopian goals of resettlement. Just as a rising tide of southeastern migrants coming into the state was expanding the demands for camps, Garst reluctantly dropped resettlement plans and focused on building a limited number of migratory camps.[80]

Of the camps built, the Arvin, Shafter, and Firebaugh camps were in cotton areas. The camps offered clean but sparse accommodations, an eagerly sought improvement over camps on private land, in shanty towns, the makeshift camps on ditch banks, or the auto parks that dotted the highways throughout the Valley. The tents originally erected were soon replaced by neat, spartan metal shelters with sanitary facilities and running water. Teams of doctors and nurses provided health care, lectured on birth control, sanitation, and child care, and tended to migrants who suffered from an interminable list of diseases including tuberculosis, rickets, and whooping cough. Moreover, the Farm Security Administration (FSA; the Resettlement Administration was rechristened as the FSA in 1937) conceived of the camps as instilling a sense of democracy in the workers. Under the direction of camp directors, the camps became experiments in cooperative self-government. Camp members established local councils, published newspapers, set up chicken coops, laid out vegetable gardens, held dances and socials, and met regularly to discuss common problems.

This publicized experiment in democracy was in some ways illusory, for the camp director had final say on camp matters. Yet more important than the freedom to elect camp leaders and discuss camp activities was the practical freedom the camp offered. Because the camp was on federal land, it was the only haven free from control by local officials and growers. Workers living in private camps felt coerced to work for low wages and could be evicted at any time.[81] Sheriffs could rout them from ditch-bank camps and shanty towns. On federal camps, they were protected from eviction. All visitors, whether growers or local sheriffs, needed permission even to enter the camp. Workers could more easily move, change jobs, or organize. It was this freedom, not democratic promises, that prompted one Arvin camp resident to say, "Now I can go any place I want to work, and if he don't pay right I can tell him nothing doing and go someplace else. In this government camp *I am a free citizen.*"[82]

Federally run housing was not necessarily incompatible with growers' desire for a resident but malleable work force. Since the nineteenth century, growers had attempted to establish a work force, "in some ways tied to the land," that would work for low wages.[83] Growers probably would have welcomed the camps if they had been under their own control. The California Department of Industrial Relations had recommended this in a 1934 report which argued that public housing

> will go far toward stabilizing agricultural labor and making it immune to the lurid appeals of the professional agitator. Its advantages will mean increased efficiency through satisfied workers, thus minimizing the economic losses that is [sic] yearly visited upon the industry through strikes, and it will be good insurance against disease, filth, poverty and petty crime.[84]

Indeed, critics of the program attacked the camps as a subsidy to California's agricultural interests. Senator Richard Russell, for example, argued that "a comprehensive [camp] program carried on over a long period of years . . . is going to be of as much a benefit to the big corporate farms by furnishing them with cheap labor as it is going to be to the people whom we are trying to help."[85]

Yet while several local growers used the camp as a source of workers, the camps' convenience in that respect was overshadowed by growers' concern that the camps would become a basis for organizing. At a conference on housing, Lee Stone, a pugnacious county health official from Madera, rhetorically questioned Drobisch, "Is it not a fact that . . . camp sites for migrants . . . are more or less going to be harboring places for promotion of subversive thought?"[86] The camps prompted the industry to dredge up their own plans for workers' camps under agricultural control. George Clements dusted off his 1929 Clearing House Plan intended to make good citizens out of farm workers who "would be to a great degree immune to radicalism," and Roy Pike, of the El Solyo ranch, presented as an alternative a series of grower-owned and -controlled camps.[87]

For all the furor surrounding them, federal camps remained largely symbolic. Due in part to congressional opposition to the Resettlement Administration as a whole, the camps received even less than the original $1.5 million set aside for the RA region in California.[88] The plan, which had envisioned housing 200,000 migratory workers, dwindled to a few model camps. And those camps housed mostly white workers. The RA, fearing racial conflicts, confined blacks to one camp in Indio. Mexican workers were generally unwelcome and stayed away. While camp managers such as Fred Ross tried to combat racism, the camps were hardly visions of interracial harmony. For all but a few cotton workers they remained an attractive but elusive alternative. Yet by remaining under the federal auspices of the RA rather than California agencies, the housing program evaded grower control until the beginning of World War

II. The camps were administered under the most progressive of New Deal agencies, and liberal camp managers allowed unions to use the camps to organize, despite government regulations to the contrary.

Agriculture's desperate economic state in the 1930s exerted greater pressure on cotton interests to support the AAA than industrial conditions placed on their industrial counterparts to unite behind the NRA. This was only in part because, as Kenneth Finegold and Theda Skocpol suggest, the organizational mechanisms that linked government and agriculture were already in place.[89] The broader questions are, why did these mechanisms exist and why did they expand in the 1930s and subsequent decades? The basis for the development and continuation of the agricultural-governmental relationship was the precarious nature of agricultural production itself. The continuous instability in cotton production, punctuated by boom-and-bust cycles, led the agricultural industry, however ambivalently, to forge increasingly strong links with the federal government in order to help stabilize production. Growers' vulnerabilities led them to accept governmental intervention in a number of areas, from the application of technology to the acquisition of workers and the regulation of growing, processing, and marketing.

As a result, agriculture continued to support the AAA while industrial support for the NRA fragmented and collapsed. This development was crucial for labor. While industrial sectors differed on programs such as social security and other legislation, agriculture was united in its determination to exclude agricultural workers from social and labor legislation. The AAA in turn helped to reinforce the organization and standardization within the industry.

By 1936 it appeared that relief would establish a minimum wage in agriculture and that federal expansion of camps would help force an improvement in working and living conditions and act as a base for organizing. But these programs were enforced at the discretion of the federal government or were administered by California and Valley agencies—all of whom were sensitive to political pressure. Growers and workers increasingly organized in response to the vicissitudes of government programs and their administration on the federal, state, and local levels. But these were shaped by economic and political forces over which workers had little power. Their relative powerlessness outside the government reduced their bargaining power within the government, and, ultimately, even the programs meant to assist them became tools of the agricultural industry.

Figure 1: Mexican cotton pickers, 1930s. Courtesy Fresno Historical Society Archives.

Figure 2: José Mendez Bañales, who was part Purépecha, exemplifies the proletarianization of many Mexican cotton workers. Originally from San Francisco Angumacutiro, Michoacán, Bañales had worked as a sharecropper in Mexico, a laborer on the Kansas railroads, and in a Los Angeles nursery before beginning to pick cotton. Photo taken in Kansas City in the 1910s or 1920s. Courtesy of Edward Bañales and Virginia Bañales Godina.

Figure 3: Roberto Castro, originally from Michoacán, worked the railroads in Louisiana, Oklahoma, and Texas before joining his father in Corcoran. He worked as a contractor on the J. G. Boswell ranch, but left the ranch to help organize the 1933 cotton strike, serving on the strike committee. Photo circa 1931 or 1932, probably en route to one of the camps around Corcoran. Courtesy Anna Castro.

Figure 4: The Corcoran strikers' camp, October 1933, where 3,000–4,000 strikers lived. Note tents and lean-tos, commonly used for shelter on ranches, and the gate erected by strikers. Makeshift streets were laid out by labor groups and named for Mexican towns and heroes. Courtesy Edward Bañales.

Figure 5: A Mexican family in their shelter at the Corcoran strikers' camp during the 1933 cotton strike. Conditions in the camp were similar to conditions on many ranches. Courtesy of the Bancroft Library.

Figure 6: The CAWIU office in the Corcoran strikers' camp during the 1933 cotton strike. Note the "100%" endorsement of the NRA, tacked above the office sign. Courtesy of the Bancroft Library.

Figure 7: Picket caravan during the 1933 cotton strike. Courtesy of the Bancroft Library.

Figure 8: Mexican strikers on the picket line, blowing bugles and shouting to get the attention of strikebreakers during the 1933 cotton strike. Courtesy of the Bancroft Library.

Figure 9: Mexican women being transported to the fields to urge strikebreakers to join the strike. Note the variation in age and in styles of dress. Courtesy of the Bancroft Library.

Figure 10: An unnamed migrant from Oklahoma who became a strike leader in the 1938 Kern County cotton strike. On his lapel are a CIO button and a button urging a no vote on Proposition 1, the anti-picketing ordinance. Courtesy of the Bancroft Library.

CHAPTER FIVE

New Migrants in the Fields

In the mid-1930s there was a dramatic change in the work force. Anglo migrants fleeing economic disaster in the Southeast and Southwest filed into California, replacing the Mexican work force which had dominated cotton since the 1910s. The Anglos were the latest in a long line of refugees from depressed cotton-growing areas, pushed out by the expansion of capitalist agriculture and the disasters that had hounded cotton farmers since the nineteenth century. They were white, and they were American citizens. Many had been small farmers, tenants, or sharecroppers. Arriving in the midst of the Depression, they had been uprooted from their communities. Contemporaries viewed them as ripe for unionization because they were white Americans. This chapter explores how their background, depressed conditions, and the federal government affected their adjustment to cotton work in California. Questions raised earlier are relevant here: How did the changing economic, social, and political parameters these new migrants encountered shape their world? To what extent did their background affect their adjustment to California cotton fields? How did the work force's changing composition affect its relation to unions, growers, and the federal and state government?

The migrant Mexican work force had begun to dwindle in 1933, and the work force declined further in 1934. This decline reversed abruptly in 1935, when over 57,000 migrants from the South and Southwest crossed the California border in search of manual employment. Their

numbers swelled to 105,185 in 1937—the peak of this migration—
before subsiding to 80,200 in 1940. Of the 400,000 who entered between 1935 and 1939, 40 percent went to the San Joaquin Valley. By
1936, Anglo migrants composed an estimated 90 percent of the work
force and supplanted the Mexican workers who had dominated the
cotton work force for almost two decades. The migrant stream of Mexican workers shrank to a trickle and then virtually disappeared in the
wave of Anglo migration.[1]

They came from the southern-plains cotton belt of Oklahoma,
Texas, Arkansas, and southern Missouri. Oklahoma alone contributed
70,857.[2] While Anglos migrated from both urban and rural areas, most
of the 190,000 who went to the San Joaquin Valley came from rural
backgrounds and had strong ties to the land. Although few had owned
the land they farmed, they had been tenants, sharecroppers, or farm
workers. Even those who had worked in their home states as construction workers, miners, or lumberjacks or on the oil rigs had been born and
raised on farms.[3] As Fred Ross, manager of the FSA camp in Arvin, said,
most of the camp inhabitants had been sharecroppers and farm workers,
and "they were all cotton . . . cotton . . . cotton and corn."[4] The new
migrants, soon to be called Okies in the scathing popular vernacular, fled
decaying prospects of cotton in other states, lured by the possibility of
work or farming in cotton in California.

This migration to California was the latest move in a decades-old
migration out of depressed cotton areas where the transition from subsistence farming to capitalist agriculture, the boom-and-bust cycles of
cotton production, and periodic natural disasters had pushed them off
the land. For over half a century, cotton farmers had been plagued by
cyclical downturns in cotton production. The quick profits to be made
in boom years encouraged the overcapitalization of land and excessive
crop loans. Hopeful of profits, farmers had expanded their production.
Yet expansion created surpluses, which depressed prices. Sharecroppers,
tenants, and small farmers were unable to repay debts. Some, facing
foreclosure, went further into debt. Others lost their land and became
farm workers. A drop in cotton prices reverberated throughout towns
that depended on the fortunes of cotton. Farmers bought less. Merchants, bankers, and other small businessmen went out of business.
People left in droves, and whole towns withered. By 1910 the regularity
of these cotton depressions had created a steady stream of migrations
from cotton areas. Some migrated north into industrial areas. Others

moved west into the plains and the southwestern states of Texas and Oklahoma to try their hand at cotton-growing again.

By the 1930s, cotton refugees who had gone to Oklahoma, Texas, and the southwestern cotton belt faced new renditions of the old dilemmas that had plagued them in the South. They planted on small farms and were harassed by droughts, falling prices, boll weevils, and continual boom-and-bust cycles. Despite an occasional good year, from the 1880s on the economic horizons of farmers steadily deteriorated. The agricultural depressions of the 1920s and 1930s told the old story of overcapitalization, bumper crops, large surplus, falling prices, and the overuse of tired land pushing workers off the land. In Oklahoma alone tenantry increased from 54.8 percent to 61.2 percent between 1910 and 1935. The Depression, soil erosion, and the droughts of 1934 and 1936 intensified the problem.[5] Even those who did not farm were affected by the general economic decline in these agricultural areas.[6]

The New Deal's attempts to rectify the situation caused by capitalist agriculture exacerbated the problems of small farmers in the southwestern cotton belt. The AAA paid more to large growers, while disproportionately cutting the acreage of small farmers. This reduced the ability of smaller farmers to compete. The AAA was partially responsible for the grim Oklahoma statistic that between 1935 and 1940, 33,274 farmers lost their farms.[7] One Oklahoma migrant family who registered at the FSA camp in Marysville, California, was described as "Wife, husband, six children . . . had a 200 acre ranch in Oklahoma. Raised cotton. . . . Were doing well and had money in the bank. Cotton acreage was cut to 30 by AAA; forced them into bankruptcy."[8]

The Agricultural Adjustment Administration inadvertently helped push tenants and sharecroppers off the land, facilitating the spread of an agricultural pattern that increasingly resembled that of California. The program paid benefits directly to landlords, not to the tenants who grew the cotton. Landowners found it more profitable to pocket the government payments, evict their tenants, and invest in tractors and machinery that reduced the need for the manual labor the tenants had provided. The use of tractors increased in the areas from which the migrants came: Oklahoma farmers had bought 60,000 tractors by 1940, six times more than they owned a decade earlier. On 141 Texas farms, growers increased their use of tractors from 25.5 percent in 1931 to 78.6 percent in 1937.[9] The *Tulsa Tribune* noted that "hundreds of thousands of tenants and sharecroppers . . . were forced off the land by

New Deal payments to big cotton plantation owners."[10] As one Texas landlord told Paul Taylor:

> I let 'em all go. In '34 I had I reckon four renters and I didn't make anything. I bought tractors on the money the government give me and got shet [rid] of my renters. You'll find it everywhere all over the country that way. I did everything the government said—except keep my renters. The renters have been having it this way ever since the government come in. They've got their choice—California or WPA.[11]

Moreover, for those who combined farming with work in the oil fields, New Deal programs undercut this work as an extra source of income. The New Deal restricted oil production, thus reducing the number of jobs in the Oklahoma oil fields. One migrant to California, who had been a farm worker and oil worker in Oklahoma, told Paul Taylor that "since the oil quota, we had no work."[12]

Government relief was a last resort. As one of the migrants put it:

> Seems to me the Government wanted people on relief or it wouldn't have passed the Agricultural Adjustment Act. . . . The farmers took the parity payments and kicked the sharecropper off, and where could they go but on relief? That's the only place they had to go. I think the AAA caused more people to go on relief than the dust or the floods.[13]

But relief offered little. By 1938, Oklahoma families were making do with $26.40 on the WPA.[14] Relief also ate away at pride in a nation that had doggedly insisted that individuals were responsible for any economic calamities that befell them. A woman cotton worker who testified at a California state hearing on cotton wages blamed the government:

> I think that the relief has ruined about the majority of working men. I said relief when maybe I should of said Government for that's where it all started when the cotton acreage was cut. That took a lot of the small farmer[s] off the farm, didn't it? They had no place to go and nothing to make a living with. They got down to where they had to sign themselves as paupers and beggars to get something to eat. When a man gets down low enough to beg like that he becomes discouraged with life and doesn't have any pride.[15]

The other option for these tenants and sharecroppers was migration. The people who traveled to California in the mid-1930s were not a settled rural peasantry. Since the nineteenth century they had been migrating in response to the ebbs and flows of cotton production; moving was a familiar, if not a welcome, way of life. In Oklahoma, tenants often moved to a different farm every year. Although most migrations occurred within relatively local circumferences inside the state, cross-state migration was not uncommon. For many families the westward trek to California was the latest in a series of moves to keep a shaky toehold in cotton farming. One family in the Arvin FSA camp had made six moves in thirteen years. In 1921 the family had moved from Arkansas to Muskogee, Oklahoma, to raise cotton and potatoes as tenant farmers. They moved again in 1922. But they lived on the margins. When sickness hit the father and children, they were too poor to hire labor and lost most of their crop. They moved in attempts to recover from the disaster, returning to Arkansas in 1926, trying again in Texas in 1928, and going back once again to Arkansas in 1932. In 1934 they gave up farming, pulled up stakes, and moved to California, where they had heard they could pick cotton.[16]

For migrants who read advertisements, saw movies, or heard glowing reports of California from friends or family, California seemed to offer an alternative to the dust storms, desolate farms, low pay, and unemployment in their home states.[17] Growers, Chambers of Commerce, and newspapers encouraged southwestern migrants to come to California. One migrant described by the FSA had come to California because he had "read such glowing accounts in the newspapers about the opportunities for a man with farming experience in California where the various Chambers of Commerce were advising agricultural laborers of the dust bowl areas that California was in need of thousands of cotton pickers. Any man that could pick cotton could make four dollars a day."[18]

Cotton-picking wages were higher in California than in other states. Wages in rural Oklahoma could dip as low as 10 cents an hour, but skilled pickers in California could make $2 to $3 a day in season. Even an average picker could make $1.50 to $2.00.[19] Perhaps even more appealing, relief benefits were higher. An unemployed California resident with a family of four could receive monthly relief of $40 to $60, whereas in rural areas of Oklahoma, Arkansas, and Texas payment was as low as $5 per family, and it seldom went higher than $15. In 1938, WPA payments in one area of rural Oklahoma were only $26.40 a month.[20]

ADJUSTMENT AND RESPONSE
TO CALIFORNIA COTTON

The lure of higher wages and relief payments turned to dross in the stunning reality of California's cotton belt. The migrants' lack of experience with other crops made it hard for them to make a living when the cotton season ended. Francisco Palomares of the Agricultural Labor

Bureau sniffed, "These people know cotton, but they cannot make a year round living in cotton alone."[21] Wages were lower than the workers had anticipated. Part of the discrepancy between promises of wages and the actual wages received resulted from variation in the estimations of how much workers could pick in a day. Growers consistently argued that workers could pick 300 to 400 pounds a day, and, based on this, estimated that workers could earn $3.00 to $4.00 a day. This, of course, assumed the exceptionally high rate of $1 per hundred pounds. While there were instances where experienced workers could pick this much in an abundant field, it was not usual. Experienced workers averaged 200 to 250 pounds a day, earning $2 to $2.50 a day.

Moreover, although many of the new migrants had picked cotton back home, a different kind of picking was demanded in California. Newcomers tended to pick not only the cotton boll but the twigs and debris around it. Supervisors on California cotton farms insisted that workers pick the cotton clean. In 1936, the *Corcoran Journal* complained of a "shortage of skilled harvest hands."[22] Experienced Mexican workers remarked on the trouble newcomers encountered in learning the California picking technique. Jessie de la Cruz worked beside new migrants who were experienced in picking Texas cotton, but "they weren't used to this kind of picking. . . . It had to be clean, no leaves, you had to leave nothing but the stalk." They picked forty-five pounds to her hundred and spent precious time going back to repick cotton lint they had left behind.[23] New migrants were initially unused to the cotton weights and measures and were unfamiliar with the devices contractors used to cheat workers. Eventually they learned how to pick the cotton, the tricks to evade contractors, and the weights—much as Mexicans had had to learn these skills earlier. But the time it took to learn the California system contributed to their initial disorientation.[24]

The higher cost of living soon ate up what difference there was between wages and relief in their home states and in California. One worker who had struggled to learn cotton picking lamented, "I finally got so that I could pick about 200 pounds, but you can't make any money out of that at a cent a pound."[25] Out of this, workers had to pay about $1.50 for a picking sack, furnish their own bedding, buy their groceries, and usually pay for gas money to get to and from the fields. Beginners could not support a family by picking cotton.[26] In 1931 a picker from the South reported that in three weeks he had made a total of $2.68.[27]

The new migrants were at a distinct disadvantage in relation to workers more familiar with the area. Finding jobs required knowledge of crops, seasons, ranches, growers, contractors, and routes. New to California, they had yet to develop stable migration patterns and learn the intricacies of farm work, the tricks and pitfalls of the contracting system, and the patterns, abilities, and eccentricities of individual contractors.

In some respects, working relations on California ranches were similar to those the new migrants had experienced in their home states. The instability of tenant life was not dissimilar to that of migratory workers. Both groups were poor, lacked enough clothing, often suffered from diseases, and generally attained only a sixth-grade education. Tenant farmers' ties to the land had been uncertain, and they had often worked for the landlord in exchange for wages.[28] In Oklahoma, for example, tenant contracts had to be renewed each year and, as a result, renters often moved annually. Despite the similarities to wage labor, however, their actual work as tenants resembled that of farmers. Tenants worked small plots of land, usually with only their own labor and that of their families. They were in charge of all steps of cotton growing and controlled the pace of their work.

In California instead of small plots, many worked on large ranches in labor crews, chopping and picking mile-long rows of seemingly endless cotton. Tenants who had tended the crop through the entire growing process were now brought in to work only for chopping and picking and then rudely let go. One boy said, "When they need us, they call us migrants. When we've picked the crop, we're bums and we got to get out."[29] Expendability replaced a sense of commitment to the crop, and impersonality replaced the personal relations with the landlord they had had as tenants.

The migrants gravitated to those areas of the San Joaquin Valley where the work and communities seemed most familiar or where they had friends or relatives. They settled in the cotton-based communities where earlier migrants had recreated aspects of southern and southwestern culture, class, and racial structures. Bakersfield in Kern County was the first stopping place after crossing the Tehachapi mountains. In Kern County, the largest cotton-growing area and an oil center, the migrants found a familiar milieu of small farmers and workers. They settled in and made their homes in migrant FSA settlements, in auto courts, and in ditch-bank camps. Squatter settlements, often called "little Oklahoma" or "little Arkansas" by the townspeople, sprouted up in Valley towns. The Kern County towns of Oildale, Arvin, Weed Patch, Buttonwillow, Wasco, and Earlimart were soon filled with new migrants and became identified as "Okie" towns. By 1944 two-thirds of Arvin's population was from Oklahoma, Texas, Arkansas, or Missouri: only 4 percent had been born in the state.[30]

The migrants' impact on local social structures is suggested by the growth of population in California counties. Certainly their impact was felt outside the Valley: Los Angeles registered a population gain of 26.1 percent. But the figures mushroomed most in Valley counties. Kern County's population jumped by 63.6 percent, from 93,185 in 1934 to 141,000 in 1938. The population of Kings County increased from 25,385 in 1930 to 31,600 in 1938. The sharpest increase was in Madera, where the population rose 400 percent from 1935 to 1937.[31]

Although they began to build social networks, newcomers lacked the intricate relationships Mexicans had built up over the years. James Gregory has pointed out that the Anglo migration was not the "atomistic dispersion of solitary families" traditionally depicted and points out that perhaps half of the migrants migrated to a community where members of their family already lived. The difference was in the extent, depth, and solidity of these networks. The decades-long pattern of local migration had loosened community ties and encouraged southwestern migrants to emphasize family more than community. Southwesterners may have migrated as families, but they did not (with some notable exceptions) migrate as parts of communities.[32] Established families did inspire the migration of family members back home, but it took time. Those new to the work could not expect to have the kinds of networks Mexicans had spent the last decade building in a better economic climate.

It is hard to disentangle the economic circumstances of the Depression from the effectiveness of social networks. The instability and insecurity of the economy chipped away at the human ties of security in families and communities. The Depression had reduced individuals' resources, diminished the job pool, and weakened community ties. In 1937 the President's Committee on Farm Tenancy concluded that the "instability and insecurity of farm families leach the binding elements of rural community life."[33]

The base of the migrants' problems was economic. Declining wages and an increasing number of workers hungry for work made jobs harder to get, work more unstable, wages lower, and migration patterns more uncertain. Migrants who had been in the state less than a year did not meet the residency requirements to receive relief. Once they did qualify for relief, they were forced to take it in order to survive. Tom Collins, manager of the Arvin FSA camp, estimated that two-thirds of the migrants in the Valley were forced to depend on relief for, on the average, an eighth of their annual income.[34]

The uncertainty of the migrants' lives was reflected in their poor health—an international marker of poverty. People living on water biscuits and water gravy, a mainstay of the diet on the Hammond ranch near Firebaugh, had little strength to fight off disease.[35] As one father said of his children, "Reckon they're not eatin' too much. Maybe it's my fault, as I've not been working. I still have $2 and we have a few more potatoes. I haven't asked for help; haven't been here a year and it's no use."[36] Okies faced the same health problems as Mexicans: rickets, tuberculosis, typhoid, smallpox, pneumonia, malaria, and parasites.[37] But poor health mirrored larger problems. One survey showed that new migrants suffered from more diseases than either long-term Anglo residents or Mexican workers, migrant or not, an indication of this new group's abject impoverishment.[38]

How the new migrants adapted, resisted, and responded depended on economic conditions and the responses they had learned over decades. Older Anglo residents in the Valley had built networks and patterns of working that differed from those of the newer migrants. The Hammett family, for example, had combined sharecropping with cotton fieldwork in California since 1925. The Depression shoved Oklahoma-based members of the Hammett extended family off their farms and shares, but it expanded W. D. Hammett's work crew, and by 1933 he was contracting a forty-five-member crew composed of friends and family, both locals and new migrants. In a decade of life in California he had gained a knowledge of the fields and jobs. The crew found work on big ranches such as the Tagus, the Boswell, and the Giffen. By 1936 they had established a pattern of work that provided a modicum of security: they picked fruit and cotton for six to eight months and lived on relief in Fresno the remaining four.[39] But new migrants, recently arrived and desperate for work, had yet to develop stable migration patterns. They relied on rumors, and more than once moved only to find they had missed the harvest, or had arrived before the crop matured, or faced hundreds like themselves also looking for work.

Such changes in migrancy were also a result of the growing pool of available workers. In the late 1930s growers no longer worried about finding enough workers for harvest: they could hire them off the road or from a nearby ditch camp. By 1937 the grower's perennial complaint of "labor shortage" had disappeared. As one small farmer said, "They came looking for it. Before we had to go looking for them."[40] As more migrants arrived in the Valley, work became increasingly uncertain, and workers were forced to migrate more. James Gregory quoted a migrant who noted, "We was out here nine years ago; then we could get a steady job. Now it seems we can't stay in one place. *We got to follow these little jobs to live.*"[41] Definite migratory patterns gave way to chaotic movements that started in rural areas, moved in many directions, and converged on areas irrespective of the number of workers required there. Random mobility exacerbated the problems of those who were part of this large labor pool, made it harder to make ends meet, increased insecurity and instability, and created a sense of personal and familial anxiety.[42]

Some responded to the problems of California by going back home. In 1936 Tom Collins reported that groups of cotton pickers were returning to Texas and Oklahoma. Edgar Crane's parents and brother returned to Texas because his brother did not like working for somebody else: "He liked to work for himself. You could do it in Texas, but you couldn't do it here."[43] The cost of living evaporated dreams of a better life in California. Some returned for good. Others went only to wait it out until the next cotton season.[44]

Many contemporary observers believed that white Americans streaming into the cotton fields would be more likely than Mexicans to join unions. Even after the predominantly Mexican strikes of 1933 and 1934, growers were at first convinced that these new migrants posed an even greater threat than their Mexican predecessors. Racism, underlined by the still firm belief that these migrants were the modern agricultural backbone to American democracy, clouded the understanding of conditions. George Clements warned a colleague in late 1936 that "white citizens are not tractable labor" and would demand "the so-called American standards of living." Clements was convinced that the new farm workers would be "the finest pabulum for unionization."[45]

Growers were not alone in this point of view. Many union supporters shared Clements's assumptions. John Steinbeck wrote a moving tract entitled "Their Blood Is Strong" that put forth the assumption that "With this new race [sic] the old methods of repression, of starvation wages, of jailing, beating and intimidation are not going to work: these are American people." Steinbeck argued that white Americans imbued with democracy and coming from communities where "democracy was not only possible but inevitable" would change the makeup of the agricultural labor system forever.[46] Yet these new migrants were, on the whole, less likely to support strikes than the Mexican work force had been. An understanding of the response by white workers points to the need again to differentiate between the new migrants and longer-term residents in the area.

Growers did complain about Anglo workers. They grumbled that the new migrants were less reliable than Mexicans and "make . . . poor fluid laborers" because of individualism and resistance to routine.[47] Edgar Crane's brother returned to Texas because "he liked to work for himself." One grower complained that Anglos did not take orders well, had to be handled carefully, and demanded respect: "I can't tell my oldest worker that he must do anything. I can *ask* him to do it and I can always get him to do it, but I cannot *tell* him or he would walk off the place."[48] One farmer preferred Mexicans because "Okies are not reliable. . . . They just act as individuals. . . . The fact that they are not organized is about the only good thing about the Okies."[49]

Resentment and exasperation exploded in individual acts of rage. In 1937 twenty-five-year-old Carl Dillard, working on the Riley ranch near Corcoran, "became enraged over an assignment of rows of cotton to be picked and struck [the supervisor] on the head" with what a newspaper surmised was the blunt end of a knife.[50] At the Slaybaugh ranch, thirty-year-old Giffen Arnold from Arkansas, after being fired by Charlie Slaybaugh, leaped into the owner's truck and furiously attacked him, almost putting out Slaybaugh's eye.[51] A Kings County undersheriff noted in 1938 that while "no labor trouble developed . . . disputes and quarrels among individuals and groups in cotton camps resulted in numerous arrests."[52]

Despair, either caused or exacerbated by the situation, was registered in newspaper accounts which indicate isolation, lack of the support of human networks, and a sense of hopelessness. Mrs. Anna Jones, from Cement, Oklahoma, twenty-eight years old, abandoned her nine-month-old daughter in a Los Angeles bus station because she and her husband could not care for the baby.[53] Twenty-four-year-old Mary Snodgrass, wife of a cotton picker on strike at the Camp West Lowe ranch, who was despondent over domestic problems, attempted suicide by drinking two glasses of kerosene.[54]

The uncertainty, tenuousness, and instability of the new migrants' lives undermined their inclination or ability to bargain with growers. By 1936, migrants new to the state, disoriented and unable to qualify for relief, were willing to take whatever wages were offered. Growers soon realized that the southwesterners were less resistant than Mexicans to wage cuts. While some growers still preferred Mexicans because, as one woman remembered, "They used to say we could produce more work than the Anglos," many preferred to hire Okies.[55] Large cotton growers, such as the owners of the Hotchkiss ranch, advertised for workers in southwestern newspapers, further fueling the migration to the western cotton fields.

The aspirations of the Anglos differed from those of Mexican workers. Many Anglo tenants and farm workers aspired to be property owners. Frustrated aspirations and immediate grievances may have led them to join strikes and unions, but the expectations that would determine their participation were different from those of Mexican workers who had no desire to have land in the Valley and who were more thoroughly proletarianized. It was true that the new migrants were, and had been, wage earners. They were clearly becoming a part of the industrial labor force. Yet they clung to the dream of being farmers and, as one organizer remarked, to "the ideology of farmers."[56] Tom Collins reported that forty families in the Arvin FSA camp were "most anxious" to get a small farm or even sharecrop.[57] The *Fresno Bee* reported that workers in Kern County in 1937 were not "strictly speaking migratory workers at all but farmers . . . who have come to California with a determination to reestablish themselves."[58] Work in Depression California eroded their illusions. Fred Ross sympathetically noted, "They were always talking about 40 acres and a mule and all that . . . [but] they weren't about to get a farm and they knew it."[59] Yet the comment by a migrant at a union meeting that "you could empty this hall in a minute if some one would say, Come with me. I'll show you how to get a piece of land" suggests the lingering persistence of this hope. The *Madera Tribune* editorialized that growers could stave off union organization by simply giving workers small farm plots.[60]

Migrants' prejudices fed California's already virulent racism. Although comments about "white trash" reverberated in heated moments in the Valley, the new migrants were white. Their attitude toward the Mexicans was probably related to attitudes toward blacks held in the South; their race put them, in their own eyes and those of the community, above Mexican and black workers. But it was not far above. Anglo migrants from relatively homogeneous areas, where racial and class or caste substructures had been relatively clearly delineated, were confronted with a society where the position of blacks and Mexicans was uncomfortably close to their own. Farm workers found themselves laboring in the fields, doing work they had stigmatized as "nigger work."

Migrants, shocked by the relative "race mixing" in California, went to some lengths to distance their views from those of what they perceived as the more racially lax Californians.[61] As James Gregory points out, the Okie culture in California was based, in part, on maintaining themselves as a bastion of Anglo supremacy against what they perceived as the relatively looser attitudes of Californians. Anglos' nervous claim to some sort of status within the white community was often based on an insistence of separation from nonwhites. Billie Pate's mother preferred to send him to a public school fifteen miles away (attended by fellow southerners but where Okies were called trash) to sending him to the ranch school with Mexican children.[62] This racism persisted to the point where the wife of an ex-union official sputtered that she refused to join the union because "you can't equalize me with no nigger."[63]

White Texans had been raised on a history of conflict with Mexicans. The seizure of half of Mexico in 1848, including what is now Texas, had created a situation characterized by economic and political subordination of Mexicans in Texas. This, in turn, led to armed conflict between Anglos and Mexicans, irredentist battles, and attempts by Texas Mexicans to retake parts of the state for Mexico in 1915. The Texas spillover of fighting during the Mexican Revolution helped keep alive an Anglo-Texan enmity with Mexicans.[64] This animosity fused racism and a strong regional identity as Texans that was based in conquest and continued conflict with Mexicans over land. When a Mexican family moved into the primarily white Arvin FSA camp, a delegation of irate Texans demanded that the manager evict the newcomers, waving the verbal saber of "Remember the Alamo! Either us or them. Can't have both of us here."[65] But racism against Mexicans was not the exclusive province of the Texans. Jessie de la Cruz remembers, "Anglos didn't want anything to do with the Mexicans [and] tried to stay away from the Mexican people. . . . They wanted their own labor camps, they didn't want to be mixed in with the Chicanos, the Mexicans."[66] Belén Flores, who pitied the "pobrecito Okies" ("poor Okies") who were in much worse shape than the Mexicans, resented the disdain with which the whites treated her and other Mexicans.[67]

Racism could subside under the influence of human contact. A man whose wife first said it "nearly killed him" to work under a black contractor later developed respect for his boss.[68] Two months after Texans had demanded the expulsion of Mexicans from the Arvin camp, the unit elected a Mexican worker to the camp council.[69] Mexicans, blacks, and

Anglos were later to work together in the Workers Alliance and in unions. Some Anglos battled racism. Yet such instances of cooperation were, on the whole, uneasy alliances. Organizers walked, straddled, crossed, and fell over the thin line drawn by the racial hostilities, insecurities, and historical enmities dividing black from white and Anglo from Mexican workers. Comments by both Anglo and Mexican workers suggest that it was only the abysmal conditions and the heat of strikes that, briefly, produced enough common ground to create a tenuous and grudging willingness to work together. If the work force had at one time been united by a common history, language, culture, and identity as Mexicans, it was now divided by historical enmities and racism.

The migrants identified with white growers as Mexicans never had. Their desire for their own land and their concomitant identification with the white farming community joined them to their grower employers, while racism divided them from other workers. Anti-communism tore many away from unions. One migrant grumbled, "We got enough troubles without going Communist."[70] Their painful desire to be respected members of an Anglo community argued against their joining organizations that denounced what they desired.

Religion influenced migrants' decisions about secular life. Migrants who came from the Bible belt were southern Baptists, Pentecostals, and members of other fundamentalist sects. That religion remained a mainstay in their lives was confirmed by the makeshift churches that cropped up in tents, ditch-bank camps, and labor camps in California. Whereas Mexicans' Catholicism survived easily outside the structure of the church, the new migrants produced a sudden flourish of formal religious activity. Pentecostal, Nazarene, and Baptist churches sprang up in towns with heavy migration, such as Corcoran and Arvin. Religion provided an emotional outlet, but also contributed to a fatalism. Disoriented, beaten down, newly arrived migrants found religion to be more of an escape from their present situation than a guide for improving it. Preachers convinced that earthly organizing undermined heavenly aspirations exhorted against unions; one Pentecostal preacher claimed, "After becoming a Christian I wouldn't even join a union. I believe it's wrong to belong to a union because you have to mix up so much with the world if you do. The Bible says we shouldn't be a striker."[71]

But the teachings of religion, when taken to heart, could conflict with conservative timidity and galvanize people against earthly barriers to the realization of their beliefs. A Pentecostal preacher whose religious conviction led him to socialism in the Oklahoma oil fields said, "The reason

that I am a Socialist, Brother, is because I *am* a Pentecostal preacher."
Critical of the "aristocratic high priests in religion today," he found it
"an all-fired shame that Man is getting preached to turn the world over
to the Devil, and then taking a small cut in the net profits." To counter
this, he had to "seek the truth, Brother, and after I have found that truth
then I've got to preach it. If that search for truth leads to economics and
politics as well as to the spirit, it makes no difference."[72] Lillian Dunn
felt no conflict between her intense religious beliefs and her work with
the Communist party: "It wasn't that I didn't believe in the Lord any-
more, but someone had to do something."[73]

ELECTORAL POLITICS

What political influences helped shape the response of these new migrants to the conditions in California? Certainly, many came from areas
where a radical tradition had flourished. In the South and Midwest the
Grangers, the Farmers Alliance, and then the Populist movement had
attracted farmers in the late nineteenth century. This movement, an
agrarian response to industrialization and entrance into the market
economy, made a futile attempt to push back capital encroachment and
restore to farmers and tenants some self-control over their economic and
political lives.[74] This was to be achieved, not through revolution or
syndicalism, but through the electoral process, by which the government
could be made more responsive to the needs of farmers. The ballot box
and electoral politics became central to the party's strategy.

The Populist impetus was taken over and transformed by socialists in
the early twentieth century. By a series of electoral upsets the socialists
threatened to "uproot the conventional pattern of local and regional
politics."[75] At the apex of the party's career, in 1912, the socialists
received their strongest national support in Oklahoma. Texas, Arkansas,
and Louisiana also made a strong show of socialist support. Each, except
Louisiana, was a major source of migrants to California in the 1930s.

Yet what influence did these movements have on migrants to California, and how did they shape their response as field workers? As James
Gregory and James Green have noted, the radical impetus behind Populism did not survive the 1920s, but descended into a neo-Populism and
movements marked by virulent racism and anti-Semitism. Why it declined has been open to conjecture: whether as a result of government
repression, vigilante terror, or economic depression, which, Green argues, undermined agrarian radicalism.[76] Both the Populists and the

socialists had appealed to small farmers and property owners. The socialist movement, which had appealed to cotton tenants in parts of Arkansas and Oklahoma, was never able to overcome their consciousness as property owners and convert it into a collective consciousness either as farmers or as potential workers. As Oscar Ameringer remarked, socialism in the southwest was more Populist than it was Marxist.[77]

Radical movements eroded under the wear of economic crisis and political repression. Continuing economic disruptions precipitated migrations that undermined the basis of resistance. As Oscar Ameringer put it, "They fought then and now they migrate." The successions of economic disasters that had plagued them had led to a despair which inhibited radical commitments. Many retreated into fatalism and faith in an afterworld. James Green quotes Floyd Murray, son of a Texas cotton tenant, who wrote a letter to President Roosevelt that reflects a fatalism and passivity untouched by radical tradition:

> I was born and raised on a tenant farm . . . in circumstances of the most stringent poverty. . . . I am fully acquainted with the hopeless misery and bleakness, the horrible mockery of civilization which is practically inevitable in this life. My own childhood and youth is a haunting nightmare, utterly devoid of a single tender memory. . . . I never at any time had even necessary clothing, not to mention such items as would allow a measure of self respect. . . . I don't know why I am not a radical today, but I am not. In those days renters accepted their lot, not through fortitude or honest patience, but through absolute inability to conceive a better plane of life as applicable to them.[78]

By the 1930s, the vestiges of the Populism and socialism that had written the organizational history in Oklahoma and Texas had faded into obscurity. The mass displacement of tenants and sharecroppers did lead to a radical revival in the 1930s. But this occurred not where socialism and agrarian movements had been strongest, but in the recently inhabited areas of the Arkansas Delta. The Southern Tenant Farmers Union (STFU), formed in Tyronza, Arkansas, in 1935, appealed, not to the small farmers who had been the backbone of the Populists, but to tenant farmers being expelled by the AAA. The STFU, faithful to its name, was a union of agricultural tenants and workers, not an organization of property-owning farmers. There were attempts to spread the union into Oklahoma and Texas. Odis Sweeden, a Cherokee from Muskogee, Oklahoma, organized several locals. But the STFU failed to find a base. Locals of the STFU remained autonomous from the national organization and declined into social organizations with few dues-paying members.[79] STFU's problems in establishing a base in these areas suggested the obstacles California unions might face in organizing new migrants.

The biggest barriers to organizing these migrants were economic depression, disruption, and the searing sense of despair and discouragement that weighed heavily on migrants in their first years.[80] This despair derived largely from their immediate circumstances. Certainly they were later to be important to labor organizing in California's plants and industries. But in the 1930s, as Fred Ross, a manager of the Arvin FSA camp used as a base by the United Cannery, Agricultural, Packing and Allied Workers of America, remarked, "You can just see the pain and agony those people had gone through to try and live. Nothing ever went right . . . it's full of failure at one place or another."[81] This despair, coupled with their disorientation and unfamiliarity with migrant field-work, made the migrants poor material for unionization at this point. That this situation was closely related to the economic conditions they came out of was underlined by one camp manager who said, "Those who have seen first hand the workings of a strike . . . are too close to the years of starvation and privation to be actively interested. Particularly is this true of refugees from the draught [sic] areas."[82] By the late 1930s few of the new migrants were socialists. Fred Ross said that he had only met three or four socialists among the residents of his camp, and those were all from the same family. Of the rest, "I never got the impression that they [the migrants] had any ideology at all. Either socialist or communist or anarchist or whatever."[83]

The older Anglo residents who had settled in the Valley in the 1920s made a more solid base for organizing. These settlements existed through the Valley, especially on the east side. Kern County was an example of an area that provided a relatively hospitable atmosphere for Anglo organizing; there, work and residency patterns facilitated community and labor support for cotton field workers. By the early 1920s, Kern County had drawn a substantial segment of people from southeastern and southwestern cotton areas. As the county grew as a center of cotton production it became a magnet for new migrants as well. Unlike many other parts of the Valley, such as Kings County, small farmers had been able to establish a base in Kern in the 1920s, and some held on, if tenuously, through the 1930s.

In Kern County traces could still be found of the progressive legacy of the southwestern states. Over the years a progressive community of oil workers, small farmers, tenants, field workers and some professionals

had grown up there. Ex-members of the IWW, possibly former bindlestiffs from the early agricultural days or workers in the oil rigs, settled
in the county, and by the 1930s they were active in unionization and
organization to obtain relief. The Communist party found enough members there to become the strongest branch in the Valley.[84] Members of
the Bakersfield branch organized among industrial workers and in the
Workers Alliance and made alliances with old Populists and ex-Wob-
blies. By 1934 a branch of the Townsend Club was meeting in Bakersfield, and by 1935 had 621 members in the chapter.[85] There was also a
chapter of the Progressive Club, which supported the Townsend Club.[86]
The Women's Christian Temperance Union, too, was active.[87]

The Socialist party was strong, and its activities were reported faithfully in the *Bakersfield Californian*. The extent of the reporting of the
socialists may magnify their importance, but they serve to indicate the
interrelationship between the issues of agricultural workers, tenants,
small farmers, and laborers in Kern County's progressive circles. The
Socialist party appealed to oil workers and proprietors. Taft, a small oil
town, boasted an active socialist local, as did the larger city of Bakersfield.[88] The Kern County socialists ran candidates for assembly and by
1933 were a part of a Socialist Federation of the San Joaquin Valley,
headquartered in Bakersfield and chaired by a local attorney, Raymond
W. Henderson.[89] Samuel White, a leading socialist in Kern County, was
a member of the oil workers' union, editor of the *Union Labor Journal*,
and member of the Kern County Labor Council.

As White's credentials suggest, the socialists took up labor issues, and
their interests extended to field workers. Local socialists took part in a
1933 California Congress of Workers and Farmers. Henderson and
White were members of the congress's executive committee, which
passed resolutions indicative of Socialist party concerns: they condemned the sales tax, voted to support farmers against mortgage fore-
closures, and agreed to aid labor unions.[90] White was a delegate to the
convention of the California State Federation of Labor.[91] The socialists
had supported workers in the 1933 cotton strike. White had asked the
Board of Supervisors to pay burial expenses for killed striker Pedro
Subia. He lambasted the board for not "making any provision to take
care of starving people," and added, "We don't want the welfare department used as a strikebreaking agency."[92] By 1936 White was secretary of the California Conference of Agricultural Workers.[93] When
Norman Thomas, socialist presidential candidate, visited Bakersfield
and Taft, he spent a day investigating the cotton strike.[94] Showing a

sympathy for small farmers as well as workers, Thomas condemned the strike as "slaves striking against slaves" before a rapt audience of 700.[95]

The Kern County Labor Council was strong, active, and peppered with progressives such as White and Clyde Champion, an agricultural organizer and member of the Communist party who ran for a seat on the labor council.[96] The council's progressive positions and its strength in this relatively industrial area were pivotal to organizing Anglo agricultural workers. The close relationship between industrial and agricultural sectors there led the council to address the problems of small farmers and agricultural workers as well as those of industrial workers. In the late 1930s the council played a pivotal role in forming and organizing agricultural unions. It provided meeting halls, gave financial support, worked with sympathetic FSA camp managers, and took a leading role in relief struggles. It was sympathetic to new migrants, as was demonstrated by an editorial in the Labor Council's journalistic organ, the *Kern County Labor Journal*, which admonished townspeople to "quit casting aspersions on the dust bowlers."[97]

This atmosphere encouraged small farmers to organize, and in 1933 they formed the original Associated Farmers, which attempted to include small farmers under the NRA legislation.[98] The original Associated Farmers had ten chapters in the Valley, in the small farmer towns of Wasco, Shafter, Delano, Buttonwillow, Earlimart, Tipton, Pixley, Tulare, Woodville, and McFarland. Small farmers also established active grange locals. Conflicts between large cotton interests and small farmers and workers flared up. Small cotton farmers, pressured by gins and land companies, devised ways to resist coercion. Grange members, for example, clandestinely bombed a water levy used by a land company to deflect water from smaller growers. In 1935 the Weed Patch Grange was the only grange in the state to oppose relief cuts.

Kern County had been the base for Anglo participation in the 1933 cotton strike. Many of these workers had had substantial experience of proletarianization. Workers in the 1933 strike tended to be settled workers who felt the grievances of 1933 as the latest in a long string of injustices. Some had been tenants or small farmers forced out of farming by the Depression. This settled group continued to be the base for local organization in the 1930s. Over time in California, their economic circumstances had seared into their brains the dilemmas, conflicts, and contradictions of capitalist agriculture. Some had participated in labor unions. Some drew on Populist and socialist models. But all these were recast in the Valley to apply to field workers as well as small farmers.

Michael Kearney, for example, a small farmer in Kings County, had been active in the American Railway Union with Eugene Debs. Attempting to escape the conflicts of the industrial world, he was brought up short by the industrial conflicts in the agricultural fields. Kearney supported the CAWIU in 1933, bailed organizers such as Lillian Dunn out of jail, and was involved in movements through the 1930s.[99] Bertha Rankin, a Kern County farmer with holdings in Weed Patch and Arvin, supported labor organization and progressive movements. Rankin donated land on which to build the controversial FSA camp at Arvin.[100] Jim White, a tenant and later a small farmer whose family had been in the Valley since the 1910s, came from a Populist farming family in Missouri. White credited his Populist background for his support of the local grange and other progressive movements in Kern County.[101] W. D. Hammett, a sharecropper since 1925 and a field worker since 1933, had experienced firsthand the descent from sharecropping to fieldwork and relief recipient and the death of hope for settling on a farm or even a share. The increasing awareness of his proletarianization no doubt helped propel Hammett into activity with the union and, later, relief organizations. It was perhaps his time in California that spurred Hammett, working on the Tagus ranch in 1933, to join Chambers and the Mexican union in the peach strike and cotton strike that followed. Hammett worked closely with Chambers but, like his Mexican counterparts, never joined the union and continued to be involved, sometimes controversially, in union and welfare organizing.[102]

Yet electoral politics was a major force in shaping the responses of small farmers and, ultimately, of white farm workers. Populists and socialists had envisioned electoral politics as the vehicle for social change. For Populists it was through electoral politics that "producers" could take over government and make it responsive to the electorate. Socialists also focused on electoral politics and had made stunning showings in the 1910s. In Oklahoma and Arkansas, radical electoral politics had been a major form of radical movements. Electoral politics were thus integrally related to the concept of radical politics. The idea of the electoral arena as a mechanism for change continued among whites who migrated to California. As such, some of the solutions they offered—while radical—were based in part on strategies of electoral politics and government participation. This dovetailed with the implicit promise by some proponents of the New Deal that the government would in fact restructure social relations.

Anglos established residency in part to qualify for relief. This stability, which might have helped create a union base, in fact created the base for participation in electoral politics. Anglos may have been ambivalent about joining a union when in doing so they identified themselves as agricultural workers instead of farmers and worked with nonwhites. They had no such compunction about voting. Voting actually affirmed their membership in a community which in other respects rejected them. No small part of their choice of participation in electoral politics was probably their memories of the Socialist and Populist emphasis on electoral politics rather than syndicalism as the appropriate means to achieve their goals. But it also resulted from the larger American ideology, which stressed participatory democracy within a homogeneous community of "producers" and shunned a Marxist or neo-Marxian analysis of class struggle, conflict, and social change. What distinguished the Anglos from other farm workers, then, was not only ethnicity but a higher incidence of citizenship, the right to vote, and a belief that voting could be of benefit to them.

Electoral politics during the New Deal galvanized the settled Anglo community in the 1930s. The 1934 gubernatorial election began to show the cleavages within the community, including the increasing divisions over the meaning of the New Deal. Socialist Upton Sinclair rattled Democratic party pundits by running on a campaign to End Poverty in California (EPIC) and winning the primary election in an upset over Democratic party candidate George Creel, NRA mediator during the cotton strike. The surprising primary results indicated support for the more radical interpretations of the New Deal. The results would leave a legacy of factional bitterness among the Democrats, as California became a political battleground between factions with differing definitions of the New Deal and between New Deal democrats and the conservative (Republican and Democrat) opposition.[103]

In a taste of what was to come, the election set off heated political divisions within Kern County. Locals of the Socialist party split: the Bakersfield party opposed Sinclair's candidacy on the Democratic ticket, the Taft socialists supported him.[104] Yet Sinclair received enthusiastic support from labor. The Taft Central Labor Council passed a resolution endorsing Sinclair; the Kern County Labor Council passed a unanimous resolution in his support and stated that his election was "essential [if] our great leader, Franklin Delano Roosevelt, [was to] have the Democratic support of California."[105] The State Federation of Labor endorsed Sinclair and his running mate, Sheridan Downey, by a vote of 400 to 2.[106]

From the beginning there had been different views of what the New Deal was or could be. If the federal government was acting on behalf of what Creel had called the "public good," then which public was being referred to and how was its good to be determined? The *Bakersfield Californian*'s editor, Alfred Harrell, supported the New Deal. As of August 1933, Harrell was a member of the federally appointed board directing the NRA in California. As an anti-communist, he endorsed the idea of labor and capital working in harmony for the common good. During the 1933 strike he had condemned Rolph and "the intolerable situation due to the coddling of reckless and communistic leadership," while congratulating farmers for "recognizing their obligation as citizens . . . [and preferring to] sacrifice their anticipated profits" in the efforts of peace.[107]

The Sinclair campaign heightened the debate over the nature of the New Deal. The *Bakersfield Californian* ran an unrelenting series of front-page anti-Sinclair editorials headlined in bold type. Harrell argued that Sinclair's candidacy was antipathetic to the New Deal and argued that those who suggested that an endorsement of Sinclair was an expression of support for New Deal policies should "realize that nothing at this time could more acutely embarrass the President . . . and render a more positive disservice than the election . . . of a lifetime Socialist, pledged to a personally written program of dreams and visions."[108] Local church pastors opposed Sinclair.[109] Merchants and employers used hard-handed political pressure to get their employees and creditors to vote for the Republican gubernatorial candidate, Frank Merriam.[110] But popular sentiment in Kern County favored Sinclair. Despite pressure, an informal preelection poll found Sinclair and Downey leading; a pro-Downey meeting was one of the largest political gatherings in the county in months.[111] Statewide, Sinclair and Downey lost, but some EPIC candidates for the state senate and assembly won. But James Gregory's discussion of the Kern County town of Arvin as a barometer of this new voting block is perhaps more to the point. That this largely Okie town cast 54 percent of their votes for Sinclair suggests the appeal Sinclair and his policies had for the new migrants.[112]

The EPIC campaign politicized a vision of the government's role on the left and encouraged electoral participation as a means of realizing these goals. Workers, organized labor, and many small farmers and tenants responded to this appeal. Sinclair's campaign added enough registered Democrats to transform California from a Republican bastion to a Democratic state. In the Valley it put growers and employers on notice that while the new migrants and local workers might not organize lasting unions, they would participate at the ballot box. The EPIC campaign was a taste of the electoral upset to come in 1938 when Culbert Olson, Sinclair-endorsed Democratic candidate for state senator from Los Angeles, was to win the election for governor of the state.[113]

Despite assumptions by John Steinbeck and George Clements that white Americans would demand better conditions and higher wages, it was, as Carey McWilliams noted, "the Mexicans, long familiar with field conditions in California, who showed the most determination to fight for a higher standard of living."[114] Although many were not citizens, were subject to deportation, and continually faced racial hostility, they were the backbone of the strongest union drives in California to that date. When Mexicans left the cotton fields, a major force for labor organizing left with them.

Mexican unionization south of the Tehachapis from 1934 to 1937 suggests the impact that Mexicans might have had on unionization in cotton areas. In 1935 and 1936 CUCOM was the most effective and active farm workers' organization in the state.[115] It worked closely with the Workers Alliance. CUCOM members also formed a major part of the left wing of the AFL's Federation of Agricultural Unions, founded in 1936, and "furnished the chief leaders and most of the rank and file membership."[116]

Since 1933, CUCOM leadership was factionalized between radicals led by William Velarde, an anarchist, and the faction led by Armando Flores and the Mexican consul, Ricardo Hill. In 1936 CUCOM split, fueled by the growing division between the AFL and those who favored a progressive unionism. Some factions joined the AFL, but the body of CUCOM was taken over by Velarde, to become one of the major unions in the state. In 1936 CUCOM led a series of strikes in Los Angeles's abundant agricultural fields, from Santa Monica on the coast south to Palos Verdes and inland to Watts, Compton, El Monte, Norwalk, and parts of the San Gabriel Valley. CUCOM built upon the bases formed, or similar to those formed, during the 1933 cotton strike: family networks, labor crews, and Mexican working-class communities. Among them were veterans of the 1933 strike wave. By 1936, CUCOM was strong enough in some areas to elicit wage hikes, contracts, and better conditions without striking. CUCOM demanded, and got, union recognition, a feat which had eluded the CAWIU. By 1937 CUCOM was one of the largest agricultural unions and was instrumental in forming the United Cannery, Agricultural, Packing, and Allied Workers of America (UCAPAWA), with which it merged.

Even in cotton areas now dominated by Anglo workers, Mexicans continued to play a role disproportionate to their numbers and were key to the expansion of UCAPAWA.[117] As Vicki Ruiz notes, Mexicans organized seven of the ten field worker locals in the San Joaquin Valley. By 1940, the predominantly Latino Local no. 250 had launched successful organizing drives among both Anglos and Mexicans in Shafter, Wasco, Earlimart, Lamont, Lindsey, and Bakersfield.[118] In Shafter, Kern County, Mexicans were crucial to organizing. When a local SERA Workers Club was formed in Shafter in 1935, meetings were held in the "Mexican town" two miles south of Shafter, and its two officers were Mexicans.[119] By 1938 Steve Rodríguez was president of the Shafter local. During the 1938 strike, meetings of whites, blacks, and Mexicans were held at the Rodríguez home in the Mexican community, and 250 workers belonged to the local.[120] Felix Rivera of Bakersfield represented the whole of the San Joaquin Valley at the UCAPAWA 1938 meeting.[121]

The change in the labor force and migration patterns requires a reassessment of the relationship between migration and agricultural organizing. Organizing migratory workers is more difficult than organizing resident workers. Migration militates against developing full-time union bases. Migrant workers have less stake in a given area or local and have no political power. Yet at the same time, as can be seen in the case of cotton, migration can contribute to the spread of grass-roots networks useful in organizing strikes. In California agriculture these networks were important, for to combat the highly organized growers, strikes had to cover the entire industry.

The increasing rates of residency of the Anglo workers did create a structural base that could have lent itself to the formation of unions or participation in electoral politics. Yet this depended on the inclinations of the workers and the strength of their organizations. Indeed, with the union, the burden of spreading organization and strikes depended increasingly on the ability of the union to communicate with workers rather than on the networks among workers formed by migration. The focus of organizing became confined within a more limited circumference. Without a strong and well-financed union and the support of workers, lack of mobility could in fact sharply limit the success of strikes.[122]

These new workers in California cotton were American citizens, from areas where social protests had focused on changing the government through the ballot. When—and if—they participated in attempts to change class relations, they did so increasingly within the arena of electoral politics. Their support for and belief in the New Deal helped in electoral politics to strengthen labor and their vote: but their reliance on electoral politics led them to place trust in a process that ultimately, because of its economic and political parameters, did not work in their favor as farm workers. Citizenship gave little respite from the conditions in the agricultural system. Citizenship and white skin would be more important in propelling the new migrants out of the fields than in changing conditions there.[123]

CHAPTER SIX

New Deal Relief Policies, Local Organizing, and Electoral Battles

Federal policies of the New Deal altered class relations in the cotton industry, precipitated organizing by both growers and workers to contest the terms of these policies, and led to increasing conflicts over their administration. In short, federal policies shifted the locus of class conflict from the fields to relief offices and legislative halls. These conflicts first focused on localized battles over the administration of New Deal policies in the Valley. But the conflict quickly escalated into the realm of California electoral politics as both sides fought out the definitions of power within the state government. This chapter focuses on federal relief policies, how those policies influenced the direction and intensity of organizing in the later 1930s, and, finally, the beginnings of electoral fights within California over the application of the policies.

CRISIS IN THE VALLEY

In 1937 and 1938 two crises, prompted in part by federal policies, hit the cotton industry. The first resulted from the 1936 Supreme Court decision to invalidate the AAA. By 1936, the cotton industry had come to rely on federally imposed crop curtailment to raise prices and prevent endemic overproduction. Yet when the Supreme Court temporarily struck down the Agricultural Adjustment Act, California growers, like their counterparts around the nation, increased production. Favorable conditions of the domestic market and expanding international markets, free from domestic regulations, encouraged increased capital investment.[1] California cotton acreage expanded by half a million acres, to reach over a million and a half acres. Growers were, as individual investors, initially optimistic: the crop was being valued at $50 million, and Valley cotton accounted for one-third of the state's entire agricultural output.[2] Yet the retrenchment of production controls recreated earlier problems of overproduction. More immediately, the expansion of cotton lured an influx of workers into California, swelling the labor pool and leading to a crisis in relief that shook the industry.

In 1937, an unprecedented 105,185 white migrants entered California. Most sought work in the Valley's cotton fields. The expansion of cotton, border and USDA officials concurred, was "one of the major reasons for the constant influx . . . of workers from Texas, Oklahoma, Arkansas, and other cotton states."[3] By the spring of 1937, cotton camps were prematurely filled to capacity with new migrants hungrily awaiting the fall cotton harvest. By July, with the other crop harvests over, an estimated 70,000 destitute workers were living in the Valley, awaiting the fall cotton harvest.[4] The situation was growing rapidly more serious as large numbers of workers continued to pour into the area in anticipation of an unprecedentedly large fall harvest.[5] Newspapers that had been oblivious to the condition of Mexican cotton workers once they left the fields now reported on the destitution of white "American" migrants on their doorsteps.

Not only were the new migrants white—which bothered local residents—but migrants posed a problem for local communities and public health and relief agencies. Harold L. Robertson, secretary of the private relief agency the Gospel of Army, reported:

> There are 70,000 homeless and jobless persons in desperate straits between Bakersfield and Stockton. Disease is frequent among them and the public health is being jeopardized. There have been three typhoid outbreaks already in one county. . . . One city found its public hospitals overflowing and set up a hospital in its high schools.[6]

This influx brought to a head the problem of how farm workers were to be supported between seasons. Mexican workers had returned to their homes outside the Valley, and any costs of caring for these workers in the off months had fallen on urban areas, not the rural communities of the San Joaquin Valley. As a result, these costs for agricultural workers had remained invisible in the Valley. But the new migrants stayed: 79 percent planned to settle permanently. Moreover, after a year's residence they would qualify for local relief.[7] The hidden cost of maintaining the

low-paid agricultural labor force in the off months was now threatened
to be borne by Valley communities. This influx coincided with a congressional cut in national relief appropriations for 1938, and it became
increasingly clear that the burden of relief would fall upon unprepared
and unwilling state and local governments.[8] California State Relief Administration director Harold Pomeroy claimed that this federal retrench-
ment would force the SRA into "bankruptcy."[9] California relief officials,
ordinarily ambivalent about federal programs, began to demand federal
relief.

ORGANIZING: WORKERS AND THE INDUSTRY

The growing conflict over relief helped activate the Workers Alliance.
The alliance had been formed in 1936, a successor to the Communist
party's Unemployed Council. The organization was formed initially to
improve conditions for workers on WPA jobs and raise relief rates for
the unemployed. Under Alex Noral, the alliance began to make inroads
into the San Joaquin Valley in the summer of 1936. The alliance assumed
a major importance in cotton areas because relief played a pivotal role
in supporting field workers during the off season, in setting a de facto
minimum wage, and in supporting or breaking strikes. Raising wages by
increasing the competing relief stipends had, in effect, become an alternative to collective bargaining. Relief programs now set wages more
effectively than the Agricultural Labor Bureau did. Because relief now
played such a prominent role in determining wages, the Workers Alliance
became, in effect, a trade union of the unemployed.

By the summer of 1937, the alliance joined with a new union,
UCAPAWA, the agricultural branch of the fledgling Congress of Industrial Organizations (CIO). From the demise of the CAWIU, in 1935, to
1937, agricultural organizing had been taken up by several groups:
left-wing locals within the AFL, many composed of old CAWIU members; CUCOM, the Mexican union; and independent Filipino unions.
Under the impetus of left-wing organizers, the AFL established 16 locals
in agricultural areas, including six cotton towns.[10] By 1936, left-wing
unionists and ethnic unions formed the Federation of Agricultural Workers of America to organize field and processing workers on an industry-
wide basis within the AFL.[11] But in 1937 these organizational efforts
became impaled on the increasingly bitter jurisdictional conflicts within
the State Federation of Labor. Reflecting the conflicts raging nationally
within the AFL, the executive council refused to support organizing into

one integrated agricultural union. By May, the national union movement had formally split, and the CIO was formed. Two months later, CIO organizers founded UCAPAWA. In California, disgruntled left-wing unionists from the AFL, CUCOM, and other ethnic unions became the organizational backbone for UCAPAWA and spearheaded organizing in the state.

In 1937, UCAPAWA seemed to offer the hope that industry-wide unionism was possible in California agriculture. The national upsurge in organizing—the wave of sitdown strikes and the CIO victory against steel—was an encouraging sign for agricultural workers.[12] UCAPAWA-CIO gave local organizers a national identification under which to organize (workers commonly referred to it as the CIO) and helped speed up the organization of agriculture.[13] UCAPAWA supported extending state and federal legislation to include farm labor and favored amending the Agricultural Adjustment Act to require participating growers to meet minimum wage and working standards as part of their contract (a demand sugar workers later obtained).[14]

Yet, ultimately, organizing field workers depended largely on local conditions, financing, and leadership. The union had only a slight impact on cotton areas in 1937. UCAPAWA had trouble establishing locals. By the spring of 1938 the union had signed up 500 members in the Valley, only 7 of whom paid the $1 fee. UCAPAWA was strongest in Kern County, where a supportive union movement had aided the organizations. Here, organizers who had worked with the CAWIU and, later, AFL locals now organized locals of UCAPAWA.[15]

It is likely that more cotton workers joined and became active in the alliance than in the union. In 1937 the alliance claimed 7,000 members in California; in 1939 it boasted 186 locals with a total membership of 42,000. While only 12,000 had paid their dues, suggesting these figures are somewhat inflated, the alliance undeniably had a substantial membership.[16] Figures are unavailable on Valley locals, but they appear to have been active. Federal FSA camps provided a base for the organization: workers in the Arvin and Shafter camps supported active alliance chapters. The alliance, recognized by the federal government as an organization of the unemployed, also cooperated briefly with the American Federation of Labor's half-hearted efforts to organize field workers.[17] In local areas the alliance worked closely with UCAPAWA; the groups organized together and participated in demonstrations at relief offices and on picket lines. Alliance members were automatically transferred to UCAPAWA during the chopping and picking seasons, returning to the alliance in off months. The two organizations formed a functionally dual-headed, year-round organization.

The increasing crisis over government policies simultaneously led to the recrudescence and growing strength of the Associated Farmers. By August, the optimistic predictions for the 1937 harvest had turned sour. The Department of Agriculture announced an impending glut on the market due to a 25 percent increase in cotton production, amounting to more than one million bales nationwide. In response, cotton prices dropped by as much as $1 to $2 a bale, to 9 cents a pound. The anticipated profits of cotton growers shrank correspondingly.[18] At the same time, the availability of relief and the expansion of cotton acreage reduced the size of the labor pool relative to the increased demand for labor.[19] With an estimated 75,000 experienced workers needed for harvest, growers felt the labor pool was too small to ensure the customary low wages, despite the continued influx of new migrants.[20] Growers battled over the wage rate at the Agricultural Labor Bureau meeting before settling on a wage of 90 cents, 10 cents less than they had paid in the 1936 season.[21]

The wage cut exacerbated the problem of getting workers. Some migrants simply pulled up stakes and went home; the *Fresno Bee* reported a "noticeable migration of cotton pickers from the Valley."[22] One small town found that within a single week 352 people had bought railroad tickets and left for Oklahoma.[23] Growers were left to compete for the remaining workers. In Kern County, where growers had already begun the harvest, destitute, out-of-state migrants were ineligible for relief and willing to work for 90 cents. Yet many were inexperienced, could not pick even 100 pounds a day, slowed the harvest, and sent growers searching frantically for more workers.[24] Some farmers reportedly were working with less than half their regular crew.[25] Capitalizing on competition among growers, the CIO promised to provide crews for growers who paid over $1 per 100 pounds. Even though Kern County sheriff Ed Champness ordered unemployed workers to pick cotton or face arrest on vagrancy charges, growers were unable to get enough workers at the low rate.[26] They began to offer $1.00 and even $1.10 and pressured finance companies to advance additional loans to cover the increase. Finally, the finance companies capitulated and advanced additional funds, advising growers to pay $1.00.[27]

This was the first time since 1933 that the Agricultural Labor Bureau had been unable to enforce wage rates. The capitulation signaled the bureau's inability to eliminate competition for labor or prevent any form of collective bargaining. The finance companies' agreement to advance loans based on a higher rate had effectively underwritten collective bargaining and, because the companies set wages through their lending policies, forced the bureau to follow suit.

The capitulation precipitated an open breach between the ALB and the industry. In Kings County, well-organized growers stuck to the rate of 90 cents. Corcoran grower, ginner, and financier J. G. Boswell condemned the wage increase and threatened to withdraw his substantial contribution to the ALB if the bureau failed to enforce uniform wage standards across the Valley. The bureau's board of directors admitted that they "found it impossible to control the situation," which forced cotton growers to seek another means of reasserting control over wages and the labor supply.[28] Their efforts led to the reappearance of the Associated Farmers.

CONFLICT OVER RELIEF

As the 1937 fall harvest approached, relief again became a major issue. Cotton growers pressured the California State Employment Service (CSES) and the State Relief Administration (SRA) to cut workers from relief rolls. By early September, the CSES publicly protested that they were "stymied by the inactivity and apparent unresponsiveness of the SRA . . . to get men off the relief rolls and into the harvest fields," and they complained that SRA requirements that growers provide workers with bedding and equipment interfered with the easy flow of workers to the fields.[29] Republican governor Frank Merriam initiated a "complete purge" of the SRA to eliminate social workers who kept workers on relief. He changed eligibility requirements, ordering that any able-bodied person, regardless of experience, be cut from SRA rolls if they refused agricultural work. Merriam's plan was dubbed the "no work, no eat" policy.[30]

Government agents laboriously went through SRA and WPA relief rolls, cutting them to the bone. In Fresno, 600 were eliminated from the SRA, and the WPA sliced their rolls in half.[31] WPA construction projects in Fresno and Tulare were closed in order to supply workers to the fields.[32] The state government agreed to shoulder the costs of blankets, utensils, and transportation required by the SRA as a condition for their releasing people to pick cotton. Workers were imported from other parts of California.[33] Although Merriam argued that the relief cuts saved tax money, the costs of equipment and transportation ate up whatever savings were made. Merriam acknowledged that the state would not save any money but promised, "We will have furnished the labor for the ranches."[34]

With the interests of the agricultural industry at stake, the SRA and local WPA offices, controlled by a local Republican administration subject to political pressures and sympathetic to agricultural interests, were transformed into labor contractors. As the Fresno WPA director openly admitted, the major intention behind the change in relief policy was to furnish labor to agriculture: "I am very definite in our program to furnish workers from our rolls whenever the need arises. When an acute shortage of agricultural labor exists we will send out as many men as we possibly can and if the need is sufficient we will shut down the entire program in order that crops may be harvested without loss."[35] A *Bakersfield Californian* editorial, applauding Merriam, pointed out the priorities of the state government, the importance of the agricultural industry to the state economy, and the unwillingness of taxpayers to shoulder the relief bill:

> One of the dangers of the relief program has been that many persons . . . began to look upon relief as a right. . . . Relief was organized as a humane project to provide sustenance during the depression. . . . That emergency has passed. [For workers to remain on relief] is an imposition on the overburdened tax payer. It is a menace to the agricultural industry.[36]

A relief agent noted that the SRA had created a reservoir of labor which intensified and perpetuated the problems of agricultural workers. Local relief projects insured that workers would stay in the area during the off months. At harvest they were thrown off relief and forced to take jobs at whatever wages were offered. Under the auspices of a malleable agency, growers found that relief could be a solution to the old problem of securing a readily available but temporary work force.

In January 1938, relief agencies reverted to the problem of caring for unemployed pickers. With 95 percent of the cotton harvested, work had all but ceased. Cotton chopping would not begin for a few months. There was little work in other crops. Workers watched grimly as the last bolls of cotton were picked or, damaged by the winter rain storms, rotted on the ground. The record number of migrants who had come to pick the crop of 1937 had no place to go and no money for gas. They stayed on the cotton ranches or in roadside camps, awaiting the next season.

By February, San Francisco newspapers estimated that 90,000 workers were left destitute.[37] Malnutrition, a chronic problem for agricultural workers, became critical for families living on a weekly income of from $2 to $5. The *People's World* reported that twenty-seven out of thirty children in the Valley suffered from some stage of malnutrition. It became harder to forage for food. People in the Arvin FSA camp were living on potatoes and onions stolen at night from nearby fields.[38]

In February and March, heavy storms flooded the San Joaquin Valley. The large ranches built on reclaimed land of the Tulare Lake Basin, on the west side of Kings and Fresno counties, were hardest hit. The Tulare Lake crested at 192 feet near Corcoran before breaking the levee and spilling over 49 square miles of land and 28,000 acres of cotton, wheat, flax, and barley land.[39] The floods increased immiseration. In a Farmersville squatters' camp, workers lived in makeshift tents suspended in a sea of mud. The west side was worse. Twelve hundred workers on Giffen's camp number 3 in the Mendotta-Firebaugh area were locked in by water and mud for over nine weeks.[40] Roads in and out of the camps became accessible only by caterpillar trucks. Hundreds of workers in west-side camps were reported "near starvation."[41] The cold, damp living conditions, lack of adequate sanitation, and lack of food contributed to an increase in disease. A smallpox epidemic broke out in Tulare and Madera counties.[42]

Widespread publicity about the 1938 flood and the plight of white Americans drew public and statewide attention to conditions in cotton camps that had existed for years.[43] Malnutrition, disease, abysmal living conditions, and annual flooding had been staples of life for cotton workers, albeit the misery was exacerbated by the Depression. Yet as long as the workers were Mexicans, little attention was paid to them except by social workers and townspeople attempting to keep the effects of these conditions from spilling over into neighboring communities. But in 1938, newspapers wrote copious articles about the "rain soaked, fog drenched army of cotton pickers [which is] today . . . scattered across the floor of the lower San Joaquin Valley matching its meager pennies against time."[44] These conditions were endemic in California agriculture, yet townspeople, relief agencies, and even growers of other crops pointed an accusing finger at the cotton industry, the largest employer and at least partially responsible for drawing destitute workers to the Valley. A Tulare grower blamed cotton as "the leading factor in bringing in . . . a class of people who do not fit into other lines of agriculture and as soon as cotton work is over are a burden on the welfare departments of the counties."[45]

Townspeople had several concerns. Part of their anxiety stemmed from the old fear, first expressed in the 1920s, that cotton cultivation would recreate poverty and the class and racial tensions that afflicted southern cotton states. Now, staring out from photographs and detailed in graphic reports of starvation, disease, and squalid camps were the ghosts of southern poverty linked to a one-crop economy based on cotton. More than Mexicans, white migrants seemed to personify the social structure of a poor-white stratum of society tied to cotton. The Kern County Department of Public Health reported that "poverty had followed cotton" into that county, where cotton production had increased 125 percent since 1936.[46] The department found a direct correlation between cotton expansion and the sudden upsurge in infant mortality, which it pointedly called a "class disease."[47]

But the primary concern was that the costs of caring for agricultural workers were now falling on local communities. The migrants who stayed sent their children to local schools, used public health facilities, and collected local relief. Kern and Madera counties, which gave more health care to workers than other counties, doubled their health and sanitation budgets between 1935 and 1940.[48] The influx of white migrants increased the school population, especially in cotton counties where the majority settled and school taxes increased more rapidly than the population: Fresno faced an increase of 134 percent, Kings County one of 282 percent, and the increases for Madera, Kern and Tulare counties fell in between.[49] As Walter Stein points out, the actual cost of taxes, when balanced against other factors, was not as burdensome as it might seem: what was important was that residents blamed tax increases on the migrants and, as a result, on cotton.[50]

Community resentment of the social costs of corporate agriculture increased. It was not surprising that a *San Francisco News* editorial rhetorically asked whether cotton growing "should not be scrapped for some less speculative and more diversified form of agriculture."[51] But these views were spreading to the Valley, where the industry had enjoyed support. Even the *Bakersfield Californian*, usually supportive of cotton interests and a defender of removing workers from relief during harvests, questioned the social costs of corporate agriculture in an article entitled "Does It Pay?":

> Every Californian must be concerned over a situation which creates an army of migrant laborers who are left without employment. A humane government cannot permit them to lack for food . . . [and] the situation . . . will become even more complex. . . . Thoughtful people will wonder . . . if it is advisable to continue an industry which attracts so many thousands . . . for what necessarily must be temporary employment. In the absence of some means of earning a livelihood later public authority must rescue them from starvation

through taxation levied upon the public as a whole, and the amount needed . . . is in excess of any profit that comes to the state through the growing
of cotton. . . . If the grower is directly advantaged, indirectly . . . all the other
permanent residents of the state are disadvantaged.[52]

By 1938, public clamor over the abject poverty of unemployed cotton
workers grew to sufficient volume to force the federal government to step
in. The FSA announced that because of the extent of "poverty and
suffering," it would care for those not covered by county or state aid.
It approved $150,000 for migrant aid, with an additional million prom-
ised to come later.[53] This shift in federal policy made relief available to
workers who had been in the state less than a year and were therefore
ineligible for public relief. In effect, the public outcry over conditions
had, indirectly, helped recreate a situation in which workers, no matter
what their residency, could fall back on relief.

Relief cases increased dramatically. The FSA supported 50,000 workers between the February floods and the October harvest.[54] The WPA
offered work relief which, in cotton areas, increased in direct proportion
to the increase in migrants. The March relief roll in Fresno increased from
an average of 500 to over 1,400.[55] Kern and Tulare counties reported
"phenomenal" rises in work relief, whereas Madera and Kings counties
showed only a slight increase. The WPA director for the Valley, Earl
Cummings, claimed the relief load had doubled and estimated that the
increase would equal that of the peak Depression years.[56]

The change in federal relief policy made relief the most pressing
political issue for growers. It increased tension between federal and state
agencies and between these agencies and the agricultural industry. The
new relief policy also posed political problems for California legislators:
migrants who qualified for federal relief would become residents eligible
for local relief after a year. State administrators were feeling increasing
public pressure to alleviate, or at least make invisible, the migrants'
condition. At the same time, tax-paying voters were increasingly dissatisfied about mounting relief and other costs.

SRA director Pomeroy opposed direct federal relief and refused to
distribute federal funds.[57] He blamed the federal government for the
local crisis, claiming that the policy subsidized workers only until they
became residents, at which point the burden for their care fell on local
areas. Pomeroy argued that the federal government should assume complete responsibility for relief: "If the federal government is going to come
into the picture, it should be on the basis of Congress making permanent
provisions for dealing with the situation."[58] In effect, Pomeroy was

proposing a federal subsidy to agriculture to maintain the low-paid seasonal work force, thus relieving local areas of the increasing tension and questioning over the costs of corporate agriculture.

Growers were less concerned about social costs than about the direct costs to them. Under the Merriam administration this had not posed a problem. The one-year residency requirement had insured that there would be a pool of workers, ineligible for relief, who would accept low wages. Relief agencies had also supplied labor and kept relief rates low enough to depress wages. The new FSA ruling threatened to remove the growers' net of impoverished workers ready to take work at low wages. Cotton growers therefore refused to cooperate, worried that acceptance of this federal relief in any form would compromise their control over the work force. The Kings County Board of Supervisors (composed of cotton and sugar beet growers and members of the Associated Farmers) refused federal aid for workers stranded on flooded cotton ranches. They were supported by the local Red Cross, the county welfare director, and health officials, all appointed by the Board of Supervisors. When the State Division of Immigration and Housing condemned the Corcoran cotton camps and urged that the 2,000 families be temporarily moved to federal camps on higher ground, growers initially refused that, too.[59] Their reluctance to comply was due, as one Associated Farmers official argued, to the concern that "if the workers are put into relief camps they might be subjected to the propaganda of radical organizers."[60] Only when the flood waters reached the knees of the workers did growers finally consent to move them to federal camps.

The conflict polarized state sentiment on the issue of relief. Growers dug in their heels, staunchly refusing federal relief, while stories of stranded white Americans drew the sympathy and support of urban-based liberal organizations. The most prominent of the sympathizers was the Simon J. Lubin Society in San Francisco, a private organization founded in 1936 and named after the progressive first commissioner of the Division of Immigration and Housing. The Lubin Society informed Californians about conditions in corporate agriculture through its news-letter, the *Rural Observer*. After the flood, the society organized the Bay Area Emergency Flood Relief Committee in conjunction with other civic, fraternal, and relief groups and sent in clothes and food to workers stranded in labor camps.[61] They gave sympathetic publicity to the plight of workers and helped bring to national consciousness the conflicts in agriculture. In April the society published John Steinbeck's evocative

pamphlet "Their Blood Is Strong," the documentary basis for his novel
The Grapes of Wrath.

Individuals such as Lucy Sullivan McWilliams, Democratic wife of a
San Francisco judge and head of the labor study group of the San
Francisco League of Women Voters, coordinated efforts to send in food,
clothes, and money. She funneled these donations to Lillian Monroe, an
ex-member of the Communist party. Monroe turned the donations over
to key workers on ranch camps. On the Giffen ranch, W. D. Hammett,
who had for years been active in the union and organizations of the
unemployed and was now a foreman, was in charge of distributing the
donations.[62] The urban-based liberal groups involved in relief issues
later became important allies in labor conflicts and urged workers to
become engaged in electoral conflicts.

ELECTORAL POLITICS

The power the state administration held over relief and the growers'
influence in Sacramento politics highlighted workers' need for advocates
to influence federal and state policies. After the SRA opposed direct
federal relief, organizers began to rally against the Merriam administration. Many workers had by now established residency, and they
registered to vote, in part because proof of residency was a prerequisite
for receiving relief. Resident workers seemed likely to become a political
force capable of changing the state and local administration.

In the spring of 1938 electoral politics began to play an increasingly
central role in the clash between cotton interests and workers. The
California liberal Democratic slate would, if elected, not only upset the
Republican party's domination of California politics but replace Republicans with New Deal administrators. Culbert Olson, the state senator from Los Angeles County, was the Democratic gubernatorial candidate. Sheridan Downey, the Democratic senatorial candidate, ran
against Associated Farmers member Philip Bancroft. Both Olson and
Downey had supported Sinclair and the EPIC campaign in 1934. In 1938
Downey endorsed the controversial Ham and Eggs plan, an offshoot of
the Townsend movement, whose labyrinthine proposal offered to give
money to all those over fifty.[63] Olson ran his campaign for governor
based on New Deal promises which were by now more progressive than
those of the retrenching New Deal administration in Washington. He
attacked the power of corporate growers and processing interests and

promised that, if elected, he would enact a Wagner Act for California and defend workers' civil liberties.[64]

Progressives, liberals, and communists (now in the popular front period) rallied behind Olson and Downey. Dorothy Ray Healey, Communist party organizer, saw the Olson election as the electoral coming together of what she called the "people's movement."[65] To the organizers who had battled recalcitrant state and local political appointees, these elections were of special interest. Not only did they raise the hope of electing a sympathetic governor, but local races also promised to replace incumbent sheriffs, relief officials, and supervisors who supported cotton interests.

Organizers turned the networks and organizations that had sent privately collected relief to farm workers into political conduits to galvanize support for the Democratic slate. Lucy Sullivan McWilliams used a private trust fund to finance a campaign to persuade resident migrants to register to vote, and the FSA supported her. She wrote to Lillian Monroe that it was "absolutely essential that work start at once. If we do not build up the Democratic Party this year we will all be under the Fascists the next year."[66]

Conservative opponents threw their considerable weight behind already well entrenched candidates and political blocs. Since 1933, cotton and related businesses had opposed New Deal social policies, such as relief and federally subsidized housing, as a threat to their interests. The Associated Farmers, agricultural organizations, and industrial interests now launched a campaign to discredit the Okies, relief, and, through them, the New Deal politics of Roosevelt and Olson. In late May of 1938, only a few months after the institution of FSA direct aid, the Committee of Sixty was formed in Kern County. Within a matter of weeks it had expanded statewide and had been renamed the California Citizens' Association (CCA).[67]

The CCA represented business interests in the state. A. C. Damion, vice president of the Bank of America and manager of its Kern County branches, was the chair. Thomas W. McManus, a real estate broker and owner of the McManus Insurance Company in Bakersfield, was appointed secretary.[68] Other members included Arthur Crites, secretary of the Kern County Mutual Building and Loan Association, and Alfred Harrell, publisher of the *Bakersfield Californian*.[69] The CCA counted among its contributors the three largest California land companies, DiGiorgio Farms, Kern County Land Company, and Miller and Lux; six banks and investment companies, including the Bank of America;

twenty-seven oil companies; and a number of other businesses and utilities.[70]

The CCA launched the propaganda fight against relief, supported by the Associated Farmers, which led the drive against unionization. The state Chamber of Commerce envisioned the battle in agriculture over relief and unionism as the spearhead of a statewide fight, warning in a confidential communication:

> If the present public sentiment continues, the results are evident, and we would point out that if agriculture goes down in this fight, all employers will go down with us. We believe that a fight should be started now in the interests of all employers but that it should be made on an agricultural front.[71]

The upcoming election, coupled with a surge in union activities, increased growers' and ginners' interest in the Associated Farmers.[72]

OLSON'S ELECTION

The Republican party, which had controlled California state politics since 1910, was divided between progressives like Hiram Johnson and old standard-bearers such as the owners of the *Los Angeles Times*. Due to the relative weakness of both major political parties, pressure groups played an unusually large role in state politics, and, as the state's largest industry, agriculture therefore exercised considerable political power. When allied with urban and industrial interests, such as the state Chamber of Commerce, processors, land companies, and Giannini's Bank of America, it formed a large part of the state's conservative lobby. Although some agriculturalists and others, such as the Bank of America, supported the New Deal, most of the conservative lobbyists, both Republicans and Democrats, opposed New Deal social programs, and they presented a relatively united opposition to labor organization.

Until the 1930s, state Democratic leadership was polarized, and the potential voting strength of the Democrats was lodged in the Progressive party. In 1932, Roosevelt gained control of the Democratic nomination through support from thirty-five counties, mostly agricultural and mountain areas. FDR won because many of the newly registered voted Democratic. The party's registration increased 219 percent between 1928 and 1936, while Republican registration fell 19 percent. Most new Democrats were in the southern counties, and the San Joaquin Valley was more strongly Democratic than the rest of the state.[73]

Sinclair's candidacy in the 1934 election had reshaped the Democratic constituency of California. The number of Democratic seats in the assembly rose from 7 in 1931 to 25 in 1933,[74] but the party's inability to heal factional disputes whose history went back to the 1920s undermined efforts to take advantage of these opportunities. By the end of Sinclair's campaign, the EPIC people controlled the Democratic party. Their voting strength rested on the newly registered Democrats in Southern California.[75]

This political realignment, begun in 1934, was briefly realized in 1938 when, as *Business Week* noted, the new migrants "may hold the balance of power" in state politics.[76] As new migrants settled in the Valley, they swelled local populations by as much as 63.6 percent, and they registered to vote.[77] From 1936 to 1938 voter registration increased 35.8 percent in Madera, 22.4 percent in Kern, 15.5 percent in Kings, 23.8 percent in Merced, 15.9 percent in Tulare, and 7.4 percent in Fresno County,[78] while Los Angeles's voting strength increased only 6.7 percent.

By the 1934 election, it was clear that the new residents could be expected to vote Democratic. Most had their roots in the traditionally Democratic South. Moreover, they tended to be liberals. In 1934 they had supported Sinclair as a vote against the Depression and, many thought, to support Roosevelt's social programs. (That Roosevelt opposed Sinclair did not undermine this belief.) Poor and displaced, they supported the New Deal because they felt Roosevelt paid attention to the problems of small farmers and the poor. The San Joaquin Valley, traditionally a stronghold of large agricultural interests, was turning not only Democratic but liberal Democratic.

In November 1938 the New Deal candidate, Culbert Olson, was elected, becoming the first Democratic governor of the state in forty years. The San Joaquin Valley vote had been decisive. Olson carried only 50.17 percent of the state vote, but 53.36 percent of the Valley voters turned out for him.[79] Philip Bancroft, the defeated senatorial incumbent, wrote, "The worst jolt the results have given me is not that of my personal defeat, but the realization that our former important farm vote is pretty much a thing of the past. The representative farmers, large and small, from one end of the State to the other were well behind me; but their efforts were neutralized by hundreds and thousands of unsuccessful farmers and farm laborers who would not even vote for Ham and Eggs."[80]

News of the Olson victory swept through the cotton belt, bringing the hope that the Olson administration might finally mean a New Deal for cotton workers. After all, Olson had pledged to improve conditions for farm workers; he had proposed a little Wagner bill for California, favored unionization, and attacked corporate growers and ginners who profited at the expense of workers. If relief had played such a crucial role in organizing, even under an adverse Republican state administration, how would a sympathetic administration pledged to supporting agricultural organizing affect the power balance between workers and the industry?

OLSON'S ADMINISTRATION

Olson and his new administration openly supported agricultural unions, arguing that unions were key to changing conditions, that "as long as workers are not organized in California agriculture, they are likely to be subjected to discriminatory treatment."[81] Moreover, Olson's appointments to the sensitive jobs of director of the State Relief Administration and of the Department of Immigration and Housing put growers on notice that the governor intended to end the state's honeymoon with agricultural interests. Olson named Dewey Anderson to replace controversial SRA director Harold Pomeroy. (In a show of colors, Pomeroy was immediately appointed executive secretary for the Associated Farmers.)[82] Olson assigned liberal lawyer Carey McWilliams to head Immigration and Housing, which regulated conditions in agricultural labor camps. In 1939 McWilliams published *Factories in the Field*, a searing historical exposé of corporate agriculture, the exploitation of agricultural labor, and the efforts that were made to suppress unionization. McWilliams had concluded that the solution to the exploitation of labor "will come only when the present wasteful, vicious, undemocratic and thoroughly antisocial system of agricultural ownership in California is abolished."[83] Growers promptly attacked him as "agricultural pest number one in California."[84]

Under Merriam's Republican administration, workers had been tossed off relief roles if they refused work at the "prevailing wages" set by growers. Dewey Anderson replaced this standard with the "fair wage policy," which allowed workers to receive relief unless they were offered work at a wage that the SRA had determined to be reasonable. In effect, that is, the SRA replaced industry-set wage scales with government-determined rates, a move with staggering implications.

Battles over relief began to be expressed as a conflict between local and state control, fueled by increasing animosity in Valley towns over the high costs of relief. The Valley press clamored for local control over relief and published a steady stream of propaganda that linked "easy" relief to "chiseling hordes of indigents."[85] A *Sacramento Bee* cartoon depicted California taxpayers as an emaciated cow, wearily pumping milk into the bowl of "liberal relief benefits" while "migrant indigent" bees swarmed in a menacing cloud overhead.[86] The California Citizens Association, the Associated Farmers, the Farm Bureau, and major cotton interests all opposed the new relief policy.

Olson also faced staunch opposition within the state apparatus. State Relief Commission chairman Archibald Young proposed that the SRA be abolished, eligibility requirements be stiffened, and relief administration be turned over to grower-dominated counties. In the San Joaquin Valley there was growing pressure to turn relief over to local county supervisors, most of whom were Republicans.[87] In March, supervisors from eight counties formally demanded that relief payments be reduced, that federal legislation be enacted to curb migration into California, and that control of relief be turned over to the counties, which were more conservative than Olson's administration. Olson vowed to veto any such attempts.

Yet the Olson administration operated within the sharply defined political parameters of California's legislative and pressure-group politics. Although Olson entered office on a wave of optimism, within weeks his program for ameliorating the agricultural ills of the state had come up against the stone wall of the conservative California legislature. Republicans controlled the state Senate. Conservative Democrats controlled the assembly. Together they created a political power bloc that opposed New Deal policies and would work to thwart Olson's attempts to change relief and housing program practices. Olson could initiate programs, propose bills, and direct administration, but the legislature could veto the bills, attack his administrative appointments, and stymie new programs.[88] In the first session the legislature attacked Olson's budget, cutting social welfare programs to the bone.

The bill for relief appropriations, the major focus of the governor's conflict with the legislature, passed the committee by only one vote and was attacked in the legislature by the Republican bloc and by conservative Democrats. They challenged the governor's right to either establish or administer relief policy, and they pushed to send control over relief back to the counties. In response, Olson threatened to exercise his veto power over any such measure.[89] After acrimonious debate, the relief bill was sharply cut, and the legislature insisted upon the humiliating provision that Dewey Anderson state in writing that it was "not the desire nor the providence of SRA to set agricultural wages."[90]

The CCA and the conservative California State Employee Association (CSEA) began a Red-baiting campaign to discredit the SRA and, by extension, Olson's social policies. They accused Olson of using the SRA to build a radical political machine. In March, two Kern County SRA employees charged that the Workers Alliance had pressured the local SRA into firing or transferring them for refusing relief to striking cotton workers in 1938. This was a continuation of an old political battle. The central figure was the local relief administrator, William Plunkert, who had himself been fired by Harold Pomeroy for lobbying against a bill that would have returned control of relief to the local areas. Pomeroy had fired liberal social workers such as Plunkert with impunity, but now the conservative CSEA demanded an investigation of the SRA.[91] Plunkert was accused of being a communist in league with the CIO, and a bipartisan legislative group called for his removal. CCA propaganda used the incident to paint the SRA as a communist organization with fiscal policies that would bankrupt the state. Bending under political pressure, Anderson fired Plunkert. By August the attacks had shifted to Anderson himself, and, under pressure from the Democratic Committee, Anderson resigned in what the *People's World* called a "purge" of the SRA.[92] Olson's surrender to his opponents undermined his already shrinking base of support. The coalition of leftists and liberals that had helped elect Olson began to fragment.[93]

In 1937 and 1938, the government policies that were changing relations in the cotton industry peaked. The influx of migrants, the expansion and then contraction of the cotton crop, and the changing policies regarding relief had led to an increase in organizing by both workers and growers. The increase in organizing had, in turn, increased the involvement of both camps in the electoral process as government policies became increasingly prominent in class relations.

CHAPTER SEVEN

End of a Hope

The Strikes of 1938 and 1939

Since 1933, workers had hoped that government policies might help them, either indirectly or directly. Despite their exclusion from national legislation, supporters believed that agricultural workers would eventually be included in federal legislation. By 1938 conditions seemed optimal. There was now a national agricultural union, allied with the CIO. Federal relief still offered some assistance, federal housing was available to at least a limited number of workers, and the camps served as a base for organizing. The cotton strikes in 1938 and 1939 raised the question as to whether cotton workers, aided by relief and an affiliated national union, could improve their situation.

KERN COUNTY COTTON STRIKE, 1938

When Congress passed a second Agricultural Adjustment Act in 1938 which reestablished production controls, growers cut cotton acreage 43 percent.[1] Yet the thousands of migrants who had been drawn to the Valley the previous year remained, glutting the labor market. Relief again became the central issue, extending into a struggle over wages during harvest. In August the Agricultural Labor Bureau slashed wages by 25 cents, offering only 75 cents to pick 100 pounds of cotton.[2] In the fall of 1938, a short-lived strike hit cotton areas of Kern County, where tensions between unemployed groups, UCAPAWA, and the California Citizens Association and the Associated Farmers had been brewing for months.

In 1937 and 1938, union organizing focused on Kern County, which had a high degree of occupational fluidity, a large number of small farmers, a community sympathetic to small farmers and workers, and the support of an institutionalized labor movement. Kern County boasted some of the strongest labor unions in the Valley. By 1938, 1,500 of Kern's 5,000 union members belonged to CIO-affiliated unions.[3] Because of the close relation between industrial and agricultural sectors, the Kern County Labor Council addressed the problems of agricultural workers and small farmers as well as industrial workers, and it was instrumental in organizing agricultural unions. In 1936, left-wing AFL unionists formed the Federation of Cannery and Agricultural Workers after meeting in the Kern County labor temple. By May, the Kern County Agricultural Union had established branches in Bakersfield, Arvin, Delano, Wasco, McFarland, and Shafter.[4]

Whether because of the close relationship between these groups or of the relatively radical views of local labor leaders, the destructive split between the AFL and the CIO had little impact in Kern County. By 1937, the State Federation of Labor had expelled the CIO, and a bitter jurisdictional battle between the AFL and CIO spilled over into field organizing, seriously impeding UCAPAWA's efforts in the state. Yet in Kern County, although the Labor Council expelled three CIO affiliates, it was done only under national pressure, and it was done with "no bitterness, with handshaking all around, and with pledges of cooperation."[5] The expulsions had little real effect. CIO unions were invited to send fraternal delegates to council meetings, and the divisions that weakened organizing in other areas were absent. The Labor Council continued to provide meeting halls, donate financial support, cooperate with sympathetic FSA camp managers, and work in relief conflicts.[6] Overall, the Kern County labor movement provided a hospitable atmosphere for organizing agricultural workers: a relatively undivided union movement, workers sympathetic to field workers, and concrete support. For example, when in April 1938, the Kern County UCAPAWA began to direct protests for unemployed on relief, the Bakersfield Central Labor Council and Labor's Non-Partisan League sent resolutions to Governor Merriam and SRA director Pomeroy, demanding that relief be given to unemployed residents.[7]

In the cotton harvest of 1938, UCAPAWA called for a rate of $1 per hundred pounds of cotton picked, twenty-five cents more than the grower's offered rate. Growers refused to negotiate with the union. The 1938 cotton strike began when three hundred workers walked off a ranch, the

Camp West Lowe near Shafter, that was owned by the Camp brothers, vociferous supporters of the Associated Farmers.[8] Kern County federal FSA camps in Shafter and Arvin had, despite federal regulations, active chapters of the Workers Alliance and UCAPAWA. When growers, fearing the stirrings in the Shafter camp, boycotted workers who lived there, the strike spread.[9] Within a week 4000 Kern county workers had walked off. The strike was initially effective, and some growers began to offer 90 cents or even $1.00.[10]

The industry, however, quickly mobilized to stop wage concessions, pressuring growers into agreeing not to pay $1.00. The industry also pressured state agencies to deny relief to strikers. SRA director Harold Pomeroy, while claiming to remain impartial, denied relief to "persons who are offered work in the cotton fields and refuse it."[11] At first the California State Employment Service, following national policy, refused to refer workers to jobs where a strike was declared. But after Pomeroy refused to recognize the strike and the State Employment Commission, in an act of legalized fiction, conveniently ruled that no strike existed, the CSES began to refer workers to boycotted ranches.[12]

Despite union appeals, the Department of Labor declined to mediate the dispute, because the Associated Farmers refused to agree to mediation. The federal government had by now ceased any mediation in agriculture. There would be no repeat of Creel's role in the 1933 strike. After the strike, the Department of Labor agreed to investigate the intimidation of workers, but government officials never attempted to use the economic lever of the AAA to force growers to mediation.

The federal foot-dragging was due in large part to the fact that the strike never reached proportions that would have forced the government to intervene. Obviously to a degree, this argument is circular: if relief had been available, the strike might have been prolonged and thus have been more effective. External conditions were more conducive to organizing that year. Strikers received an unprecedented amount of support. Urban-based groups such as the John Steinbeck Committee to Aid Agricultural Organization, International Labor Defense, and urban chapters of UCAPAWA sent support, publicized the strike, and sent representatives to the Valley.[13] By now UCAPAWA locals had a backbone of long-term residents in FSA camps and in the communities.[14] Some were Anglos, but a disproportionate number were Mexicans.

CIO locals varied considerably depending on the particular dynamics of the local workers, their background, and their relations with growers and the surrounding community. Which workers joined the union? Probably a core group were resident Anglos who had been in the area at least two years or more. Residents were more connected to local communities and, because they were eligible for relief, had the resources to remain in the area. Yet in the pivotal Local no. 42 in Shafter, over half of the union members were Mexican, black, or Filipino.[15]

The Shafter local was founded in 1938, as part of a mass organizing drive in the San Joaquin Valley. Moving outward from the CIO base in Bakersfield, three unpaid UCAPAWA organizers set up camp committees in Kern County, including one in Shafter. One of these organizers, Tony Angus, claimed to have enrolled 284 members in the Shafter local.[16]

By the strike in the fall of 1938, the base for the Shafter UCAPAWA local came from two sources: the FSA camp of Anglo migrants, and the established Mexican community to the south of the town. The FSA camp residents had been active in efforts to maintain relief in face of CCA attacks. Workers in the Shafter camp had circulated a petition demanding federal relief be continued; in the process they enrolled members in the CIO.[17] During the strike Mexicans, Anglos and blacks met in the Rodríguez home nightly for strike meetings.[18] M. M. Rodríguez, a resident of Shafter, was elected treasurer of the union and was reportedly the leader of agricultural workers' Local no. 250.[19] Two hundred and fifty members of the union were listed, but meetings of over two thousand were sometimes reported, when a home-made public address system blared the proceeding to anxious strikers waiting outside the house.[20]

Union efforts were aided by the Workers Alliance, urban-based liberal groups, the San Joaquin Council of Agricultural Workers, which operated as an umbrella group for organizations seeking to aid agricultural workers, and ethnic organizations such as El Congreso del Pueblo de Habla Española (the Congress of Spanish Speaking People).[21] Yet there was not enough support for the strike among the desperate workers who composed the swollen labor pool. As Clyde Champion had told a reporter several months before, the "huge surplus of labor constitutes a menace to all labor. Those who are on the verge of starvation will not refuse work anywhere at any price and when these people are refused relief they serve to drive down the low wage scale already in existence."[22] Most of them were too disoriented and poor to join the union, let alone support what must have seemed a quixotic strike. The arrest of thirty-nine strikers broke morale of strikers and drew the strike to a close. The strike failed to spread beyond Kern County, and it had collapsed within two weeks.[23]

The defeat of the 1938 strike and pressure from growers ultimately ended the camp's life as a union base. By March 1939, the camp council, under pressure from the Associated Farmers, local farmers, and the local press, voted to suspend CIO meetings at the camp temporarily and to discontinue the CIO-written articles in the paper, pending investigation of the camp.[24] By September 1939, the manager of the Shafter FSA camp reported there was no noticeable union activity in the camp, although the Shafter union remained in existence.[25]

By November 1939, the UCAPAWA base had shifted to the Mexican colony. As union organizing in Shafter was grinding to a halt, organizers from the local based in the Shafter Mexican colony attended a conference of UCAPAWA locals in the San Joaquin Valley. At that conference it was decided that all field workers' locals had to be self-supporting and that the number of paid organizers in the Valley would be cut from five to three. The Shafter local called a meeting of workers on the Hoover ranch to vote for representation by UCAPAWA and was active in the letter-writing campaign to encourage Governor Olson to continue with the wage-rate hearings.[26]

COTTON: TEST CASE FOR OLSON'S POLICIES

Perhaps more important than its outbreak in the strike was the expansion of the conflict over relief and unions into state electoral politics. Throughout the 1938 strike, organizers stressed the importance of the upcoming gubernatorial, senatorial, and local elections. Olson's victory occurred at an important point, and for a short time it seemed that it might alter the state's role in class relations.

In 1939, cotton became the test case for Olson's agricultural policies. The struggle took place within the government itself, between state and local offices over administration of relief and over wages and conditions in the cotton camps. This division paralleled the growing conflict between the Associated Farmers and their allies and the Workers Alliance, UCAPAWA, and their supporters. The 1939 battle focused on Madera, a town of 8,000 that was the hub of California's fourth largest cotton area. The Madera conflict had its genesis in four years of changing social, economic, and political conditions in the community.

Madera's development was paradigmatic of the changes in cotton. In 1937 alone cotton acreage in Madera had expanded 300 percent. The population increased 400 percent between 1935 and 1937. The influx of white migrants reflected in these figures dramatically changed the social and political composition of the area. Voter registration increased by 35.8 percent. The number of school children mushroomed, hospital loads soared, and demands for housing and relief intensified. Older residents became increasingly concerned about escalating taxes, social changes in the community, and shifting configurations of political power. The extent and rapidity of change in Madera contributed to polarization of the community and encouraged both growers and workers to organize to protect their position.[27] Madera became a test case for the Associated Farmers, the Workers Alliance, and the union.

The cotton industry formed a solid phalanx with local press, law enforcement officials, local SRA branches, the Board of Supervisors, and the district attorney in opposition to the Workers Alliance and UCA-PAWA. In early 1937, prominent Madera businessmen with ties to the cotton industry, the Agricultural Labor Bureau, the press, and relief administration formed a branch of the Associated Farmers which, as the statewide organization, represented not only growers but the political and economic establishment. Among its members were Ben Hayes, a large cotton grower and secretary of the ALB; Dr. Lee Stone, the pugnacious director of the county health department; and Ted Marston, business manager and part owner of the community newspaper, the *Madera Daily Tribune*. The organization developed such close ties with local branches of law enforcement that the Associated Farmers, the district attorney, and the Board of Supervisors collaborated in writing the local anti-picketing ordinance. The ordinance, similar to those passed in many communities since the 1933 strikes, provided a legal mechanism against union organization by making it a misdemeanor to use violence, intimidation, threats, or even profanity to induce laborers to stop work.[28]

While UCAPAWA was concentrating on organizing processing industries and was smarting under the bitter jurisdictional fight with the AFL, CIO activities in Madera focused on organizing field workers. Organizing in cotton took place against the backdrop of a 1938 strike by 650 sheep shearers, most of whom were Mexicans. Both the union and the Associated Farmers saw the shearers' strike as a test case, and it had been the site of intense organizing by both groups. Although the strike was defeated, it brought workers into the CIO. Gene Luna, one of the sheep shearers who also picked cotton, learned from the strike that he could have "a piece of the pie," making him willing to join the cotton strike in 1939.[29]

The increasing strength of the Associated Farmers and the upsurge in the local population spurred the organization of the Workers Alliance in Madera among the growing number of residents on relief. Since the union's formation in 1937, UCAPAWA locals had dwindled steadily.

But the alliance had grown, until by 1939 it had 186 local units in California with a claimed membership of 42,000 and a paid membership of 12,000. By the fall of 1939, one branch of the Workers Alliance was active in the town of Madera, and in January, workers in Chowchilla formed a local with 175 members.[30]

In this already tense atmosphere, Olson's relief policy sparked a conflict during the 1939 cotton chopping season. When the ALB set a 20 cent wage for cotton chopping, dissatisfied workers and the Workers Alliance asked Olson to hold hearings to establish a "fair wage" that would set the level below which workers could not be refused relief if they would not take jobs in cotton. The Workers Alliance discouraged workers from taking jobs (while not calling a strike), picketed local SRA offices, and held mass meetings. Meanwhile, the Madera relief office began to deny relief to those who refused work at prevailing wages. Under pressure, Olson appointed a committee to determine a fair wage for chopping cotton, which would be used as a benchmark to determine the SRA scale. The commission, headed by Carey McWilliams, denounced the ALB rate as not even "subsistence" and recommended 27.5 cents or $1.25 an acre, the same amount suggested by the alliance. With this ruling, the state government effectively created a higher wage in cotton, guaranteeing that workers who refused low wages would still receive relief.[31]

The *Madera Tribune* lambasted the relief ruling as "illustrat-[ing] . . . just what business has suffered under the New Deal and what the state administration is preparing for it."[32] The Executive Committee of the Associated Farmers warned that "an arbitrary wage fixed by some governmental agency would be disastrous" because it would put California growers at a competitive disadvantage that would drive them out of business.[33] Under pressure from the Madera Associated Farmers, local SRA and WPA administrators refused to comply with the state administration and continued to force clients off relief rolls to work in the fields. By August the California SRA was in open conflict with the Madera office.[34]

The conditions in local cotton camps became an issue of contention between state and local government. As head of Housing and Immigration, Carey McWilliams had begun to enforce long-ignored housing codes for cotton camps. McWilliams posed an interesting argument for increased government regulation. He pointed out that not only did laws on the books need to be enforced, but because government relief payments often went to pay rent on cotton camps, the government was indirectly subsidizing growers for housing and thus claimed a direct interest in their conditions. McWilliams tripled camp inspections and directed the glare of publicity to the camps by describing bad camps on the radio and organizing tours of social workers through Valley labor camps.[35]

Dr. Lee Stone, director of the Madera County health unit, reacted with anger, claiming that the county's 2,600 cabins complied with state regulations and refusing to enforce housing laws. The Workers Alliance accused Stone of violating state law, began a letter-writing campaign detailing rotted housing, unsanitary conditions, and the contagious diseases that flourished in the camps. McWilliams pressed on with camp inspections, threatening to close those found not to be in compliance. Stone in turn attacked McWilliams as an ally of the CIO and the Workers Alliance.[36]

COTTON AND THE NATIONAL SCOPE

By 1939, the low wages and poor conditions in cotton had become a national concern through the publication of exposés, novels, and photographs of dispossessed white farmers, sharecroppers, and field workers. In 1939 several widely read books were published: the evocative *You Have Seen Their Faces*, featuring photographs by Margaret Bourke White and a text by Erskine Caldwell; *An American Exodus*, with Dorothea Lange's photographs of dispossessed sharecroppers and migrant workers and text by her husband, Paul Taylor; and James Agee and Walker Evans's *Let Us Now Praise Famous Men*. The moving photographs by Bourke White, Lange, and Evans and the lucid prose of Agee and Taylor personalized the poverty of southern and southwestern cotton tenants and sharecroppers, many of whom now worked in California's cotton fields. Housing and Immigration director Carey McWilliams published *Factories in the Field*, a searing exposé of California's farm labor system culled from a series of articles McWilliams had written throughout the 1930s. But perhaps the most powerful of them all was John Steinbeck's *The Grapes of Wrath*.

The Grapes of Wrath was based on a series of newspaper articles Steinbeck had written on white migrant agricultural workers, reissued by the Simon J. Lubin Society as a pamphlet in 1938, "Their Blood Is Strong." The pamphlet and book were based on Steinbeck's own experience with migrants, and they captured the human toll of economic deprivation, disorganization, and disruption of these white agricultural workers as no statistical report had done. The book indicted the agricultural industry as a whole. But as it had been for decades in the South, cotton labor was a major focus in the book. When Rose-of-Sharon's baby died in a cotton picker's shack made from a converted freight car, the onus was symbolically laid squarely on the cotton industry. Drawing upon the historical sense of an earlier cotton-labor system and its literary critics, reviewers compared the book to Harriet Beecher Stowe's *Uncle Tom's Cabin*. *The Grapes of Wrath* became a best-seller, and by 1940 the film version, which starred Henry Fonda, became a major box office attraction, even in towns of the San Joaquin Valley.[37] Hoping to avoid a repetition of the social impact of *Uncle Tom's Cabin*, growers launched a cathartic but largely ineffective counterattack against the novel. Kern County supervisors banned the book from school and public libraries. W. B. Camp led the Associated Farmers in an unsuccessful attempt to institute a statewide ban and officiated over a public book-burning. The furor over the novel brought workers' conditions to national attention: *Life* published a photo essay on the labor camps, and the *New York Times* ran a series of articles.

Although these works brought considerable ideological gains, they were based on preconceptions lodged in the American psyche that were ultimately to push migrant field workers to the margins of written history. These works, like many other studies, focused on the growth of capitalist agriculture and the related decline of the family farm and were concerned about the implications for American culture, the political economy, and the agrarian dream. The questions asked about capitalist agriculture were shaped by broader questions about American democracy, as measured against a mythologized past. Farm workers, rather than being seen as individuals in their own right, were depicted as primarily the degraded result of the family farm's demise. As was the case with histories of unskilled workers in industry, these writings on farm people became molded by the pressing conditions of their lives, and the wretchedness of their condition became confused with their social world. Pictured as victims of a brutal system, they emerged as powerless, passive, and, ultimately, outside the flow of history.[38]

The political response to the growing concern over white migrants focused less on migrant farm workers than on the problems of the small family farmer who had been forced to migrate. The lingering pressure to include farm workers in federal legislation was dimming. Instead Congress focused on investigating the conditions of migrants and, as an offshoot of the migrant problem, the problems they encountered as agricultural laborers.

California's congressional delegation urged the federal government to send relief to help the immediate situation but was deeply divided on the appropriate long-term governmental solution. Republicans argued that the federal government should return migrants to their home states and refuse relief to nonresidents. Liberal representatives Jerry Voorhis and John Tolan pointed out that the problem, being national, defied localized solutions. They linked migrancy to southern economic stagnation and requested an investigation into interstate migration. Voorhis argued that the government should help migrants adjust to their new homes; Tolan introduced a resolution to investigate the problems of interstate migration. Roosevelt, increasingly wary of touching the politically sensitive issue of farm labor, refused a congressional delegation that asked him to have federal agencies tackle the problem. As a result, the issue languished in Congress, where even the California delegation itself was unable to agree on anything beyond the Tolan investigation. When Congress adjourned in August, no action had been taken.[39]

Elements in Congress still favored extending collective bargaining rights to farm workers. But by now the strategy had shifted from including farm workers in bills to launching investigations into conditions. In late 1938 Senator La Follette's Committee to Investigate the Violations of Free Speech and the Rights of Labor expanded to investigate the Associated Farmers of California, and it conducted hearings into the cotton industry's economic structure, the Agricultural Labor Bureau's activities, and the cotton strikes. The committee's political agenda was to argue for the need for federal protection of agricultural unionization and to include agricultural workers in federal social legislation. The hearings turned the national spotlight on labor relations in California agriculture, but by the time legislation might have been proposed, the impending war in Europe was overshadowing concerns about farm workers. As a result, although there was more national awareness of the problems of farm workers in 1939, no substantial change would be made in their situation.

THE STRIKE

By the fall of 1939, federal policies had helped improve the structural position of the cotton industry in relation to workers. AAA payments and government policies were credited with increasing net profits by 4 cents a pound.[40] Eighty percent of the California crop was sold to Japan and India and was thus free from the restrictions of the Domestic Allotment

Plan.[41] Cotton prices had increased to $13.00 a bale, with further increases expected due to the impending war. Federal payments—over $950,000 to Madera county growers alone—provided a more stable economic base for the industry.[42] Growers who had initially resisted federal encroachment were now endorsing the program. As Madera grower Ben Hayes testified in May, "In the past few years . . . the cotton business was generally a profitable business. Our costs were much higher than they are today. I think that we can all agree that the control program . . . was a good program."[43]

The federal relief program had contributed to the disorganization of labor by indirectly subsidizing the larger growers, who received a major share of the AAA payments. These growers, who supported the ALB and the Associated Farmers, hired the most workers and thus had a greater stake in labor issues. In Madera, members of the Associated Farmers received the highest government subsidies. Twenty-three members of the Associated Farmers (2 percent of Madera growers) grew crops on 56 percent of the total land planted and received total benefits of $274,995.[44]

Whereas growers and the Associated Farmers were in a stronger position than before, UCAPAWA was facing problems. The union seemed to have its greatest potential strength in California, which had a history of farm labor organizing. Yet UCAPAWA activity, despite its national affiliation and membership, compared poorly with that of the CAWIU. By 1939 organizing had dwindled from its high in 1933 and 1934, when 23 strikes involving a total of 42,000 workers, mostly Mexicans, had broken out in the state. While in 1937 agricultural strikes increased in other states, only 16, involving only 4,000 workers altogether, broke out in California. In 1938 there were 12 California strikes. California's UCAPAWA had only fifteen locals: six of these were inactive. Only in 1939, with the cotton strike, did California's agricultural strikes begin to reach the level and magnitude of the earlier conflicts.[45]

While it could be argued that these figures might indicate more substantial organizing which did not necessitate strikes, that does not appear to be the case. In November 1938, after a series of unsuccessful strikes which drained the union's treasury, UCAPAWA leadership decided to forbid unauthorized strikes and to focus organizing among workers in the processing, packing, and canning industries.[46] They reasoned that if they were able to organize these better-paid and more stable workers, they could use them as a base from which to extend organizing to field workers. Although the shift in focus was intended only as a temporary expedient, the union never again expended as much effort on field workers as they had done earlier.[47]

Processing workers organized by the CIO did help in organizing field workers that year. Highly skilled black cotton-compress and gin operatives, originally from Oklahoma, Louisiana, and Arkansas, became critical to CIO organizing in processing plants. In April 1937, the Los Angeles International Longshoremen's and Warehousemen's Union (CIO) local at San Pedro organized black workers into a union at an Anderson Clayton cotton compress plant and won a wage increase.[48] The CIO was simultaneously organizing Clayton plants in Phoenix and Fresno. Workers had already organized plants in Bakersfield and Fresno, and negotiations were underway for wage increases by 1938.[49] These gin and compress workers formed an additional base for UCAPAWA organizing in the Valley and had taken part in demonstrations against relief policy in Kern County.

Yet in 1939, CIO unionization in processing was foundering on the bitter conflict between the AFL and the CIO. The AFL's ability to control the processing, shipping, and packing industries undercut the degree to which UCAPAWA could use those workers to help organize in the fields. By November 1938 the union changed the tactics it used in agricultural organizing. Worried about the debilitating effects of unsuccessful strikes, the executive board ruled that the union would oppose strikes except as a last resort.

Following the direction of labor unionism as a whole, UCAPAWA began to rely increasingly on federal intervention. Olson's political victory had revived hopes that unionization, even in agriculture, could proceed with help from a benevolent administration. UCAPAWA's increasing dependence on the government to set wages and their reluctance to resort to strikes were understandable: strikes drained their very limited funds. Yet government policy and its administration depended on the economic and political configuration of power.[50] Agricultural workers were more vulnerable than their industrial counterparts, and they faced an industry that was better organized and more powerful as a body than the nonagricultural sector. As the strike of 1939 was to show, government intervention meant little without the political clout to support sympathetic policies and their administration.

In August, UCAPAWA and the Workers Alliance set their wage demands. Arguing that the impending war was raising cotton prices, they demanded that $1.25 be paid for each hundred pounds picked and that wages be placed on a sliding scale, to be renegotiated if cotton prices increased.[51] The ALB set the rate at 80 cents, and local SRA offices refused to comply with the state relief policy, cut people off relief, and ordered them to work at prevailing wages.

The legislature had defeated Olson's earlier efforts to set up a permanent wage-rate board to set relief rates. But in late September, hoping to stave off a strike, Olson called for a wage hearing and appointed a wage board. The specific appointments to the wage board suggested that Olson was bowing to political pressures: while there were representatives of agriculture and the state, and some sympathetic to labor, there were no union representatives on the board. Board members R. L. Adams, agricultural professor at the University of California, Berkeley; Ray Wiser of the California Farm Bureau Federation; and George Sehlmeyer, master of the California Grange (representing the interests of some of the small farmers) all supported cotton growers' interests. William B. Parker, director of the state Department of Agriculture, sat on the board; Carey McWilliams, Mrs. H. E. Erdman, and H. C. Carrasco were appointed to represent labor's interests. The Agricultural Labor Bureau and the Associated Farmers refused Olson's invitation to participate, on the grounds that the state had no legal right to set wages.[52] But Olson did not ask the CIO, which requested representation, to sit on the board.[53]

The union was relying on the wage hearings to set relief rates. Given that government rulings had changed wages, this was understandable. Yet the extent to which the union relied on the government detracted from the sense of urgency needed to induce the government to act in workers' favor. Rather than encouraging strikes that could exert pressure on the board, union organizers actively attempted to stop workers from striking before the board issued a ruling.[54]

The board's recommendation was divided. Three members favored keeping workers on relief unless $1.25 was offered. Four suggested a flexible relief formula by which workers would not be forced to work for less than the SRA rate. But the board refused to set wage guidelines. Without these guidelines, their unanimous agreement that wage rates should be determined by "genuine collective bargaining" was gratuitous.[55] The majority opinion opposed state or union intervention and urged that wages be set "by agreement between individual growers and their respective workers, and not by opposing groups with divergent interests."[56] The union's faith in the government had proved to be misplaced.

The board's refusal to set a rate indicated the government's sensitivity to political pressure. In effect, the board put relief policy back into the hands of local relief offices, and, consequently, under the influence of growers. They signaled that the union could not count on the state government for substantial support. And they precipitated a strike.

With no other recourse left, the union called a strike. The strike spread sporadically around the Valley; walkouts were reported in Corcoran, Visalia, Poplar, Fresno, Kern County, and Madera. Yet the strike was limited. Organizers reported that they had little trouble in getting workers to walk out, yet as soon as the CIO truck disappeared over the horizon, they returned to work. Fresno reported only one ranch on strike. In Kern County strikes were called on 150 ranches, but strikebreakers were found and picking continued.[57] In Corcoran, where growers were paying 90 cents, estimates of the number on strike varied between 400 and 1,000.[58] Although strikes were called on 274 ranches outside Madera, the strike failed to slow production decisively. Estimates varied, yet even the highest figures calculated that less than a third of the work force went on strike.[59]

Why did the strike fail to spread more widely? In many areas workers had simply not heard of the strike. In others, workers walked out, but only temporarily. The memory of last year's failed strike was still fresh in workers' minds. The competition for work was still fierce, and workers new to the area and hungry for work were available to replace strikers. Some workers were indifferent or hostile to a union. The work force lacked the ethnic cohesiveness that had reinforced Mexicans' sense of class in opposition to growers. Many workers still maintained a "small farmer" mentality. Yet by 1939 there were signs that after a few seasons in California agriculture some new migrants became more receptive to workers' organizations. In areas where strong local organizations existed, such as the FSA camps and Madera, more workers supported the strikes than in areas without organizations. Yet the overall difficulties in organizing were greater in 1939 than they had been in 1933.

UCAPAWA, like the CAWIU before it, was underfunded and understaffed. Their increasing national resources were, by 1938, devoted to organizing processing workers, not field workers. The union refused to sanction wildcat strikes, which were common in agriculture. And in November 1939 the number of Valley field organizers was cut to three. Thus the union failed to provide substantially greater support than the CAWIU had supplied earlier. In 1939, as in 1933, most activities still depended on local leaders and the informal networks of workers themselves. But these were weaker than they had been in 1933. The social networks Mexicans had built over the years had helped precipitate, spread, and maintain the strike in 1933. But the new migrants were only beginning to develop such networks locally and had developed nothing comparable on a more extended basis. In part, this was because they were new to the area. In part it was because instead of continuing to migrate, they became residents. Residency provided their base for developing social networks, communities, and long-term social and labor organizations. The Workers Alliance, for example, depended on residents as members. But agricultural workers were part of a common labor pool and worked on farms spread over several hundred miles; without a strong union, the decline in migrancy thus reduced workers' contact with each other and limited their connection with organizing efforts.

In this same period the industry had enhanced its statewide coordination. The expanded, better financed, and more sophisticated Associated Farmers was a growing impediment to unions. Stable union organization demanded strong local bases, yet the union had to cover several hundred miles and stop production among enough growers to force the entire industry—not just one or several growers or even a particular region—to agree to a rate change. This necessitated industry-wide strikes, which became increasingly difficult to conduct. As workers became more localized, the responsibility for spreading organization and strikes shifted to the union, whose resources were inadequate for the job. Recognizing this dilemma, in 1939 the union concentrated on selected areas: Arvin, Corcoran, Pixley, Visalia, and Madera. Even then, only in Madera did the strike come close to a complete walkout.

The Madera UCAPAWA and the Workers Alliance held mass meetings, but after growers refused to meet with union representatives, spontaneous strikes began. Growers evicted strikers from labor camps, and the evicted strikers congregated in the city park near the center of town. By 10 October, 60 to 75 percent of the pickers had joined the strike.[60]

UCAPAWA received support from other unionists, from liberals, and from Mexican organizations. UCAPAWA organizers from Los Angeles, San Francisco, and Seattle came to help.[61] Black compress workers, recently organized by UCAPAWA, sent representatives, and two compress workers, Elijah Samuel Moore and Jesse McHenry, became strike leaders.[62] The support from the compress workers indicates that the idea of using processing workers as a basis for organizing was a practical one. These workers were more geographically stable, more economically secure, and covered by the National Labor Relations Act.

Mexicans continued to play a role disproportionate to their numbers. Some of the 650 Mexican sheep shearers who had participated in the 1938 strike now joined the cotton strike.[63] Mexicans remained crucial to UCAPAWA activities in the Valley. They had founded seven of the ten field workers' locals and were a substantial part of the membership in Shafter, Wasco, Earlimart, Lamont, Lindsey, and Bakersfield locals.[64] Mexican CIO organizers, among them Stephen Rodríguez, secretary of the Shafter UCAPAWA local, went to Madera.

Mexican organizations worked in tandem with UCAPAWA, Labor's Non-Partisan League, and the Workers Alliance. Josefina Fierro de Bright and Eduardo Quevedo, leaders of the statewide Congreso del Pueblo de Habla Española brought clothes and food to strikers. Leaders of the Congress worked with left-wing allies: Quevedo was also an officer of the Los Angeles Workers Alliance. The Congress had been formed in Los Angeles in 1937 as a statewide federation of organizations that included unions, social clubs, and mutual aid societies. The organization drew upon Mexican leftists involved in earlier conflicts. By 1939, they had established branches in the cotton towns of Shafter, Bakersfield, and Tulare. The Tulare branch had over five hundred members.

More research needs to be done on the Congress, but there is evidence that it played multiple roles.[65] It may have served as an organizational center for the informal social networks that had been employed in the 1933 strike. It supported unions and organized workers. A member of the Tulare branch of the Congress testified before the state cotton-wage hearings in 1939, "We bargain collectively for our members and we . . . defend our members if they are thrown in jail because they take active part in trouble."[66]

UCAPAWA used tactics reminiscent of the strike of 1933. The union began mass picketing of cotton ranches. Long caravans of cars carrying placards wound along the roads by the fields. To prevent more walkouts, on October 12 the sheriff arrested 142 strikers for violating the anti-picketing ordinance. Yet the arrests backfired. The Madera jail, too small to hold 142 people, abutted the park and became a visible symbol of repression. The next day the district attorney ordered strikers released because the arrests "seemed to be inflaming the situation."[67] He was right. After the arrests, picket lines doubled in size. Workers formed new caravans. By October 15, 90 percent of the workers were on strike. Picking had ended.

Faced with the largest strike since 1933 and with local officials helpless to stop it, the Associated Farmers tapped the state organization for help. Stuart Strathman, field secretary for the Associated Farmers, had been traveling through strike areas and went to Madera. Publicly he urged a "cool, calm and thorough" legal offensive. He was franker in a private letter to Harold Pomeroy, ex-SRA chief who was now secretary of the Associated Farmers: "Well, we really have a tough situation on our hands. . . . Although to date we have played it down for obvious reasons . . . we plan to break loose on it now as we can not hurt the situation by giving it the works in certain areas."[68]

Strathman brought to bear the more sophisticated anti-union techniques growers had developed since 1933. Strathman and the Madera Associated Farmers formed the Growers Emergency Committee, which became the organizational nucleus for anti-union activity. Fourteen members of the committee belonged to the Associated Farmers and nine sat on its board of directors. Together they organized a joint strategy to stop the strike. The group drew up blacklists, giving organizers' names along with their license plate numbers. They established a telephone network to summon growers to a ranch quickly if pickets appeared. They urged growers to evict strikers and to stay on their property to prevent strikers from coming onto their land.[69]

They had learned the importance of public opinion. A public relations committee was appointed to "give the papers the facts."[70] The so-called Emergency Committee ran a full-page ad in the *Madera Daily Tribune*, addressed to "our fellow citizens," which denounced the strike as "an attempt to disrupt the community, [and] overthrow law and order." No longer limiting themselves to Red-baiting, the Associated Farmers' rhetoric had become broader based, if not more sophisticated. Invoking the shibboleth of American property rights, growers presented themselves as paragons of democracy fighting the threat of communism and Hitler. The committee claimed that Madera had "been invaded by two organizations organized and headed by nationally known Communists . . . following in the footsteps of Hitler and Stalin, using Communistic propaganda and Hitleristic tactics in an effort to dominate an entire community whose prosperity is founded upon the American right of the farmers to grow, harvest and transport their crops. . . . It is the duty of all right thinking people . . . [to] reestablish true American thinking and dedicate ourselves anew to the principles of democracy."[71] In a sense they were presenting themselves as fighting against some form of agricultural fifth column.

Yet behind their rhetoric lay a well-organized vigilantism. An editorial in the same edition of the paper emphasized the threat of physical violence against strikers. Calling on the history of the decline of the IWW, the editorial warned, "There is one little item that has been overlooked by the radicals. . . . The history of the rise and cause for the sudden crackup of the IWW. History always repeats itself. Enthusiasm for the IWW ceased when a few were found suspended from a bridge. . . . So far as anyone has ever been able to prove, they were letting themselves down with ropes, the ropes slipped and became tangled about their heads."[72] The threat was not hyperbole. Organizers were threatened with lynching, and one black organizer narrowly escaped being lynched by fifteen white men.[73] Armed growers patrolled the countryside looking for strike leaders and the American Civil Liberties Union attorney A. L. Wirin, threatening to roll them in melted tar and cotton.[74] One thousand ranchers armed with shotguns and rifles marched on the strikers' camp in the Madera park. In a situation described as "extremely tense," they were dissuaded from a bloody confrontation by the sheriff and Strathman, who preferred less public confrontations. Growers, meeting later with Strathman, punctuated his speech with threats to tar and feather strikers and "string up the speaker [for the strikers]."[75]

Some strikers urged that they arm themselves for self-defense, especially the Mexican and black workers who were the butt of the threats. Yet the union continued to rely on government protection, appealing to Governor Olson to send in the California Highway Patrol. Only after the armed confrontation did Olson send in twenty-four men from the Highway Patrol and Colonel Charles Henderson of the California National Guard. Vowing to "protect the civil rights and liberties of all individuals, including the lowliest transient agricultural laborer," Olson also ordered George Kidwell, director of Industrial Relations, in to mediate the strike.[76]

But the assistance was too little and came too late. The strike had gone beyond mediation. Moreover, the Associated Farmers and growers flaunted state directives with impunity, issuing an ultimatum to Henderson: either imprison strikers and end picketing or, as one grower said, "If you do not stop it, we will. We will be the law."[77] According to evidence presented at a government hearing, Henderson made no effort to stop growers but, according to growers, made a private agreement with them to stop picketing and arrest "agitators." Whether or not the agreement was made, neither Henderson nor the state administration tried to stop the Associated Farmers. Strike leaders were arrested for violating the anti-picketing ordinance, and A. L. Wirin, attorney for the American Civil Liberties Union, was forcibly escorted across the state line. Picketing stopped.[78] Wirin's appeal to Olson to send in the militia went unanswered.[79]

Unchallenged by the state government, approximately two hundred growers armed with pick handles, clubs, and rubber hoses attacked a car caravan of pickets and then the three hundred unarmed strikers in the city park. The California Highway Patrol ended the hand-to-hand fighting by shooting tear gas into the park to rout the strikers and surrounding the park with guards to prevent them from returning.[80] This action in effect ended the strike. The International Legal Defense condemned the Highway Patrol, brought in ostensibly to protect the civil rights of workers, for doing nothing but, as the *People's World* described it, "helping deputies throw tear gas into the crowd of strikers to prevent them from fighting back in self defense." The paper rhetorically asked whether there was any difference between Olson and ex-governor Merriam "with regard to the constitutional rights and liberties."[81]

The tentative and overdue steps taken to mediate (but not arbitrate) the strike were meaningless without the power of enforcement. With the strike effectively squashed, the Growers Emergency Committee rejected mediation offers by Industrial Relations. This time, unlike 1933, the federal government did not intervene. Although an NLRB conciliator was in Fresno, reportedly waiting to intervene if state efforts failed, the federal government did not become involved, and the conciliator left the area. But the strike was over. Rains, not the union or the state or federal government, pushed some growers to offer $1.00. Within six days picking had returned to normal.

The union's faith in the Olson administration had failed to take into account the rocky shores of state politics. Olson verbally supported the strike, but political pressures were taking a toll. His beleaguered administration had been defeated in legislation, and was accused of being sympathetic to communists. One member of the strikers' committee said Olson refused to take a stand because of rumored communist involvement in the strike.[82] The news coming out of Madera of communist involvement, the increasing right-wing tenor of the state and the nation, and the relative sophistication of Associated Farmers propaganda no doubt contributed to Olson's immobility. But the importance of agricultural interests to the state economy and within the political structure and administration undoubtedly also contributed to Olson's decision not to protect farm workers, much as it had dissuaded FDR and Congress from supporting farm workers on a national scale.

The defeat of the largest agricultural strike since 1933 signaled the end of an era. Considering the proportion of Mexicans to Anglos in the overall field worker population by the late 1930s, Mexicans played a disproportionate role in union locals, as they had in the California representation in the formation of UCAPAWA. Yet what gains had been made had been overshadowed by the growing organizational strength of the cotton industry, buttressed as it was by federal programs. The defeat signaled UCAPAWA's failure in the fields and ended hopes that a New Deal, either federal or state, would help cotton workers.

CHAPTER EIGHT

"Down the valleys wild . . . "

Conclusion

Driving south through the San Joaquin Valley on Highway 5, RVs and cars returning from vacations in the Sierra Nevada, San Francisco, and the growing metropolitan areas of the Valley sail past big rigs transporting goods to market in Los Angeles. Urban areas infringe on crop land, pushing out those crops no longer able to compete with urban sprawl. International competition has changed the patterns of crops. Farms have gone the way of manufacturing, and some crop production has fled to Mexico, Costa Rica, and other countries with less stringent environmental laws and cheaper labor. A dwindling water table contaminated by years of pesticide and chemical use brings periodic bouts of the flu-like symptoms of pesticide poisonings to towns, and rashes of cancer deaths that allegedly are due to the pesticides. The Valley remains a cornucopia of California agriculture, but it is a modern, streamlined, mechanized, technology-dependent agriculture in which the organization laid down in the 1920s and 1930s has reached a high gloss. The complex interplay among the government, the agricultural industry, and workers persists in a different time and under changed economic conditions.

Yet the pentimento of the past can be discerned beneath the colors of the present. The voices of workers carry faint reverberations of earlier days. Cotton work is now mechanized, but the work force in other crops is still drawn from an international pool of marginal workers. Settled Chicanos form the most stable element of the work force, augmented by an underclass of workers who migrate, sleep in the fields on plastic sheeting, are subject to periodic deportations, and are paid rock-bottom wages. The glaring disparity between the bounty of the agricultural industry and the conditions and wages of its workers remains as painfully evident as it has been for the past hundred years. These workers are from a variety of areas: Central Americans who fled wars in their homelands, immigrants from such traditional sources of labor as the Mexican states of Michoacán, Guerrero, and Jalisco, and newer migrants such as Mixtec Indians from the southern Mexican state of Oaxaca.

The United Farm Workers (UFW), organized in 1965 and a potent force in mobilizing the Chicano movement of the 1960s and 1970s, has moved from its Valley birthplace in Delano to the town of Keene in the Tehachapi mountains. Some suggest that the UFW move from the fields to the mountains is symbolic. The UFW has been plagued by internal divisions, stung by memories of earlier ambivalent positions toward immigrant workers, and exhausted by the wear and tear of struggling to survive economic depression, international competition, and declining public and governmental support. Cesar Chavez has died.

Yet the spirit of the early UFW—and, indeed, of the strikes of the 1930s—lives on. It has moved with its ex-field organizers to work in urban areas. It has been sparked among new immigrants in the fields who are organizing unions based on social and familial networks. Mixtecs are organizing campesinos, and a resurgence of transnational migration has led to transnational organizations such as the Frente Mixteco-Zapoteco Binacional (Mixtec-Zapotec Binational Front). In September 1993 the Frente signed a pact with the UFW. In April 1994, the spirit of the early UFW was revived as union leaders and members walked the 330-mile pilgrimage from Delano to Sacramento in commemoration of the march 28 years ago, and returned to organizing workers in the fields. Activity in California fields resonates across the border in Mexico.

Ironically, it is perhaps in these fields, even more than in factories, that one can view and evaluate the complex interrelation between human agency and the economic and political structures within which people live and work. The agricultural industry remains powerful, technologically sophisticated, and the beneficiary of government support. The labor system remains structurally similar to the one introduced a hundred years ago. But in the 1990s, the conditions historically faced by agricultural workers—of being drawn from a pool of mobile, expendable workers— have been extended to factory workers, teachers, engineers, and other professionals. With the restructuring of the U.S. economy, the questions raised in agriculture have become more obviously and personally pertinent to the rest of society.

In viewing a section of California's agricultural cornucopia, it becomes clear that struggles of people cannot be fully appreciated or understood without an elucidation of the points of view of all the protagonists. Central to understanding the world of the workers is an understanding of the constraints, needs, and limitations of the cotton industry itself. Leaving aside moral pronouncements about growers, the capital structure of the cotton industry itself imposed requirements on growers and workers alike. Economic competition and the cutthroat world of profit margins chartered the course for California agriculture and was early on embedded in land values, taxes, and profits, and especially in the treatment and payment of workers. California agriculturalists were businessmen who did apply solutions from industry to agriculture, to the degree possible given the nature of production. Company towns, permanent workers, improved housing, and increasing specialization, regulation, and economically mandated cooperation became part of the cotton industry's lexicon.

Yet the business solutions applied in other industries were limited in agriculture by the nature of production there: agricultural production, unlike production by machines, is regulated by seasons, harvests, and weather. Nature, that unreliable and capricious dictator of production, shaped the labor demands, formed the industry's harsh and unambivalent opposition to unions or any possible interference during the slim window of harvest time, and pressured the industry to organize to increase control over labor sources and activities. The effort to regulate production and labor united the industry in a way that never occurred within the manufacturing sector as a whole. It created a unity among cotton interests that served them well in their relations with the state and local governments and with unions and the work force. Ultimately, the economic structure of the industry formed the basis for its political response to New Deal economic and social programs.

Workers came to labor within the parameters formed by these economic and political constraints. Because of the considerable power of the agricultural industry, the wretchedness of working conditions tended to be confused with the social world of the workers. Studies lumped agricultural workers together as "victims" of the system. Yet the image of the victim, with its implied absence of will, choice, or creativity devalues the intangible human elements workers brought with them: their work culture, national consciousness, experiences as workers, history of organizing and struggles. All of these components, which themselves differed over time and among the workers, together shaped how they would respond.

The history of agricultural unionization must take these factors into account. Organizing agricultural workers into unions was and remains formidable because of the impermanence of migrancy, wage levels too low to support a union, the seasonality of work, and the power of the agricultural industry. Yet these factors do not make up a sufficient explanation. In spite of these formidable restrictions, farm workers have organized to improve conditions. How they have organized, the alliances they have formed, and the outcomes have depended on the economic period and on the intersection of culture and consciousness with the historical experiences that molded them. The image of the victim is belied by the persistent and creative struggles waged in the agricultural fields. As the case of cotton suggests, it is not merely the structural relation of immigrants or workers to the state or the industry that shapes their response, but the intersection between these structural factors and the unquantifiable aspect of human agency.

The importance of these human factors was evident in the 1930s. Mexicans faced structural barriers higher than those presented to their Anglo counterparts. Mexicans without documents were subject to deportation, and many were swept up in the raids of the early 1930s, which decimated Mexican communities. They faced discrimination based on their skin color and nationality. Yet this vulnerable position in relation to the society and the government did not create a numbing paralysis like that Manuel Castells describes among immigrant workers. Mexican workers of this period, despite their vulnerability and their exclusion from the mainstream labor movement, waged the largest agricultural strikes in what was the agricultural precursor to the industrial strikes that erupted among manufacturing workers.

They used social structures and organizations they created. How they formed and used these structures depended on consciousness, identity, and culture as well as economic and political conditions. Social networks mediated between the demands of the work place and the needs of workers and were crucial to their adaptation. They formed the basis for migration and a nucleus for communities, and they provided the social insurance needed to survive. Within these overlapping circles of networks, women formed their own networks. The agricultural labor system relied upon and reinforced the Mexicans' social networks. Fostered by the piece-rate system, social networks formed the basis for work crews and were crucial to the growers' ability to recruit, train, organize, and, at times, control workers. Yet these same networks underlay workers' adaptation and resistance. Those same work crews formed the base for spontaneous strikes and, eventually, for union organizing.

The historical period intensified the conditions they found in the cotton fields. Mexicans of the 1920s and 1930s lived within historical memory of the takeover of the southwestern United States from Mexico, and many had personal memories of the Mexican Revolution. Historical conflicts with Anglo-Americans and Anglo business investments and appropriation of Mexican labor on both sides of the border taught them that the labor sucked from them in the cotton fields was related to a broader usurpation of Mexican land and labor. These historical conflicts created a consciousness in which class, racial, and national identities intersected. This reinforced a sense of being Mexican and of separation from the Anglo community. Their tenacious refusal to relinquish Mexican citizenship challenges assertions that this generation longed for assimilation into the Anglo world. Racism, their undocumented status, and poverty made them vulnerable, but also brought them together and intensified their need to rely on one another.

The pragmatic calculations they made to survive—through adapting, resisting, or both—contributed to apparently contradictory views of Mexican workers as either passive and malleable workers or militant, well-organized, and intense strikers. The group economy disciplined people to act in accordance with group needs. Yet while social pressure could act as a conservative force, it could also escalate the problems and grievances of one worker into that of the entire crew or community. As adaptability was necessary to survival, so was resistance.

It is often argued that the determining factors for the failure of unionization among agricultural workers were the patterns of migration, the uncertainty of obtaining work, and the low wages. This is all true, yet the issue of human agency in cotton indicates the need to revise our concepts of the relation between agricultural unions and workers. The CAWIU of the early 1930s was crucial to crystallizing demands and linking together disparate social networks and spontaneous individual strikes. But the question of who formulated strategic and tactical decisions and how they were implemented has been debated. This study counters the perception that the 1933 cotton strike was led by Communist party directorates based in San Francisco. Pat Chambers, an organizer with Wobbly inclinations, had serious disagreements with democratic centralism. Field organizers were often more pragmatic than party directives. Despite Communist party pronouncements, both Chambers and Caroline Decker openly acknowledged that the union "followed the workers" and that strikes were erupting "union or no union."[1] The union utilized inherent organization among workers. Union organizers depended on social and familial networks among workers. Day-to-day leadership arose from these networks. The earlier affiliation with the Partido Liberal Mexicano and the Industrial Workers of the World set a precedent for transethnic cooperation among progressives and laid the basis for a particular openness to the CAWIU.

There are several reasons for the persistent notion that Anglo organizers molded passive Mexican workers. Reporters were disinclined to understand, let alone deal with, Spanish-speaking Mexican workers and instead focused on Anglo organizers and the "communist menace." Government officials and growers apparently could more easily understand and deal with an English-speaking cadre, even if communist, than they could fathom the contradiction between their perceptions of "passive Mexican workers" and the tumultuous strike wave of 1933–1934. CAWIU literature frequently presented the union as the major force, rather than as one crucial force among many. And English-speaking writers, even if sympathetic, rarely spoke with Mexican workers in depth.

Economic conditions and cultural factors influenced the response of Anglo workers as well. That these were hardly strict cultural determinants is clear from the difference in response between the more settled Anglo community and the new migrants in the mid-1930s. Those who had been in the area for a generation or so had established residency and developed a sense of community with interlinked family, social, and work relations that helped lay the basis for unionization in Kern County. Small farmers subject to the whims of larger cotton interests or driven into the ranks of farm workers had developed a strong sense of class distinctions in the Valley and in the agricultural industry. These settled working-class communities helped foster active UCAPAWA branches and galvanized support from related CIO unions. The social capital Mexicans carried into the migrant streams, these Anglos developed in residence.

Anglo migrants who came to the Valley in the 1930s lacked this stability of social structure and economy. The Depression thrust them into an engorged labor pool which they augmented and which diminished their already meager wages. As James Gregory has shown, many came out with or to see family members, and they, too, relied on social and familial networks for work and support. But the hardships of the

Depression undermined the stability of these networks and their ability to mediate for the new migrants.

Their contemporaries, including Carey McWilliams and John Steinbeck, believed that these white American citizens would be more likely to organize into unions than the Mexicans would. The case in cotton suggests that this belief was mistaken. For Anglos as for Mexicans, historical circumstances intersected with class, racial, and national identities. Many new Anglo migrants harbored hopes of recreating a life as small farmers or sharecroppers. Few achieved this goal, and with the onslaught of World War II, the opening of munitions plants, and the urbanization of much of this work force, this idea of returning to rural life faded. Yet in the late 1930s this desire for land and the belief that their sojourn in agriculture was temporary militated against organizing. Although they too depended heavily on family connections, they subscribed to the mainstream American faith in rugged individualism. The myth of upward mobility and the perceived value of citizenship and white skin kept alive the hope they could leave the ranks of farm labor. Their more favored position within society as whites and citizens made it harder to develop a consciousness as workers. Thus those very aspects one might argue made them less vulnerable to oppression militated against achieving solidarity as a group. Even though they were called "Okies" and treated not much better than their non-white co-workers, racism and a desire to be accepted members of the white community made it hard to establish common ground with Mexicans.

The organizational legacy of Anglos dovetailed with the message and aspirations of the New Deal. The Populist and socialist legacies had depended heavily on electoral politics and emphasized taking over the government by means of the ballot box. What set new Anglo migrants apart from Mexicans was not only their white skins, citizenship, and access to the vote, but their belief that electoral politics could do them more good than unions. The case of these new migrants suggests that residency provided only one (albeit an important) precondition for the development of permanent unions. It was the meaning attached to residency that made the difference. For the new Anglo migrants, residency meant access to relief checks and the possibility of becoming part of a community, and was to prove a stronger basis for electoral politics than for union organizing.

The neo-Populist predilections of Anglo migrants coincided with the changing role of government intervention and shifting union expectations of the state. The CAWIU position toward government intervention had seemed ambivalent. Party directives and union slogans in the pre–Popular Front period lambasted the government and the New Deal. Yet field organizers believed that beneficial New Deal programs offered one of the best vehicles for organizing agricultural workers. CAWIU organizers met eagerly with government officials, argued in the cotton hearings that the government should maintain the principles of the NRA, and strategically used government relief to sustain strikes and organizing.

UCAPAWA augmented this pragmatic approach toward the government. At its founding in 1937, conditions for federal intervention seemed heavily weighted in their favor. Ex-members of the CAWIU and Communist party members had dissolved separate unions and were working within the established unions. Massive strikes had spread to industry, and UCAPAWA now belonged to a national federation, the CIO. The NLRB had been passed and, with some reservations, implemented. And a growing number of farm workers were establishing residence in Valley communities. Yet the conditions in California, which seemed optimal for developing the strongest UCAPAWA branches, failed to create a lasting union or one as successful in its strikes as the CAWIU had been.

Ironically, the breadth of UCAPAWA aspirations may have undercut its strength in California's fields. In an effort to conserve scarce resources, UCAPAWA sublimated organizing field workers to organizing processing workers, who, organizers felt, would provide a more stable basis for unionization. The idea was not unsound. Processing workers did help support field workers. But the plan foundered on the bitter rivalry between the AFL and the CIO, and by discouraging field workers' strikes the union undercut one of its few tools to pressure growers and government agencies. UCAPAWA organizer Ted Rasmussen reflected on his disillusionment with relying on government intervention: "It is the process of collective bargaining which produces fundamental social readjustments. Favorable laws help, but there has never been a law that couldn't be interpreted two ways and there has certainly never been a law affecting the conditions of labor which took any effect until backed by the organizational strength of labor itself."[2]

Mexicans' experiences and organizational traditions gave them ambivalent expectations of the role of government. Mexicans were well aware of the possibilities to labor being extended by the New Deal. Section 7a was a potent force in mobilizing farm workers. Workers appealed to the NLRB and held meetings to discuss the implications of New Deal programs. Small merchants worked to support Roosevelt. But the long-standing acrimonious relationship between Mexicans and the

U.S. government was not easily forgotten. Promises for future change were outweighed by the raw experience of forced deportation, low wages, exploitation, and racism. Overall, Mexicans placed more trust in bread-and-butter organizing than in the electoral process. They remained disproportionately active in the agricultural union movement of the 1930s. Even by the mid-1930s, as Mexicans were being repatriated and were largely replaced in the cotton fields by Anglo migrants, they remained crucial to the labor movement. The Mexican union CUCOM was the most influential of California's agricultural unions until UCAPAWA was formed. Within UCAPAWA, Mexicans and Mexican organizations such as El Congreso del Pueblo de Habla Española, played a major role in organizing field workers.

Ultimately, the failure of UCAPAWA and agricultural unionization rested largely on the growing disparity between the power of the cotton industry and that of its workers. This brings us to the role of federal intervention and the nature of the New Deal. The underlying dynamics that propelled the New Deal's involvement in agriculture were similar to those active in industry. Ideologically, state programs and policies in both sectors were based on the justification that they were acting to produce a harmonious relation among labor, capital, and the state. In practice, in both agriculture and manufacturing the state acted in response to class pressures, which ultimately led to the inclusion of industrial workers in this New Deal arrangement of democratic capitalism and to the exclusion and more visible repression of farm workers. Yet this dichotomy was not due simply to the mobilization and countermobilization of interest groups. The outcome was also dependent on systemic and structural issues inherent in each sector. In manufacturing, where capital was more divided than in agriculture and workers had developed stronger organizations, industrial workers were able to wrest at least limited concessions from the federal government in the 1930s. But in agriculture, as the case of cotton shows, the large interests were more unified, the workers more divided and weaker.

The economic structure shaped each sector's relation to the government. In manufacturing, divisions between segments of capital were replicated in competing federal programs, would eventually lead to the NRA's demise, and worked to the advantage of labor. The unity of the agricultural industry was propelled by the high degree of instability that plagued production. But the production demands that forced agriculture to organize meant that they presented a unified front to the federal government. Agriculture thus managed to maintain control of the AAA, and agriculturalists were able to use New Deal programs to enhance their power over labor.[3] This unity of purpose contributed to the organization of the Associated Farmers as an increasingly sophisticated means to combat unionization and exert control over smaller farmers and the work force.

The economic importance of agriculture and the relative unity of the industry contributed to the political power of the farm bloc. This power, along with the weakness of agricultural labor and business and consumer aversion to higher food prices, resulted in the exclusion of agricultural workers from the federal legislation that aided laborers in industry. State involvement on their behalf was granted to agricultural laborers, not as workers, but as destitute farmers or homeless migrants.

Exclusion from state programs reinforced field workers' political powerlessness in relation to the cotton industry and in relation to workers in manufacturing. In turn, each exclusion from federal legislation such as the NLRA or social security increased the powerlessness of agricultural workers and further undermined their ability to influence government programs or policy. It was a vicious cycle of impotence. At the same time, the AAA and growers' responsibility for the administration of federal programs strengthened industrial interests. Eventually they were able to control federal programs created to aid workers. Relief briefly held out the hope of establishing a minimum wage in cotton but ultimately became a means of providing cheap labor for growers.

Although the New Deal improved conditions for industrial workers, its programs had the opposite effect on agricultural workers. In 1940 the La Follette Committee found that "the developments in California agriculture . . . [have] greatly increased the insecurity of agricultural workers and have reduced their average incomes from levels that at best were inadequate."[4] The destabilization of workers in the 1930s was the first phase of a process that helped destroy domestically based work crews, depressed wages, and undercut unionization for two decades.

Agricultural workers were and are paid abominable wages and work under horrible conditions. But the system does not define the worker, despite the fact that it creates the parameters within which he or she works. When the situation is examined closely, the oppressive weight of the industry is seen to throw into greater relief the kinds of struggles farm workers have waged to improve conditions in the fields. This closer look counters their deprecation in the public mind and in literature as passive subjects of history. It also illuminates the exceptional lengths to which agricultural workers have been willing to go to change an unrelenting system of abuse. It shows, not a discontinuity of union efforts, but a remarkable continuity of struggle by the disparate peoples who have worked the fields. La lucha continúa.

Tables

TABLE 1. COTTON ACREAGE HARVESTED, SAN JOAQUIN VALLEY, 1919–1942

	Fresno	Kern	Kings	Madera	Merced	Tulare	Total
1919	5,551	1,296	640	39		297	7,823
1920	3,000	10,000	3,000	2,200		2,800	21,000
1921	1,600	1,200	200			500	3,500
1922	400	1,800	—	—		300	2,500
1923	1,500	5,800	1,000	—		700	9,000
1924	7,600	16,000	6,500	1,400	400	5,900	37,800
1925	17,000	32,800	12,000	11,000	5,400	17,200	95,400
1926	20,700	33,000	11,200	9,800	9,300	22,300	106,300
1927	9,500	28,800	9,000	7,600	5,000	18,900	78,800
1928	27,100	41,000	15,100	15,300	17,200	34,600	150,300
1929	57,000	52,400	22,000	26,000	23,000	67,000	247,400
1930	47,000	51,300	25,200	26,800	21,600	63,400	235,300
1931	34,100	45,000	17,700	18,900	12,100	43,800	171,600
1932	23,000	35,000	9,000	12,800	10,300	25,400	115,500
1933	42,300	56,600	17,100	22,500	13,680	43,000	195,180
1934	46,600	51,900	22,800	22,100	11,804	49,000	204,204
1935	46,700	48,200	17,700	24,300	10,300	55,800	339,900
1936	81,000	78,000	37,200	44,800	22,100	81,800	344,900
1937	134,700	121,800	66,300	77,300	44,100	143,400	587,600
1938	82,380	68,360	24,260	45,900	22,550	83,940	327,390
1939	73,850	67,200	32,180	44,900	19,590	79,170	316,890

SOURCE: California Department of Agriculture, Bureau of Agricultural Statistics, California Crop and Livestock Reporting Service, *California Cotton—Acreage, Yield, Production: State Data 1910–1965, County Data 1919–1965* (Sacramento: Government Printing Office), June 1966.

TABLE 2.   COTTON YIELD PER ACRE, 1934–1939 (POUNDS)

County	1934	1935	1936	1937	1938	1939
Kern	654	646	648	631	700	776
Tulare	556	514	631	614	619	639
Fresno	531	493	552	556	574	642
Madera	547	502	543	543	574	596
Kings	628	553	612	614	532	669
Merced	485	400	473	468	513	512

TABLE 3.   AVERAGE COTTON ACREAGE
PER COTTON-PRODUCING FARM,
UNITED STATES AND SELECTED STATES,
1924–1959

	1924	1929	1939
United States	20.3	21.8	14.3
California	61.1	69.8	59.5
Texas	40.6	42.6	29.7
Mississippi	13.4	14.2	9.4
South Carolina	12.8	15.0	10.5

SOURCE: United States Department of Commerce, Bureau of the Census, *United States Census of Agriculture: 1959, Vol. II, General Report: Statistics by Subjects* (Washington, D.C.: U.S. Government Printing Office, 1962), 829, 835.

TABLE 4. COTTON FARMS BY ACREAGE,
IN CALIFORNIA AND IN PRINCIPAL
COTTON-PRODUCING COUNTIES OF
THE SAN JOAQUIN VALLEY, 1930

	Under 20	20–49	50–99	100–174	175–259	260–499	500 and over
California	5.5	35.6	23.7	16.8	5.5	7.3	5.6
Kern	8.8	44.0	22.5	13.7	2.7	4.8	3.6
Kings	3.7	15.7	20.4	14.9	12.9	16.5	15.8
Tulare	3.3	28.3	27.4	19.8	7.6	8.5	5.1
Fresno	3.9	39.4	21.7	13.6	5.0	6.7	9.7
Madera	1.8	41.3	28.6	14.5	4.6	4.6	4.6
Merced	8.3	40.8	15.0	15.0	4.2	5.8	10.0
San Joaquin Valley	5.2	36.9	24.1	15.8	5.3	6.8	5.9

SOURCE: Paul Taylor, "Producing California Cotton: Statistical Analysis of Cotton Production" (typescript), Paul Taylor Collection, Bancroft Library.

TABLE 5. AVERAGE WAGE FOR PICKING
100 POUNDS OF COTTON, BY YEAR

	California	United States
1925	1.55	1.27
1926	1.65	1.55
1927	1.25	1.12
1928	1.00	1.10
1929	1.45	1.06
1930	0.75	0.63
1931	0.50	0.41
1932	0.45	0.42
1933	0.60/0.75	0.53
1934	1.00	0.60
1935	0.90	0.58
1936	1.00	0.69
1937	0.90	0.69
1938	0.75	0.57
1939	0.80	0.58

SOURCES: La Follette Committee, *Hearings*, part 51, 18837; *Pacific Rural Press*, 2 April 1947; Stuart Jamieson, *Labor Unionism in American Agriculture*, United States Department of Labor, Bureau of Labor Statistics, Bulletin 836 (Washington, D.C.: U.S. Government Printing Office, 1945), 176; Farm Placement Service, Reports, Wage Chart by Crops, 1938 (typescript), Giannini Library, University of California, Berkeley.

TABLE 6. WAGES AS A PERCENTAGE OF
GROSS FARM INCOME, BY YEAR

	California	United States		California	United States
1924	24.5%	9.4%	1932	20.1%	8.9%
1925	23.3	9.3	1933	15.2	6.9
1926	23.4	10.1	1934	13.2	6.5
1927	23.4	10.0	1935	13.7	5.9
1928	22.8	10.0	1936	13.8	—
1929	20.0	10.0	1937	13.1	—
1930	24.7	10.7	1938	16.6	—
1931	24.1	10.5			

SOURCE: "Gross Income, Wage Expenditures and Percent of Gross Income Spent on Wages, United States and California Agriculture, 1924–1938" (typescript), Carey McWilliams Collection, Special Collections, UCLA.

TABLE 7.   AAA PAYMENTS FOR
SELECTED COTTON GROWERS AND
COMPANIES, 1938 (BY AMOUNT
OF PAYMENT RECEIVED)

No. of AAA	Area	Farms	Payment ($)
Hotchkiss Estate Co. and family holdings	Firebaugh	7	72,876
J. G. Boswell*	Corcoran	13	53,822
Russell Giffen	Mendotta	1	43,427
Camp West Lowe*	Shafter	1	30,263
California Lands Inc. (Bank of America)**	various	450	22,013
Merritt and Ocheltree*	Madera	1	17,523
Marvin C. Baker*	Madera	1	14,211
Camp and O'Hanneson	Shafter	1	13,681
Camp West Lowe Ginning Co.*	Shafter	4	13,226
Peterson Farms	Corcoran	1	9,716
J. W. Guiberson	Corcoran	1	9,186
R. L. Poythress	Madera	1	9,023
Banducci*	Bakersfield	2	8,791
Jess Hanson	Corcoran	1	8,301
F. A. Yearont*	Fresno	1	7,585
Harp Brothers	Corcoran	1	7,074
Hugh Jewett	Bakersfield	1	6,092
J. B. Boyette	Corcoran	1	4,758
O. L. Baker*	Madera	1	4,470
H. L. Pomeroy* and H. S. Jewett*	Bakersfield	1	4,456
J. M. Hansen	Corcoran		4,428
Glenn-Harp-Boswell*	Corcoran	1	4,090
Forrest Frick*	Bakersfield	1	3,672
K. S. Battelle	Corcoran	4	2,650
Hayes Brothers	Madera	1	2,094
San Joaquin Cotton Oil Co.*	Bakersfield	1	1,557
Claire Guiberson	Corcoran	1	1,507
Ray Monroe***	Tulare	1	1,486

(Continued)

No. of AAA	Area	Farms	Payment($)
L. W. Frick*	Bakersfield	2	1,445
Edgar Combs	Bakersfield	1	1,411

*Active member of Associated Farmers.
**This includes the total landholdings of the company. It overstates their investment in cotton, but as cotton was their largest earning crop it suggests the total payments going to an influential factor in the cotton industry.
***Receiver for Farmers Cotton Gin.

SOURCE: La Follette Committee, *Hearings*, part 50, 18824, and part 62, 22889, "1938 Agricultural Conservation Program—Names and Addresses of Persons Receiving Net Payments in Excess of $1,000 (as of November 30, 1939)." A total of 1,230 payees received $3,604,981. I cite the *Hearings* because a check with the National Archives and other federal, state, and county sources indicates that the records of AAA payments to individual growers have been destroyed. This table is impressionistic in that it lists prominent cotton growers and most probably excludes others. It does, however, give a sense of the size of payments received by large growers and some of the prominent members of the cotton industry. Several people had multiple holdings.

TABLE 8. MEXICAN POPULATION IN COTTON-GROWING COUNTIES, CALIFORNIA, 1910–1930

County	1910	1920	1930 (estimated)
Fresno	615	3,1652	16,255
Kern	1,492	1,856	2,307
Kings	278	689	1,707
Los Angeles	11,278	33,644	95,953
Madera	412	387	363
Tulare	560	1,746	5,444
Merced	355	658	1,219
Imperial	1,464	6,414	28,157

SOURCE: C. C. Young, *Mexicans in California: Report of Governor C. C. Young's Mexican Fact-Finding Committee* (Sacramento: California State Printing Office, 1930), 46, 54–59.

TABLE 9.   LEGAL MEXICAN ENTRANTS,
1920–1929

Year	Immigrants	Non-Immigrants	Total
1920	51,042	17,350	63,392
1921	29,603	17,191	46,794
1922	18,246	12,049	30,295
1923	62,709	13,279	75,988
1924	87,648	18,139	105,787
1925	32,378	17,351	49,729
1926	42,638	17,147	59,785
1927	66,766	13,873	80,639
1928	57,765	3,857	61,622
1929	38,980	3,405	42,385

SOURCE: *Annual Reports* of the Commissioner General of Immigration, quoted in Cardoso, *Mexican Emigration*, 91–95. See also Paul Taylor, *Mexican Labor in the United States: Migration Statistics*, University of California Publications in Economics 12, no. 3 (Berkeley: University of California Press, 1934).

TABLE 10.   AVERAGE WAGE RATE
FOR PICKING 100 POUNDS OF COTTON
IN SELECTED STATES

	1932	1934	1936	1938
California	$ .45	.90	1.00	.75
Arkansas	.44	.60	.75	.60
Texas	.45	.60	.65	.55
Oklahoma	.48	.75	.75	.70
Missouri	.52	.80	.95	.75

SOURCE: California State Chamber of Commerce, *Migrants: A National Problem and Its Impact on California*, reprinted in Tolan Committee, *Hearings*, part 6, 2785.

TABLE 11. INCREASE IN RELIEF
CLIENTS, 1937–1938

	1937	1938
California	154,000	164,500
Tulare	6,325	16,880
Fresno	1,446	6,581
Madera	29	249
Kern	793	1,062
Merced	198	387

SOURCE: *Fresno Bee*, 24 May 1938.

Proposal of the Associated Farmers

The full proposal of the original Associated Farmers, which was composed of small farmers, both gives a taste of their demands and, through their language, suggests some of the ideological roots of their movement. Since this has not, to my knowledge, been discussed in other works, the full proposal is included here:

We propose the elimination from the farmers' consul of the petty politician and the paid organizer.

We demand the exclusion of the pseudo-farmer, who merely farms the farmers, who derives his principal income from other sources, who is the friend of traders and financiers, and who, by intimidation, intrigue or politics assumes a leadership to which he is not entitled.

We propose that from among his own number he choose leaders who with no thought of profit or personal glory will carry his declaration of participation into the state and national councils and convince the industrial world of the farmer's fitness to rule his own domain.

Specifically and locally we propose the following definite program for immediate prosecution:

1. ESTABLISHMENT OF CONTACT WITH SECRETARY WALLACE, ADMINISTRATOR PEETE [*SIC*], ADMINISTRATOR JOHNSON AND OTHER GOVERNMENTAL FORCES, WITH A VIEW TO THE FOLLOWING.

a. Co-ordination of our units with the Farm Credit Administration to provide financing for future crops.

b. Elimination of cotton brokers, cotton processors, oil mill, warehouse and press operators from the position of intermediate toll-takers between us and government funds.

c. The condemnation of future seed contracts involving discounts. The enactment of state or federal laws or rules either restoring competitive bidding in the seed market or establishment of code prices.

d. A convention of representative farmers of the San Joaquin Valley to [illegible] state a code of fair business practices for submission to the NRA. [illegible] with such problems as wages, . . . , seed marketing or price [illegible]

e. Appointment of a commission to investigate and report upon the activities of the California Cotton Cooperative Association with a view to the utilization of its resources and machinery as a universal market medium, such committee to suggest needed reform or change in methods deemed proper.

f. The selection of a representative or representatives to proceed to Washington under the sponsorship of Senator McAdoo, to meet Secretary Wallace and submit the following proposal with proper arguments in its support:

We submit that for the convenience of processors and producers some form of futures trading in commodities be maintained.

We affirm that the present machinery governing such trading is inadequate, insufficient and tends to encourage hysterical and unwarranted fluctuations at the will of a few super manipulators.

We suggest that the following system governing future trading be set up by Secretary Wallace under the authority granted by the NRA:

1. That to the agricultural extension service be delegated the duty of inspection as hereinafter specified and be empowered to exact from the producer a suitable fee to cover actual costs.

2. That the producer be granted the right to sell to any trader, factor, broker, processor or person for future delivery a certain part of his prospective crop subject to the following procedure [procedure follows]

3. We suggest that the secretary of agriculture forbid the sale by any trader, factor or broker of any commodity for future delivery, except in such amounts as may be indicated by certificates of probable production in his possession, duly signed by its producer and countersigned by the inspection service, and that he provide, under penalty, for transfer of such certificate within 24 hours of consummation of such sale.

Source: *Bakersfield Californian*, 7 and 9 September 1933

Notes

INTRODUCTION

1. The San Joaquin Valley proper is two hundred and sixty miles in length, from the end of the grapevine in the southern Valley to Lodi in the north. David Lantis, Rodney Steiner, and Arthur Carinen, *California: The Pacific Connection* (Chico: Creekside Press, 1989), 330, 369.

2. Frederick J. Turner, *The Significance of the Frontier in American History*, ed. Harold P. Simonson (New York: Frederick Ungar, 1963). For Turner's influence on other historians, see Richard Hofstadter, *The Progressive Historians: Turner, Beard, Parrington* (New York: Knopf, 1968); and Gerald D. Nash, *Creating the West: Historical Interpretations 1890–1990* (Albuquerque: University of New Mexico Press, 1991); Henry Nash Smith, *Virgin Land: The American West as Symbol and Myth* (New York: Random House, 1950).

3. Even a recent study of California farm workers suggests that the "cohering force" for farm workers was not culture, social ties, families, or organization but a "kinship of powerlessness." Cletus Daniel, *Bitter Harvest: A History of California Farmworkers, 1870–1941* (Ithaca: Cornell University Press, 1981), 71.

Among those who have dealt with workers and labor organization in California agriculture are: Paul Taylor, *Mexican Labor in the United States: The Imperial Valley*, University of California Publications in Economics 6, no. 1 (Berkeley: University of California Press, 1930); Sam Kushner, *Long Road to Delano* (New York: International Publishers, 1975); Joan London, *So Shall Ye Reap* (New York: Thomas Y. Crowell, 1970). More have dealt with the United Farm Workers (UFW), but many of these have been journalistic, such as Peter Matthiessen, *Sal Si Puedes: Cesar Chavez and the New American Revolution* (New York: Delta, 1969); Eugene Nelson, *Huelga: The First Hundred Days of the Great Delano Grape Strike* (Delano: Farm Worker Press, 1966); John

Gregory Dunne, *Delano* (New York: Farrar, Straus and Giroux, 1971). Ernesto Galarza, in his excellent study of union struggles in which he was an organizer, goes closer to the heart of the interaction between workers and agriculture. Ernesto Galarza, *Farmworkers and Agri-Business in California, 1947–1960* (Notre Dame: University of Notre Dame Press, 1977). Several unpublished works deal with agricultural workers: see especially Margaret Rose, "Women in the United Farmworkers: A Study of Chicana and Mexicana Participation in a Labor Union, 1950–1980," Ph.D. diss., University of California, Los Angeles, 1988.

Carey McWilliams's and Paul Taylor's books on agribusiness, although generally superb, tend to ignore the internal organization of workers. See Carey McWilliams, *Factories in the Field: The Story of Migratory Farm Labor in California* (orig. published 1939; Hamden: Archon Books, 1969); Carey McWilliams, *Ill Fares the Land: Migrants and Migratory Labor in the United States* (New York: Barnes and Noble, 1967). Paul Taylor, *Labor on the Land: Collected Writings, 1930–1979* (New York: Arno Press, 1981); Paul Taylor, *On the Ground in the Thirties* (Salt Lake City: Peregrine Books, 1983).

Films have reinforced the victim image of farm workers. The first, and the one that most seared the consciousness of the United States, was Edward R. Murrow's *Harvest of Shame*. This has been followed by a series of white papers by NBC, each a depressing update on Murrow's piece. One of the few exceptions to the cinematic depiction of farm workers as victims was *Alambrista*, directed by Robert Young.

4. The subject of culture has been extensively explored by historians, anthropologists, sociologists, and theorists of popular culture. Anthropological works that have made major contributions to the debate are: Clifford Geertz, "Deep Play: Notes on the Balinese Cockfight," *Daedalus* 101, no. 1 (Winter 1972); and James Clifford and George Marcus, eds., *Writing Culture: The Poetics and Politics of Ethnography* (Berkeley: University of California Press, 1986). For Marxist discussions, see: Raymond Williams, "Base and Superstructure in Marxist Cultural Theory," in Raymond Williams, *Problems in Materialism and Culture* (London: Verso Press, 1980); and Tim Patterson, "Notes on the Historical Application of Marxist Cultural Theory," *Science and Society* no. 39 (1975). Ernesto Laclau and Chantal Mouffe, *Hegemony and Socialist Strategy: Towards a Radical Democratic Politics* (London: Verso, 1985).

Some works that have explored theoretical aspects of Mexican culture are: Juan Gómez-Quiñones, *On Culture*, Popular Series no. 1 (Los Angeles: UCLA, Chicano Studies Research Center Publications, 1977); José Limón, *Mexican Ballads, Chicano Poems: History and Influence in Mexican-American Social Poetry* (Berkeley: University of California Press, 1992); Hector Calderón and José D. Saldívar, eds., *Criticism in the Borderlands: Studies in Chicano Literature and Ideology* (Durham: Duke University Press, 1991); Teresa McKenna, *Parto de Palabra: Studies on Chicano Literature in Process* (Austin: University of Texas Press, in press); Ramón Saldívar, *Chicano Narrative: The Dialectics of Difference* (Madison: University of Wisconsin Press, 1990); Gloria Anzaldua, *Borderlands/La Frontera: The New Mestiza* (San Francisco: Spinsters/Aunt Lute Book

Co., 1987); Maria Herrera-Sobek, *The Mexican Corrido: A Feminist Analysis* (Bloomington: Indiana University Press, 1990).

5. Edward P. Thompson, *The Making of the English Working Class* (New York: Vintage, 1963); Herbert Gutman, *Work, Culture and Society in Industrial America* (New York: Vintage Books, 1976); David Montgomery, *Workers' Control in America* (Cambridge: Cambridge University Press, 1979).

6. Pegge Pascoe, "Western Women at the Cultural Crossroads," in Patricia Nelson Limerick, Clyde Milner, and Charles Rankin, eds., *Trails: Toward a New Western History* (Lawrence: University Press of Kansas, 1991); Richard White, *"It's Your Misfortune and None of My Own": A New History of the American West* (Norman: University of Oklahoma Press, 1991); Patricia Nelson Limerick, *The Legacy of Conquest: The Unbroken Past of the American West* (New York: W. W. Norton, 1987). See also the selection of articles in Limerick, Milner, and Rankin, *Trails*. For good examples of this new western history see: Richard White, "The Winning of the West: The Expansion of the Western Sioux in the Eighteenth and Nineteenth Centuries," *Journal of American History* 65 (September 1978); Thomas D. Hall, *Social Change in the Southwest, 1350–1880* (Lawrence: University Press of Kansas, 1989); articles in Lillian Schlissel, Vicki L. Ruiz, and Janice Monk, eds., *Western Women: Their Land, Their Lives* (Albuquerque: University of New Mexico Press, 1988); Ramón Gutiérrez, *When Jesus Came, the Corn Mothers Went Away: Marriage, Sexuality and Power in New Mexico, 1500–1846* (Stanford: Stanford University Press, 1991). For intriguing discussions on the nature of the ideology of the West, see Richard Drinnon, *Facing West: The Metaphysics of Indian Hating and Empire Building* (New York: Schocken Books, 1990); Richard Slotkin, *The Fatal Environment: The Myth of the Frontier in the Age of Industrialization, 1800–1890* (Middletown: Wesleyan University Press, 1985); and Richard Slotkin, *Gunfighter Nation: The Myth of the Frontier in Twentieth Century America* (New York: Atheneum, 1992).

7. See Jean-Paul Sartre, *Search for a Method* (New York: Vintage Books, 1963); Thompson, *The Making of the English Working Class*; Edward P. Thompson, *The Poverty of Theory and Other Essays* (New York: Monthly Review Press, 1978); Perry Anderson, *Arguments Within English Marxism* (London: Verso Press, 1980).

8. "Men make their own history, but they do not make it just as they please; they do not make it under circumstances chosen by themselves, but under circumstances directly encountered, given, and transmitted from the past." Karl Marx, *The Eighteenth Brumaire of Louis Bonaparte* (Moscow: Progress Publishers, 1934), 2.

9. Sartre, *Search for a Method*, 87.

10. Georg Lukács, *History and Class Consciousness: Studies in Marxist Dialectics* (Cambridge, Mass.: MIT Press, 1971); Antonio Gramsci, *Selections from the Prison Notebooks* (New York: International Publishers, 1971); Herbert Marcuse, *Eros and Civilization* (New York: Vintage Books, 1962); Martin Jay, *The Dialectical Imagination: A History of the Frankfurt School and the Institute of Social Research, 1923–1950* (Boston: Little, Brown, 1973).

11. Thompson, *The Making of the English Working Class*, 9–11.

12. Anderson, *Arguments Within English Marxism*, 43. These arguments also challenged the preoccupation with class consciousness. As Anderson pointed out, "Classes have frequently existed whose members did not 'identify their antagonistic interests' in any process of common clarification or struggle," and "it is probable that for most of historical time this was the rule rather than the exception." Anderson, *Arguments Within English Marxism*, 40. Eric Hobsbawm also makes this point in "Notes on Class Consciousness," in his *Workers: Worlds of Labour* (New York: Pantheon Books, 1984).

Following the publication of Thompson's *The Poverty of Theory* and Anderson's *Arguments Within English Marxism* were a flurry of articles, many of which appeared in *History Workshop* and contributed to the debate. Among them were: Richard Johnson, "Critique: Edward Thompson, Eugene Genovese, and Socialist-Humanist History," *History Workshop* 6 (Autumn 1978); Keith McClelland, "Some Comments on Richard Johnson, 'Edward Thompson, Eugene Genovese, and Socialist-Humanist History,'" *History Workshop* 7 (Spring 1979); Simon Clarke, "Socialist Humanism and the Critique of Economism," *History Workshop* 8 (Autumn 1979); and the articles that appeared on debates related to *The Poverty of Theory* in Raphael Samuel, ed., *People's History and Socialist Theory*, History Workshop Series (London: Routledge and Kegan Paul, 1981), 365–408. In the late 1970s and early 1980s there were more direct attacks on Thompson's formulation of class. Eugene Genovese and Elizabeth Fox Genovese argued that a focus on workers outside the larger context led to inaccurate conclusions and "history with the politics left out." Eugene Genovese and Elizabeth Fox Genovese, "The Political Crisis of Social History: A Marxian Perspective," *Journal of Social History* 10 (Winter 1976).

13. Some of the works I have found particularly interesting are: Michael Burawoy, "The Functions and Reproduction of Migrant Labor: Comparative Material from Southern Africa and the United States," *American Journal of Sociology* 81, no. 5 (March 1976); Stephen Castles and Godula Kosack, *Immigrant Workers and Class Structure in Western Europe* (London: Oxford University Press, 1973); James O'Conner and the Santa Cruz Collective, "The Global Migration of Labor and Capital," in Antonio Ríos Bustamante, ed., *Mexican Immigrant Workers in the United States* (Los Angeles: UCLA, Chicano Studies Research Center, 1981); Alejandro Portes, "Toward a Structural Analysis of Illegal (Undocumented) Immigration," *International Migration Review* 12, no. 4 (Winter 1978); James D. Cockcroft, *Outlaws in the Promised Land: Mexican Immigrant Workers and America's Future* (New York: Grove Press, 1986); Jorge Bustamante, "The Historical Context of Undocumented Mexican Immigration to the United States," in A. Ríos Bustamante, *Mexican Immigrant Workers in the United States*; Jorge Bustamante, "Commodity Migrants: Structural Analysis of Mexican Immigration to the United States," in *Views Across the Border: The United States and Mexico*, ed. S. R. Ross (Albuquerque: University of New Mexico Press, 1978); Wayne Cornelius, "Mexican Immigrants and Southern California: A Summary of Current Knowledge," Working Papers in U.S.-Mexican Studies 36 (La Jolla, Calif.: Center for U.S.-Mexican Studies, 1982); Wayne Cornelius, "Implementation and Impacts of the U.S. Immigration Reform and Control Act of 1986: A Comparative Study of Mexican and Asian

Immigrants, Their Employers and Sending Communities" (La Jolla, Calif.: Center for U.S.-Mexican Studies, n.d. [typescript]); Alejandro Portes and Robert Bach, *Latin Journey: Cuban and Mexican Immigrants in the United States* (Berkeley: University of California Press, 1985); James Stuart and Michael Kearney, "Causes and Effects of Agricultural Labor Migration from the Mixteca of Oaxaca to California," Working Papers in U.S.-Mexican Studies 28 (La Jolla, Calif.: Center for U.S.-Mexican Studies, 1981; Angelina Casillas Moreno, *La mujer en dos comunidades de emigrantes (Chihuahua)* (México D.F.: Secretaría de Educación Pública, 1986).

14. Manuel Castells, "Immigrant Workers and Class Struggles in Advanced Capitalism: The Western European Experience," *Politics and Society 5*, no. 1 (1975), 51.

15. By this I mean that immigrants' relations to the state depend on their legal status, whether or not they are undocumented, whether they have special privileges (as Cuban refugees do) or lack them (as Haitians do). The economic class of the immigrants also has a direct bearing on how they are treated by the state.

16. Several articles focus on the people who migrate and on the effect of immigration on the immigrants themselves: see Lourdes Arguelles, "Undocumented Female Labor in the Southwestern United States," in *Between Borders: Essays on Mexicana/Chicana History*, ed. Adelaida Del Castillo (Encino, Calif.: Floricanto Press, 1990); Arizona Farm Workers organizers Lupe Sánchez and Jésus Romo, "Organizing Mexican Undocumented Farm Workers on Both Sides of the Border," Working Papers in U.S.-Mexican Studies 27 (La Jolla, Calif.: Center for U.S.-Mexican Studies, 1981); Magdalena Mora, "The Tolteca Strike: Mexican Women and the Struggle for Union Representation," in A. Ríos Bustamante, *Mexican Immigrant Workers in the United States*; Mario Vásquez, "Immigrant Workers and the Apparel Manufacturing Industry in Southern California, in A. Ríos Bustamante, *Mexican Immigrant Workers in the United States*; Arturo Santamaría Gómez, *La izquierda norteamericana y los trabajadores indocumentados* (n.p.: Universidad Autónoma de Sinaloa, Ediciones de Cultura Popular, S.A., 1988).

Al Gedicks's article compares Swedish and Finnish participation in labor conflicts, looking at the background of each in their country of origin as well as the conditions in the United States. Such an approach places culture within the larger economic and social framework. Al Gedicks, "The Social Origins of Radicalism Among Finnish Immigrants in Midwest Mining Communities," URPE *Review of Radical Political Economics* 8, no. 3 (Fall 1976).

17. David Brody, *Steelworkers in America: The Nonunion Era* (New York: Harper and Row, 1960). Brody has also raised questions about the problem of the cultural approach and the "jettisoning" of class. David Brody, "The Old Labor History and the New," in Daniel J. Leab, ed., *The Labor History Reader* (Chicago: University of Illinois Press, 1985), 14.

18. J. Carroll Moody, "Introduction," *Perspectives on American Labor History: The Problems of Synthesis*, ed. J. Carroll Moody and Alice Kessler-Harris (DeKalb: Northern Illinois University Press, 1990), xii.

19. David Montgomery, *The Fall of the House of Labor: The Workplace, the State and American Labor Activism, 1865–1925* (Cambridge: Cambridge Uni-

versity Press, 1987); David Montgomery, *Beyond Equality: Labor and the Radical Republicans, 1862–1872* (New York: Knopf, 1967); David Montgomery, *Workers' Control in America.*

20. This is true not only of the United States but of other countries as well. See David Goodman and Michael Redclift, *From Peasant to Proletarian: Capitalist Development and Agrarian Transitions* (Oxford: Basil Blackwell, 1981).

21. Some of these are: Emilio Zamora, *The World of the Mexican Worker in Texas During the Early 1900s* (College Station: Texas A and M Press, 1993); David Montejano, *Anglos and Mexicans in the Making of Texas, 1836–1986* (Austin: University of Texas Press, 1987); Juan Gómez-Quiñones, "The First Steps: Chicano Labor Conflict and Organizing, 1900–1920," *Aztlán* 3, no. 1 (Spring 1972); Juan Gómez-Quiñones, *Development of the Mexican Working Class North of the Rio Bravo,* Popular Series 2 (Los Angeles: UCLA, Chicano Studies Research Center Publications, 1982); Juan Gómez-Quiñones, *On Culture,* Popular Series 1 (Los Angeles: UCLA, Chicano Studies Research Center, 1977); Ricardo Romo, *East Los Angeles: History of a Barrio* (Austin: University of Texas Press, 1983); Mario García, *Desert Immigrants: The Mexicans of El Paso, 1880–1920* (New Haven: Yale University Press, 1981); George Sánchez's dissertation tackles the development of the community in Los Angeles from a different perspective (George Sánchez, "Becoming Mexican American: Ethnicity and Acculturation in Chicano Los Angeles, 1900–1943," Ph.D. diss., Stanford University, 1989), as does his book, *Becoming Mexican American: Ethnicity, Culture and Identity in Chicano Los Angeles, 1900–1945* (New York: Oxford University Press, 1993). For older but still illuminating work on Mexican immigrants, see Manuel Gamio's books written in the 1920s, *Mexican Immigration to the United States: A Study of Human Migration and Adjustment* (New York: Dover Press, 1971) and *The Life Story of the Mexican Immigrant: Autobiographic Documents* (New York: Dover Press, 1971).

22. Sarah Deutsch, *No Separate Refuge: Culture, Class and Gender on an Anglo-Hispanic Frontier in the American Southwest, 1880–1940* (Oxford: Oxford University Press, 1987). For the activities of women in labor, see Vicki Ruiz, *Cannery Women, Cannery Lives: Mexican Women, Unionization and the California Food Processing Industry, 1930–1950* (Albuquerque: University of New Mexico Press, 1987); Antonia Castañeda, "The Political Economy of Nineteenth Century Stereotypes of Californianas," in Del Castillo, ed., *Between Borders;* Antonia Castañeda, "Presidarias y Pobladoras: Spanish-Mexican Women in Frontier Monterey, Alta California, 1770–1821," Ph.D. diss., Stanford University, 1990; Deena J. González, "The Spanish-Mexican Women of Santa Fe: Patterns of Their Resistance and Accommodation, 1820–1880," Ph.D. diss., University of California, Berkeley, 1985. See also Del Castillo, ed., *Between Borders;* Emma Pérez, "'A la Mujer': A Critique of the Partido Liberal Mexicano's Gender Ideology on Women," in Del Castillo, ed., *Between Borders;* Clementina Durón, "Mexican Women and Labor Conflict in Los Angeles: The ILGWU Dressmakers' Strike of 1933," *Aztlán* 15, no. 1 (Spring 1984); Emilio Zamora, "Sara Estela Ramírez: una rosa roja en el movimiento," in Magdalena Mora and Adelaida R. Del Castillo, eds., *Mexican Women in the United States: Struggles Past and Present,* Occasional Paper 2 (Los Angeles: UCLA, Chicano

Studies Research Center Publications, 1980); Devra Anne Weber, "Raíz Fuerte: Oral History and Mexicana Farmworkers," *Oral History Review* 17, no. 2 (Fall 1989); Devra Anne Weber, "Mexican Women on Strike: Memory, History and Oral Narratives," in Adelaida R. Del Castillo, ed., *Between Borders*. For a current discussion of Chicanas in agricultural labor see Patricia Zavella, *Women's Work and Chicano Families: Cannery Workers of the Santa Clara Valley* (Ithaca: Cornell University Press, 1987).

23. Douglas Massey, Rafael Alarcón, Jorge Durand, and Humberto González, *Return to Aztlan: The Social Process of International Migration from Western Mexico* (Berkeley: University of California Press, 1987); Lilia Moreno and Omar Fonseca, *Jaripo: Pueblo de migrantes* (Jiquilpan, Michoacán: Centro de Estudios de la Revolución Mexicana "Lázaro Cárdenas," 1984); Gustavo López Castro, *La casa dividida: Un estudio de caso sobre la migración a Estados Unidos en un pueblo michoacano* (Zamora, Michoacán: Colegio de Michoacán,1986); Merilee S. Grindle, *Searching for Rural Development: Labor Migration and Employment in Mexico* (Ithaca: Cornell University Press, 1988). For labor organizing across the border, see Paco Ignacio Taibo II, *Bolshevikis: Historia narrativa de los orígenes del comunismo en México, 1919–1925* (México D.F.: Editorial Joaquín Mortiz, 1986); Javier Torres Parés, *La Revolución sin frontera: El partido Liberal Mexicano y las relaciones entre el movimiento obrero de México y el de Estados Unidos, 1900–1923* (México D.F.: Facultad de Filosofía y Letras Universidad Nacional Autónoma de México, 1990); Jorge Durand, *Los obreros de Río Grande* (Zamora: Colegio de Michoacán, 1986); Roger Rouse, "Mexican Migration to the United States: Family Relations in the Development of a Transnational Migrant Circuit," Ph.D. diss., Stanford University, 1989; Paul Friedrich, *Agrarian Revolt in a Mexican Village* (Englewood Cliffs, N.J.: Prentice-Hall, 1970); Peter Baird and Ed McCaughan, *Beyond the Border: Mexico and the U.S. Today* (New York: North American Congress on Latin America, 1979); Robert C. Smith, "'Los ausentes siempre presentes': The Imagining, Making and Politics of a Transnational Community Between New York City and Ticuani, Puebla," Columbia University, Institute for Latin American and Iberian Studies, Working Papers on Latin America (typescript), October 1992.

24. García, *Desert Immigrants*, 109. He has made the same point in "La Familia: The Mexican Immigrant Family, 1900–1930," in Mario Barrera and Alberto Camarillo, eds., *Work, Family, Sex Roles, Language* (Berkeley: Tonatiuh-Quinto Sol International, 1980).

25. Many Chicano historians disagree with García's analysis. For a discussion on labor conflicts and militancy among Mexicans, see: Gómez-Quiñones, "The First Steps"; Gómez-Quiñones, *Development of the Mexican Working Class North of the Rio Bravo*; Gómez-Quiñones, *On Culture*; Zamora, *The World of the Mexican Worker in Texas*. The special issue on Labor History and the Chicano of *Aztlán* 6, no. 2 (Summer 1975), contains a number of articles on Mexican labor activity. See especially Luis Leobardo Arroyo, "Notes on Past, Present and Future Directions of Chicano Labor Studies," and Víctor B. Nelson-Cisneros, "La clase trabajadora en Tejas, 1920–1940."

26. The term *sin fronteras* was politicized in the 1970s by the Los Angeles-based Mexican organization Centro de Acción Social Autónomo-HGT. While it

was applied to a later group of Mexican immigrants, the perception of being without borders characterized 1920s immigrants as well.

27. An excellent discussion of immigrants and their families is found in John Bodnar, *The Transplanted: A History of Immigrants in Urban America* (Bloomington: Indiana University Press, 1985). Other excellent historical works on immigrant families and labor include: Louise A. Tilly and Joan W. Scott, *Women, Work, and Family* (New York: Holt, Rinehart and Winston, 1978); Tamara K. Hareven, *Family Time and Industrial Time* (Cambridge: Cambridge University Press, 1982); Virginia Yans-McLaughlin, *Family and Community: Italian Immigrants in Buffalo, 1880–1930* (Ithaca: Cornell University Press, 1977); Richard Griswold del Castillo, *La Familia: Chicano Families in the Urban Southwest, 1848 to the Present* (Notre Dame: University of Notre Dame Press, 1984). Several books discuss the importance of these social and familial networks in current Mexican immigration. One of the best is Massey, Alarcón, Durand, and González, *Return to Aztlan*.

28. Bodnar, *The Transplanted*, 71.

29. Zamora, *The World of the Mexican Worker in Texas*, 86. Zamora devotes a chapter to the concept of mutuality as an expression of "Mexicanist" political culture and its relation to voluntary organizations that developed in Texas. See also Gómez-Quiñones, *On Culture*.

30. Daniel, *Bitter Harvest*; Ramon Chacon, "Labor Unrest and Industrialized Agriculture in California: The Case of the 1933 San Joaquin Valley Cotton Strike," *Social Science Quarterly* 65, no. 2 (June 1984).

31. Oral histories of Mexican women challenge stereotypes, suggesting that the dislocation and disruptions caused by migration changed families and gender relations. Much remains to be learned about women in Mexico, but it is possible that shifting gender relations seen in the 1920s and 1930s were not divergences from practices in Mexico but partially responses to the upheavals in their lives and partially a continuation of unexplored older patterns in private relations which, although perhaps proscribed by tradition, were alive and well. Weber, "Raíz Fuerte."

32. Walter J. Stein, *California and the Dust Bowl Migration* (Westport, Conn.: Greenwood Press, 1973).

33. James Gregory, *American Exodus: The Dust Bowl Migration and Okie Culture in California* (New York: Oxford University Press, 1989), 158–64.

34. Some of the germane works on the New Deal are: David Montgomery, "American Workers and the New Deal Formula," in his *Workers' Control in America*; Christopher L. Tomlins, *The State and the Unions: Labor Relations, Law, and the Organized Labor Movement in America, 1880–1960* (Cambridge: Cambridge University Press, 1985); Thomas Ferguson, "From Normalcy to New Deal: Industrial Structure, Party Competition and American Public Policy in the Great Depression," *International Organization* 38, no. 1 (Winter 1984); Peter Gourevitch, "Breaking with Orthodoxy: The Politics of Economic Policy Responses to the Depression of the 1930s," *International Organization* 38, no. 1 (Winter 1984); Jill Quadagno, "Welfare Capitalism and the Social Security Act of 1935," *American Sociological Review* 49 (October 1984); Theda Skocpol and Kenneth Finegold, "State Capacity and Economic Intervention in the Early New

Deal," *Political Science Quarterly* 97, no. 2 (Summer 1982); Kenneth Finegold, "Agriculture, State, Party, and Economic Crisis: American Farm Policy and the Great Depression," paper presented at the Conference on the Political Economy of Food and Agriculture in Advanced Industrial Societies, Rural Sociological Society Annual Meeting, Guelph, Ontario, 19–23 August 1981; Stanley Vittoz, *New Deal Labor Policy and the American Industrial Economy* (Chapel Hill: University of North Carolina Press, 1987); William E. Forbath, *Law and the Shaping of the American Labor Movement* (Cambridge: Harvard University Press, 1989).

For a discussion of the New Deal in agriculture, see: Linda Majka and Theo Majka, *Farmworkers, Agribusiness and the State* (Philadelphia: Temple University Press, 1982); Daniel, *Bitter Harvest*; Kenneth Finegold and Theda Skocpol, "Capitalists, Farmers and Workers in the New Deal—The Ironies of Government Intervention: A Comparison of the Agricultural Adjustment Act and the National Industrial Recovery Act," paper presented at Session on Class Coalitions and Institutions in American Politics, Conference Group on the Political Economy of Advanced Industrial Societies, at the Annual Meeting of the American Political Science Association, Washington, D.C., 31 August 1980.

35. Arthur M. Schlesinger, *The Age of Roosevelt*, 3 vols. (New York: Houghton Mifflin, 1957–1960).

36. Tomlins, *The State and the Unions*.

37. Montgomery, "American Workers and the New Deal Formula"; see also David Milton, *The Politics of U.S. Labor: From the Great Depression to the New Deal* (New York: Monthly Review Press, 1982).

38. Tomlins, *The State and the Unions*, 147.

39. Linda Majka and Theo Majka, *Farmworkers, Agribusiness and the State* (Philadelphia: Temple University Press, 1982); Theda Skocpol and Kenneth Finegold, "State Capacity and Economic Intervention in the Early New Deal," *Political Science Quarterly* 97, no. 2 (Summer 1982). An intriguing and useful discussion of the federal government's role in relation to agricultural labor in California is Daniel, *Bitter Harvest*.

40. For an excellent discussion of the definition of the state, see Nora Hamilton, *The Limits of State Autonomy: Post-Revolutionary Mexico* (Princeton, N.J.: Princeton University Press, 1982), 3–39.

The debate over the state is a long one. I rely on several writers for a general framework of the concept of the state. Alan Wolfe, in *The Limits of Legitimacy: Political Contradictions of Contemporary Capitalism* (New York: Free Press, 1977), expands the analysis to include the contradiction between the state's function of assisting in the accumulation of capital and the ideological foundations of the liberal state. Wolfe's argument about the "franchisement" or granting state power to elements of capital is insightful. He treats hegemony, however, as a somewhat mechanistic process of legitimation for accumulation. Erik Olin Wright argues in *Class, Crisis, and the State* (London: New Left Books, 1978) that the contradictions in the state are not defined by legitimation constraints, but are a result of class conflict itself. Historically, capital has been dominant. Samuel Bowles and Herbert Gintis, in "The Crisis of Liberal Democratic Capitalism," *Politics and Society* 11, no. 1 (1982), argue that the functions of accumulation and legitimation may conflict and that the tenets of liberal democratic society are in fact in basic conflict with what is necessary for capitalist accumulation. Martin Carnoy, *The State and Political Theory* (Princeton: Princeton University Press, 1984). Discussions with Charles Noble on the state were invaluable to my work.

41. The ultimate shape of these results from conflicts fought out both within the state and in the society as a whole. These in turn affect class relations, which, again, affect relations within the state.

CHAPTER ONE. "WE ARE PRODUCING A PRODUCT TO SELL . . .": THE BUSINESS OF COTTON

1. *The California Farmer*, 25 May 1854, quoted in Varden Fuller, "The Supply of Agricultural Labor as a Factor in the Evolution of Farm Organization in California," United States Congress, Senate, Subcommittee of the Committee of Education and Labor, *Hearings* of Senate Resolution 266, Violations of Free Speech and the Rights of Labor (hereafter cited as La Follette Committee, *Hearings*) (Washington, D.C.: U.S. Government Printing Office, 1939), part 54, Exhibit 8762-A, 19802.

2. *Transactions* of the California State Agricultural Society (1877), 88, quoted in Bryan T. Johns, "Field Workers in California Cotton," Master's thesis, University of California, Berkeley, 1948, 17.

3. O. F. Cook, "Cotton Farming in the Southwest" (USDA Miscellaneous Circular, 132, 1913), quoted in John Turner, *White Gold Comes to California* (Bakersfield, Calif.: California Planting Cotton Seed Distributors, 1981), 33.

4. Turner, *White Gold*, 35.

5. A note on nomenclature: Different varieties of cotton are called strains. The cotton lint, or white part of the plant, is called the staple. These come in two lengths, long staple and short staple. Short-staple cotton was most typically grown in the South. In the United States, the only source to that date for long staple was the islands off the Carolina coast.

6. James H. Street, *The New Revolution in the Cotton Economy: Mechanization and Its Consequences* (Chapel Hill: University of North Carolina Press, 1957), 39.

7. *California Cotton Journal*, November 1925.

8. In 1924 Guiberson farmed six hundred acres of cotton. Power, plowing, cultivation, irrigation, seed, and planting cost $23.72 an acre; labor, the most expensive item, cost him $25 a bale and accounted for over half of his costs; ginning the cotton cost $7 a bale. The total costs per acre came to $55.

Guiberson harvested 632 bales. Three hundred bales were sold in the fall for 26 cents a pound, grossing $44,490.42; the remaining bales sold in mid-December at 26 1/4 cents a pound, grossing $45,501.75; the cotton seed sold for $30 a ton, grossing approximately $150 an acre. The total gross was $89,992.17. Guiberson's net profit was $57,000 for the year, or $95 an acre.

9. Stanley Pratt, "Building the Cotton Industry of the San Joaquin Valley on a Sound Conservative Basis," *California Cotton Journal*, November 1925.

10. *California Cotton Journal*, December 1925. For the development of California cotton see Turner, *White Gold*; Steven John Zimrick, "The Changing Organization of Agriculture in the Southern San Joaquin Valley, California," Ph.D. diss., Louisiana State University, 1976.

11. *California Cotton Journal*, January 1924.

12. *Corcoran Journal*, 24 March 1924.

13. Turner, *White Gold*, 46.

14. *California Cotton Journal*, November 1925.

15. *Pacific Rural Press*, 21 November 1925.

16. *Pacific Rural Press*, 1 August 1925; letter from Brodie Hamilton of Anderson Clayton to Mr. A. P. Giannini of the Bank of Italy, 1 November 1932, Bank of America Archives, San Francisco, California.

17. Cotton seed was used in making margarine, Crisco, and other food products.

18. Japan was the largest buyer of California cotton. *Corcoran Journal*, December 1925. For the method of processing see Zimrick, "The Changing Organization of Agriculture."

19. Alfred D. Chandler, *The Visible Hand: The Managerial Revolution in American Business* (Cambridge, Mass.: Belknap Press, 1977), 19–23, 214.

20. Bank of Italy memo, 31 March 1924; memo to the executive committee on Cotton Business in California of the Bank of Italy by Paul Dietrich, vice president of the Bank of Italy, 26 March 1925, Bank of America Archives. Anderson Clayton was an international firm. In 1921 Anderson Clayton began to invest in Mexican cotton production.

21. *California Cotton Journal*, January 1926.

22. *California Cotton Journal*, December 1925.

23. Interest rates varied, ranging from 6 1/2 cents, charged by grower-organized cooperatives, to 20 cents on the dollar, charged by the finance companies. Letter from L. B. Mallory, Department of Division of Labor Statistics and Law Enforcement, to Louis Bloch, 3 April 1930, California State Archives.

24. *Pacific Rural Press*, 2 January 1926.

25. *Corcoran Journal*, 24 March 1924.

26. Paul S. Taylor and Tom Vasey, "Contemporary Background of California Farm Labor," *Rural Sociology* 1, no. 4 (December 1936), 403. In 1930 California had 133 large-scale farms while Mississippi had only 22. According to the 1930 census, by 1930 40 percent of cotton farms were less than 50 acres. Two-thirds were less than 100 acres. And 5 percent had more than 500 acres. By county, Kern claimed 821 that were primarily cotton farms; Kings, 108; and Tulare, 727. Kern had 256 farms that grew cotton but were not classified as cotton farms; Kings, 51; and Tulare, 260. Altogether 2,163 farms raised cotton. Of these, 77 percent were specialized enough to be called cotton farms; that is, more than 40 percent of their income came from cotton. See "Cotton Farms by County," U.S. Department of Commerce, *15th Census of the United States*, 1930, vol. 3, part 3, quoted in Zimrick, "The Changing Organization of Agriculture," 165.

27. *Corcoran Journal*, 24 March 1924.

28. *Fresno Morning Republican*, 11 October 1926. The Tagus was owned by Pacific States Corp.

29. *Pacific Rural Press*, 23 February 1928 and 12 March 1932.

30. *Pacific Rural Press*, 1 May 1926.

31. *Pacific Rural Press*, 2 April 1927.

32. *Corcoran Journal*, 16 May 1924; *Fresno Morning Republican*, 13 September 1925; *Pacific Rural Press*, 18 August 1925; Bryan T. Johns, "Field Workers in California Cotton," Master's thesis, University of California, Berkeley, 1948, 36; Turner, *White Gold*, 74.

33. Mr. W. B. B., Walter Goldschmidt Wasco field notes, quoted in Gregory, *American Exodus*, 9.

34. Interview with Jim White [pseud.], Bakersfield, California, 1 January 1982. All interviews cited are by author unless otherwise noted.

35. H. R. Philbrick, "Legislative Investigative Report," 28 December 1938, UCLA, Special Collections.

36. See Michael Paul Rogin and John L. Shover, *Political Change in California: Critical Elections and Social Movements, 1890–1966* (Westport, Conn.: Greenwood Press, 1970).

37. Gerald Nash, *State Government and Economic Development: A History of Administrative Policies in California, 1849–1933* (Berkeley: University of California Press, 1964), 241. Farm leaders who served on the committee were: Alex Johnson, secretary of the California Farm Bureau Federation; Ray Wiser, president of the Farm Bureau; A. C. Harrison, president of the Farm Bureau from 1923 to 1925; S. Parker Friselle, member of the Agricultural Labor Bureau and president of the Associated Farmers from 1934 to 1935; Roy Pike of the Associated Farmers; R. V. Garrod, president of the Farmers' Union; and Ralph Taylor, executive secretary of the Agricultural Council of California. No member of the grange belonged to the committee from 1929 to 1941. Clarke A. Chambers, *California Farm Organizations* (Berkeley: University of California Press, 1952), 53–60.

38. In the early 1920s, initial enthusiasm for the crop led to planting in unsuitable areas. Whole areas planted hundreds of acres to the crop and erected gins, only to go out of business within a few years. The issue was where cotton would grow well enough to become commercially viable. Cotton became geographically specialized and production began to focus on Kern and Kings counties, with portions of Tulare, Madera, and Merced. *Fresno Morning Republican*, 28 February 1925.

39. Merchants began to push growers to grow short-staple cotton, which could be sold more easily on the national market. It is likely that it would not have been successful in California, because it was not as profitable as the long-staple Acala. It is probable that growers would simply have switched to other crops. For this reason the attempts by some merchants to introduce other strains into the state was seen as a serious threat to the long-term viability of cotton in California.

40. *Pacific Rural Press*, 21 November 1925; *Fresno Morning Republican*, 8 October 1925.

41. For information on this law see Johns, "Field Workers in California Cotton," 47; Wofford B. Camp, "Cotton, Irrigation, and the AAA," an oral history conducted by Willa Baum, 1962–1966, University of California, Berkeley, Regional Oral History Office, 73.

42. Wofford B. Camp, "Cotton, Irrigation, and the AAA," 73.

43. Some men who wrote the bill formed the California Cotton Planting Seed Distributors in 1925. Stanley Pratt later borrowed money from the Bank of Italy (of which Camp became an agent) to develop Producers Cotton Oil Company. See Camp interview, 164.

44. *California Cotton Journal*, May 1926.

45. They ginned 150,000 bales in California. La Follette Committee, *Hearings*, part 51, 18579; notes on gin companies in the Paul Taylor Collection, Bancroft Library, Berkeley, California.

46. *Pacific Rural Press*, 11 January 1930; La Follette Committee, *Hearings*, part 51, 18580.

47. *Fresno Morning Republican*, 27 January 1926; *California Cotton Journal*, January, February, May 1926.

48. Los Tulares, "The Tagus Ranch," *Quarterly Bulletin of the Tulare Historical Society* 125, no. 2 (December 1979); *Fresno Morning Republican*, 18 September 1926; *California Cotton Journal*, September 1927.

49. *Corcoran Journal*, 8 August 1924; *Pacific Rural Press*, 11 August 1925; *Fresno Morning Republican*, 5 September 1926.

50. In Corcoran, for example, by 1921 two growers joined promoter, banker, and grower J. W. Guiberson and formed the Corcoran Cotton Growers Association to borrow money to build a local cotton gin. Many of these cooperatives were local enterprises. The Cotton Growers Cooperative Association was organized around Oroville to finance growers and thus avoid the ginners' and merchants' finance companies. The Merced County Cotton Growers Association was organized to market the cotton of Merced growers. *Pacific Rural Press*, 18 July 1925; 12 January and 2 February 1928.

51. *Pacific Rural Press*, 25 July 1928.

52. This is the tip of the iceberg of a series of mergers, consolidations, and conflicts between cotton interests who wanted to consolidate their control. A major fight centered around the efforts of the Anderson Clayton Company to maintain control over the California cotton market in the face of mounting competition. In 1927 Pacific Cottonseed Products, Inc., run by the Bencini interests, oversaw the merger of three oil companies. At the same time the Bencini interests financed the new California Cotton Cooperative Association in an effort to undercut Anderson Clayton. By 1930, the association broke with Bencini to receive funding from the Farm Board. Anderson Clayton, which felt threatened by the cooperative movement, attempted to undercut the association by offering comparable rates. *California Cotton Journal*, April, August, September 1927; *Pacific Rural Press*, 19 September 1927 and 8 February 1930.

53. These included Anderson Clayton, McFadden and Sons, the locally based Producers Cotton Oil Company, and the California Cotton Growers

Association. Marquis James and Bessie R. James, *Biography of a Bank: The Story of Bank of America* (New York: Harper and Row, 1954), 261–62.

54. The bank's personnel were also individually involved with the cotton industry. In 1928, Harvey M. Kilburn, superintendent of the Herbert Hoover Ranch, president of the California Cotton Growers Association, and supporter of the One Variety Act, joined the Bank of Italy to supervise farming in the San Joaquin Valley. In 1928 W. B. Camp became chief appraiser and then head of the California Lands, Inc., the land-owning subsidiary of Bank of Italy. Turner, *White Gold*, 72; memo to the executive committee on Cotton Business in California by Paul Dietrich, vice president of the Bank of Italy, 26 March 1925; *San Francisco Examiner*, 17 December 1929; *San Francisco Chronicle*, 24 November 1928; Al Haase, "The Cotton Boom 1920–1930" (typescript), Bank of America Archives; Bank of America Report, 31 March 1924, Bank of America Archives.

55. *Pacific Rural Press*, 8 February 1930.

56. Letter from L. B. Mallory, Division of Labor Statistics and Law Enforcement, 3 April 1930, California State Archives, Sacramento.

57. La Follette Committee, *Hearings*, part 51, 18580–82, 18576. One newspaper article claimed that in 1930, 95 percent of the state's crop was financed by three firms. This seems inordinately high, although they probably did finance more of the crop in the 1920s than in the 1930s. *Pacific Rural Press*, 8 February 1930.

58. For more on financing, see James and James, *Biography of a Bank*; McWilliams, *Factories in the Field*, p. 42; Haase, "The Cotton Boom 1920–1930."

59. *California Cotton Journal*, November 1925.

60. *California Cotton Journal*, December 1925. They urged the Department of Agriculture to sponsor a study into pests that afflicted cotton, in order to stave off any damaging blights. And they recommended that a committee investigate the possibility of developing local textile mills to create a local market for the crop.

61. *Pacific Rural Press*, 4 September 1925.

62. Demand for labor in truck crops increased 50 percent; in fruit and nuts, 30 percent. Mark Reisler, *By the Sweat of Their Brow: Mexican Immigrant Labor in the United States, 1900–1940* (Westport, Conn.: Greenwood Press, 1976), 134.

63. Imperial Valley cotton farmers also employed Hindus, Japanese, Filipinos, Koreans, and Native Americans from Arizona. The primary labor force, however, was Anglo and Mexican. Taylor, *Mexican Labor in the United States: The Imperial Valley*, 5–18.

64. Fuller, "The Supply of Agricultural Labor," part 54, 19862. By 1933, 75–95 percent of the work force was Mexican; Paul S. Taylor and Clark Kerr, "Documentary History of the Strike of Cotton Pickers in California, 1933," in Taylor, *On the Ground in the Thirties*; Federal Writers Project, "Labor in California Cotton Fields" (typescript), Oakland, California, n.d. but circa 1939.

65. Letter from George Clements to Governor C. C. Young re the Box bill, 28 December 1927; Charles A. Thompson, "What of the Bracero? The Forgotten Alternative in Our Immigration Policy," *Survey* 54 (June 1925), 291.

66. *Pacific Rural Press*, 17 December 1928, quoted in McWilliams, *Factories in the Field*, 126.

67. McWilliams, *Factories in the Field*, 9; Reisler, *By the Sweat of Their Brow*, 13; *Fresno Morning Republican* of 26 February 1926 reported on undocumented Mexicans living in the Fresno area. The paper estimated that there were 8,000 undocumented Mexicans living in Fresno alone. There were also undocumented Mexicans living in Mexican settlements in Firebaugh, Madera, Merced, and other towns.

68. *California Cotton Journal*, December 1925, 30. The Bloch report noted that the Imperial Valley suffered from a labor shortage in part because Mexican workers, enticed by higher wages in the San Joaquin Valley cotton fields, were migrating north rather than staying in the Imperial Valley. This is correct. Yet San Joaquin Valley growers saw immigration legislation and its enforcement as a major threat to their labor force. Louis Bloch, "Report on the Mexican Labor Situation in the Imperial Valley," Twenty-second Biennial Report of the Bureau of Labor Statistics of the State of California, 1925–1926 (Sacramento: California Bureau of Labor Statistics, 1926).

69. *Pacific Rural Press*, 21 November 1925.

70. *California Cotton Journal*, December 1925.

71. Quoted in the *California Cotton Journal*, February 1925.

72. *California Cotton Journal*, February 1926, 13.

73. This includes losses due to higher wages and cotton left unpicked or rotting due to a shortage of labor. Both figures come from Francisco Palomares, manager of the Agricultural Labor Bureau. The high quote is from a mimeographed letter from Palomares dated 1 March 1927. The more conservative estimate is from the pamphlet by the Agricultural Labor Bureau of the San Joaquin Valley, "The Agricultural Labor Bureau of the San Joaquin Valley: Its Aims and Purposes," n.d. but circa 1926. The two sources agreed that cotton suffered the heaviest losses of the season, followed by raisins. The loss for the entire agricultural industry was estimated at between $2,000,000 and $3,000,000.

74. Fuller, "The Supply of Agricultural Labor," 19866. Fuller's argument has been summarized by Carey McWilliams in *California: The Great Exception* (New York: A. A. Wyn, 1949), 152–53.

75. Beverly Tangari, "Federal Legislation as an Expression of United States Policy Toward Agricultural Labor," Ph.D. diss., University of California, Berkeley, 1967, 366.

76. *Fresno Morning Republican*, 21 and 31 January 1926.

77. A survey in 1938 of eight Kern County cotton farms showed that agricultural labor for cultivating, planting, and chopping accounted for 22 percent of the cost of the crop. Picking costs were an additional 22 percent. The cost for labor was six times the national average. On top of this, by 1938 the costs for labor were probably lower than in the mid-1920s, both in individual rates and in the amount of labor needed on the ranches. United States Congress, Senate, *Report, Violations of Free Speech and the Rights of Labor*, Report no. 1150, 77th Congress, 2d session (Washington, D.C.: U.S. Government Printing Office, 1942) (hereafter referred to as La Follette Committee, *Report*), 497.

78. In 1926 the subscriptions to the Agricultural Labor Bureau totaled $15,000, broken down as follows: cotton, $5,000; dried fruit, $3,700; general industry and utilities, $3,700; fresh fruit, $1,200; Chambers of Commerce, $700; Farm Bureaus, $700.

By 1930, the subscriptions of cotton interests had increased, both absolutely and relatively. Banks had also begun to contribute directly to the organization. The overall decline in subscriptions, as well as the decline by other interests, probably reflects the effects of the Depression and, possibly, complacency: the amounts of the contributions would increase following the strikes of 1933 and 1934. The 1930 total was $9,731, broken down as follows: cotton, $6,678; public utilities, $950; Farm Bureaus, $500; Chambers of Commerce, $500; fresh fruit, $483; dried fruit, $318; banks, $150; misc., $150.

The sources of the bureau's income from 1932 and 1938 reflect the preponderance of cotton money in the bureau, as well as the increase in other sectors' contributions. Seventeen cotton gin companies contributed $29,605.50; 26 fresh and dried fruit companies, $6,522.50; Chambers of Commerce, $5,275.00; Farm Bureaus, $1,962.50; and 31 other companies, including utilities, land, and insurance companies, $9,129.75, for a total of $52,495.25. "The Agricultural Labor Bureau of the San Joaquin Valley, Inc.: Its Aims and Purposes," printed in La Follette Committee, *Hearings*, part 51, 18818, 18821–27.

79. Chambers of Commerce of the United States, Proceedings of the Agricultural Conference of Chambers and Associations of Commerce (Fresno, 26–27 March 1926), 43. Parker Friselle also was sent to Congress "to get us Mexicans and keep them out of our schools and out of our social problems." Quoted in McWilliams, *Factories in the Field*, 125, from *Pacific Rural Press*, 13 February 1926.

80. *Fresno Morning Republican*, 24 January 1926.

81. The Spanish-speaking Palomares, who came from an old Los Angeles upper-class Mexican family that had been in the state for generations, acted as a glorified labor contractor. Married to an Anglo woman, he had worked for and socialized with Anglo-American agriculturalists. Like many other upper-class Mexicans, Palomares had a paternalistic disdain for the Mexican workers he recruited. Yet at the same time he was painfully aware of the racism directed at them and, by implication, at all Mexicans, including himself. By 1939 he was secretary of the Associated Farmers in Fresno, lived in Los Angeles, and was a registered Democrat.

Information on Palomares comes from the Second Annual Conference of the Farm Placement Supervisors, Chambers of Commerce, Los Angeles, USES Department of Labor, in George Clements Collection, Special Collections, UCLA, box 66; Testimony of E. J. Walker, manager of the Arizona Cotton Growers Association, La Follette Committee, *Report*, 148; La Follette Committee, *Hearings*, 18613, 22553; interview by Paul S. Taylor with Francisco Palomares, Paul Taylor Collection, Bancroft Library, University of California, Berkeley.

82. Letter to the board of directors of the Agricultural Labor Bureau from F. Palomares, manager of the Agricultural Labor Bureau of the San Joaquin Valley, 1 March 1927, in George Clements Collection, Special Collections,

UCLA. Palomares's letters to the board detail his efforts to supply and distribute labor to Valley growers. See also letters of 1 and 3 March 1930.

83. Flyer from the cotton file, Industrial Relations: Labor Law Enforcement, "Cotton Conference," 24 April 1930, California State Archives, Sacramento.

84. Report on the Agricultural Labor Bureau by Francisco Palomares to the board of directors, 1 March 1927, George Clements Collection, Special Collections, UCLA.

85. La Follette Committee, *Hearings*, 18635.

86. Even that difference in pay was not significant. The two ranches were smaller, employing between them only thirty-five workers. The going wage rate that year was $1.50: these farmers were paying $1.60 and $1.65. California, Department of Industrial Relations, Division of Housing and Sanitation, "Mexican Survey: Cotton, 1930," typescript of inspections in cotton camps conducted in September and October 1928, California State Archives, Sacramento.

87. For a good discussion of the wage system see Clark Kerr, "Industrial Relations in Large-Scale Cotton Farming," *Proceedings of the Nineteenth Annual Conference of the Pacific Coast Economic Association*, Palo Alto: Pacific Coast Economic Association, 1940. For the demise of the ALB's power in cotton see Galarza, *Farmworkers and Agri-Business in California*, 121-27.

88. Report on the Agricultural Labor Bureau by Francisco Palomares to the Board of Directors, 1 March 1927.

89. Interview with Luis Lima, Brawley, California, June 1982.

90. Stuart Jamieson, *Labor Unionism in American Agriculture*, United States Department of Labor, Bureau of Labor Statistics, Bulletin 836 (Washington, D.C.: United States Government Printing Office, 1945), 6.

91. Fuller, "The Supply of Agricultural Labor," 19865.

92. Interview with Eduardo López, Calexico, California, 21 May 1982.

93. Carey McWilliams made this point: La Follette *Hearings*, part 59, p. 21887. Post–World War I use of cars extended the area of migration and led to development of auto camps.

94. Annual Report of the Commission of Immigration and Housing of California, January 1925 (typescript), Bancroft library.

95. Although the parallel is apt, the differences should also be noted. Some workers used the camp as a home base and migrated to other ranches. Therefore even these camp residents were not there at all times. Also, there was no political apparatus in the camps. Yet growers did have an inordinate influence on local town and county politics, from which Mexican workers were completely excluded. Most Mexican workers were disenfranchised, either by formal residency in other areas or by the fact that a large number were not United States citizens. Few Mexicans voted. Carey McWilliams testified that the cotton labor camps "afford a close parallel to company towns in the mining industry": La Follette Committee, *Hearings*, part 59, 21889. The La Follette Committee, *Report*, 353, made the same point, saying that "employers' reason for providing housing was to have a labor supply readily available to be able to better control it."

96. Department of Industrial Relations, Division of Housing and Sanitation, "Mexican Cotton, 1930." By 1939 cotton ranches had more camps than ranches

devoted to other crops: 32 percent of the state's total agricultural labor camps (140 in Kern county, 94 in Tulare, 74 in Fresno, 80 in Madera, 42 in Merced, and 31 in Kings). Carey McWilliams's testimony, La Follette Committee, *Hearings*, part 59, 21887. On the east side of the Valley, in Madera, where ranches tended to be smaller, another investigator found 40 Mexicans living on the Hayes brothers' ranch and 66 Mexicans and their children, 82 people altogether, living on Marvin Baker's ranch. In 1925, 150 pickers lived in the largest of Corcoran's labor camps. *Fresno Morning Republican*, 11 October 1925.

97. Rev. George L. Cady, "Report of Commission on International and Interracial Factors in the Problems of Mexicans in the United States," n.d. but circa 1920s, George Clements Collection, Special Collections, UCLA; "The School Follows the Child," *Survey*, 1 September 1931.

98. "The School Follows the Child." The point was reiterated in many articles, e.g., *Fresno Morning Republican*, November 29, 1927.

99. The ranch owner erected the building and provided minimum equipment such as tables and benches. The trustees of the school provided the teacher. "The School Follows the Child."

100. *California Cotton Journal*, January 1927. In the beginning of the 1933 season, a school was opened on the Boswell ranch for 51 pupils, one on the Harry Glenn ranch for 42, and one on the Helms ranch for 30. Each had one Anglo teacher. *Corcoran Journal*, 6 October 1933. In 1926 Fresno County established temporary schools to teach 400 children of migrant workers. *Fresno Morning Republican*, 24 September 1926.

101. Interviews with Edward Bañales, Corcoran, California, January 1982 and May 1983. *Fresno Morning Republican*, 12 February 1926; Parker Friselle testimony, United States Congress, House, Committee on Immigration and Naturalization, *Hearings on Seasonal Agricultural Laborers from Mexico,* 69th Congress, 1st session (Washington, D.C.: Government Printing Office, 1926), 5–27.

102. Interview with Carlos Torres, Tulare, California, January 1982.

103. Interviews with Edward Bañales; interview with Carlos Torres; interview with Pat Chambers, Wilmington, California, 11 May 1982; interview with Luis Sálazar, Tulare, California, March 1982; interview with José Callejo [pseud.], Visalia, California, 23 January 1982. See also *Labor Herald*, 24 March 1938.

104. Interviews with Edward Bañales.

105. For a description of labor relations, see Kerr, "Industrial Relations in Large-Scale Cotton Farming," 89.

106. Paul Taylor in La Follette, *Hearings*, 19993.

107. Kerr, "Industrial Relations in Large Scale Cotton Farming" 90; interview with Roberto Castro, Corcoran, California, January 1982; interviews with Edward Bañales; interview with Narciso Vidaurri, Hanford, California, January 1982.

108. Southern-born Colonel Boswell sent flowers and greeting cards to workers for birthdays, and condolences to the family of workers who died. Boswell used paternalism to stave off strikes: following a strike in 1938, Boswell formed a baseball team on the camp, outfitted it with snappy uniforms, and called it the

Boswell Bears. Interview with Magdalena Gómez [pseud.], Los Angeles, California, February 1982.

109. This point should not be overemphasized. Obviously, paternalism was more ingrown and pervasive in company towns in mill areas, for example, where the work force was more permanent. Yet the interest some growers displayed in forms of paternalism or welfare capitalism indicates that they felt some need to come to terms with the needs of workers and suggests a degree of interplay between grower and workers that has been largely ignored in writings about capitalist agriculture. Cletus Daniel gives an evocative description of growers' attempts to stabilize the labor force. Daniel, *Bitter Harvest*, 40–70.

110. Parker Friselle's address at the statewide cotton conference in Los Angeles, reported in *California Cotton Journal*, February 1926, 26.

111. Hulett C. Merritt had, with his father, bought large holdings in the iron mines and organized the Lake Superior Consolidated Iron Mines, a Merritt-Rockefeller syndicate, which Hulett directed at the age of twenty-one. These underground and open-pit iron mines fed the steel mills of Youngstown, Gary, Cleveland, and Pittsburgh. Merritt sold his interest in the mines and in the Mesabi and Northern Railroad, which he had also helped create, to the United States Steel Corporation. By 1941 Merritt not only controlled the Tagus Ranch Co., Tagus Ranch Stores, and Tagus Cotton Gins, but he also boasted an impressive list of holdings and positions in numerous gas and power companies, iron companies, and other business enterprises. His son, Hulett C. Merritt Jr., was the general manager of the Tagus ranch and president of the Tagus Oil Company and the Tagus Cotton Gins, Inc. Hulett Merritt Jr. was also an influential peach grower and by 1941 held, among other titles, the directorship of the Agricultural Labor Bureau of the San Joaquin Valley. Russell Holmes Fletcher, ed., *Who's Who in California*, vol. 1 (Los Angeles: Who's Who Publication Co., 1941), 621; Melvyn Dubofsky, *We Shall Be All: A History of the I.W.W.* (New York: Quadrangle Books, 1969), 319–21.

112. Robert N. McLean, *That Mexican: As He Really Is, North and South of the Rio Grande* (New York: Fleming H. Revell Company, 1928), 176.

113. Merritt at the Second Annual Conference of Farm Placement Supervisors, 23–25 September 1938, George Clements Collection, Special Collections, UCLA, box 66.

CHAPTER TWO. *SIN FRONTERAS*: MEXICAN WORKERS

1. From 1900 to 1930 an estimated 10 percent of Mexico's population came to the United States, over a million people in all. By 1930 an estimated 1.5 million Mexicans resided in the United States. George Sánchez, "Go After the Women: Americanization and the Mexican Immigrant Woman, 1915–1929," working paper, series 6 (Stanford: Stanford Center for Chicano Research, n.d.); Ricardo Romo, "Responses to Mexican Immigration, 1910–1930," *Aztlán* 6, no. 2 (1975), 173.

The number of foreign-born Mexicans expanded. In 1929, Governor C. C. Young's Fact-Finding Committee noted substantial increases in population in the cotton counties, the Imperial Valley, and Los Angeles. See Appendix A.

2. In 1922, 63 percent of the immigrants passing through Los Angeles came from Chihuahua, Durango, Jalisco, and Zacatecas. By 1926, 67 percent came from the western states of Baja California, Sinaloa, and Sonora; 19 percent from the central states of Aguascalientes, Guanajuato, Jalisco, Michoacán, and Zacatecas; 5 percent from the northeastern border states of Chihuahua, Coahuila, Durango, and Nuevo León; and 9 percent from other states in Mexico. In the Imperial Valley, half of the Mexican population originated in the western Mexican states, one-third in the central plateau, and the remainder in the northeastern border states and other parts of Mexico. Anne Christine Lofstedt, "A Study of the Mexican Population in Pasadena, California," Master's thesis, University of Southern California, 1922, 3, quoted in Romo, *East Los Angeles*, 32. The sources of immigration may have shifted because experienced cotton pickers from the states of Baja California who had until the mid-1920s gravitated to the Imperial Valley may have then begun to work in the San Joaquin Valley, where wages were higher.

3. Lawrence A. Cardoso, *Mexican Emigration to the United States, 1897–1931* (Tucson: University of Arizona Press, 1971), 7; James D. Cockcroft, *Intellectual Precursors of the Mexican Revolution, 1900–1913* (Austin: University of Texas Press, 1977), 29; Frank Tannenbaum, *The Mexican Agrarian Revolution* (New York: Macmillan and Co., 1929), 32; Gamio, *Mexican Immigration to the United States*, 13.

4. Rodney Anderson, *Outcasts in Their Own Land: Mexican Industrial Workers, 1906–1911* (DeKalb: Northern Illinois University Press, 1976), 46–49.

5. Anderson, *Outcasts in Their Own Land*, 48.

6. Gómez-Quiñones, *Development of the Mexican Working Class*, 18; Anderson, *Outcasts in Their Own Land*, 41; Centro de Estudios Históricos del Movimiento Obrero Mexicano, *Historia Obrera* 2, no. 5 (June 1975).

7. Interview with Juana Padilla, Brawley, California, June 1982. Please note that the Spanish in interview quotations is often colloquial and, while it may not conform to formal grammar at all times, does reflect the speaker's usage.

8. Anderson, *Outcasts in Their Own Land*, 50–52; Dirk Raat, *Revoltosos: Mexico's Rebels and the United States, 1903–1923* (College Station: Texas A and M University Press, 1981), 65–91.

9. Interviews with Edward Bañales. Purépechas are commonly called Tarascan Indians. They call themselves Purépechas, however: Tarascan is the name given them by the Spaniards.

10. López Castro, *La casa dividida*, 33–37. The Hacienda la Guarucha, established in the eighteenth century, was one of the largest in the area and by the twentieth century covered 35,000 hectares. See also Grindle, *Searching for Rural Development*.

11. Interviews with Edward Bañales. Conditions on haciendas were also discussed in interviews with Roberto Castro, Corcoran, California, January 1982; Vicente Trinidad Navarro, Ancihuácuaro, Michoacán, Mexico, 17 July 1987; Juan Gutiérrez-García, Ancihuácuaro, Michoacán, 18 July 1987; Tannenbaum, *The Mexican Agrarian Revolution*; Grindle, *Searching for Rural Development*, 117–25.

12. Grindle, *Searching for Rural Development*, 117. Vicente Trinidad Navarro remembers that local priests, often relatives of the landowners, worked closely with the hacendados and landowners. Priests reported confessions of workers to landowners and threatened people with hell if they defied the wishes of the landowners. As a result, despite their continuing faith in Catholicism, Navarro and others developed an antipathy to clerics. Interviews with Vicente Trinidad Navarro and Juan Gutiérrez-García.

13. Grindle, *Searching for Rural Development*, 118–20; Alvaro S. Ochoa, "Arrieros, braceros, y migrantes de Oeste Michoacán (1849–1911)," El Colegio de Michoacán, n.d.; López Castro, *La casa dividida*, 16–24.

14. For many, the journey was long and arduous. Vicente Trinidad Navarro remembered that his grandfather had taken three months to travel by burro from rancho Ancihuácuaro in northern Michoacán to the United States. In 1906 three workers from the Tierras Blancas hacienda in Michoacán took six months to walk to the border near Tijuana. López Castro, *La casa dividida*, 36; interview with Vicente Trinidad Navarro.

15. López Castro, *La casa dividida*, 32.

16. Cardoso, *Mexican Emigration*, 27, 73–75.

17. Ibid., 11.

18. Cardoso, *Mexican Emigration*, 11, 42; Marjorie Ruth Clark, *Organized Labor in Mexico* (Chapel Hill: University of North Carolina Press, 1934), 9.

19. Cardoso, *Mexican Emigration*, 22.

20. Writing on November 10, 1910: "En general, se calcula que la emigración tanto a los Estados Unidos de America, como en los [estados] de Veracruz, Campeche y Tabasco, ha causado en el Distrito un descenso de más de 5,000 habitantes." ("In general, it is calculated that because of emigration to the United States and also to Veracruz, Campeche, and Tabasco, this district has lost more than 5,000 inhabitants." Ochoa, "Arrieros."

21. Romo, *East Los Angeles*, 42.

Both Mexican and United States figures on Mexican immigration are inexact. This was due in part to poor methods of recording immigration flows and to the different methods used by the Mexican and the U.S. governments. It was also due to the large number of people who crossed into the United States without documents and, thus, without being recorded. In 1924 Secretary of Labor James Davis testified that he believed that between 200,000 and 300,000 Mexicans had entered the United States without papers in the period between 1921 and 1924.

22. Interview with Eduardo López, Calexico, California, 21 May 1982.

23. Gamio, *Life Story of the Mexican Immigrant*, 104–9, 149–50; interview with Roberto Castro; interview with Ray Magaña, Corcoran, California, 17 September 1982; interview with Juana Padilla; interview with Jessie de la Cruz, Fresno, California, 1 June 1981.

24. Romo, *East Los Angeles*, 80.

25. Romo, *East Los Angeles*, 6–88, discusses the formation of Los Angeles and its outlying barrios. Young, *Mexicans in California*, 46; "Labor and Social Conditions of Mexicans in California," *Monthly Labor Review* 32 (January 1931); Ricardo Romo, "Work and Restlessness: Occupational and Spatial Mo-

bility Among Mexicanos in Los Angeles, 1918–1928," *Pacific Historical Review* 46, no. 2 (May 1977); Carey McWilliams, *North from Mexico: The Spanish-Speaking People of the United States* (Westport, Conn.: Greenwood Press, 1948), 169.

26. Taylor, *Mexican Labor in the United States: The Imperial Valley*, 28.

27. Interview with Ray Magaña.

28. To inquiries sent to California realty boards, twenty-four out of forty-seven replied that they had segregated districts composed of Mexicans. Other districts inserted racially restrictive clauses in real estate contracts. Young, *Mexicans in California*, 176–77.

29. C. C. Young's report stated that in a number of counties some schools had a Mexican population of over 90 percent. Los Angeles had ten such schools, Orange County fourteen, Imperial County eight, and San Bernardino six-teen. Young, *Mexicans in California*, 176–78. Cf. Romo, *East Los Angeles*, 84–85, 139–40, 168; Taylor, *Mexican Labor in the United States: The Imperial Valley*, 67.

30. For unincorporated Los Angeles County, in 1916 there were 70 infant deaths per 1,000 births among Anglos and 285 among Mexicans. By 1927, deaths had decreased to 45 per 1,000 for Anglos and 97 for Mexicans; by 1929 the rate was 40 for Anglos and 105 for Mexicans. Young, *Mexicans in California*, 184.

31. In 1928 Mexicans in the county of Los Angeles were an estimated 11 percent of the county population. In the city of Los Angeles, they were about 10 percent. Young, *Mexicans in California*, 177.

32. The remainder who were born in the United States were primarily children of immigrants or migrants from other parts of the Southwest. Young, *Mexicans in California*, 72.

33. Ibid., 72. The rate of naturalization declined again from 1920 to 1930. Young, *Mexicans in California*, 61–74. In a 1923 study, 25 percent of the Mexicans questioned felt they had to remain politically loyal to Mexico. Most flatly refused to answer the question of national affiliation. Evangeline Hymer, "A Study of the Social Attitudes of Adult Mexican Immigrants in Los Angeles and Vicinity," Master's thesis, University of Southern California, 1923, 51.

34. Gamio, *Mexican Immigration to the United States*, 128.

35. Mary Cecilia Lanigan, "Second Generation Mexicans in Belvedere," Master's thesis, University of Southern California, 1932, 80–81.

36. Jay S. Stowell, *The Near Side of the Mexican Question* (New York: George H. Doran Co., 1921), quoted in Sánchez, "Go After the Women," 102.

37. Interview with Guillermo Martínez, Los Angeles, California, 21 April 1971.

38. Gamio, *Mexican Immigration to the United States*, 129.

39. Helen Walker Douglas, "The Conflict of Cultures in First Generation Mexicans in Santa Ana, California," Master's thesis, University of Southern California, 1928; Clara Gertrude Smith, "The Development of the Mexican People in the Community of Watts, California," Master's thesis, University of Southern California, 1933, 59.

40. Gamio, *Mexican Immigration to the United States*, 94–95. The verse "Defense of the Emigrants," a seeming rejoinder, explains the economic reasons people remained in the United States, while reaffirming their loyalty to Mexico:

> Many people have said
> That we are not patriotic
> Because we go to serve
> For the accursed "patolas."
> But let them give us jobs
> And pay us decent wages.
> Not one Mexican then
> Will go to foreign lands.
> We're anxious to return again
> To our adored country,
> But what can we do about it,
> If the country is ruined?

J. Frank Dobie, *Puro Mexicano* (Dallas: Southern Methodist University Press, 1935), 249.

41. Contractors, for example, exploited workers. Mexicans swindled Mexicans, and Mexican lawyers were known to swindle their Mexican clients. Enrique Bravo, Mexican consul based in Fresno, complained to his superiors about conflicts within the Fresno Mexican community. Letter from E. Bravo to Sr. Manuel Otalora, Oficial Cónsul de México en Los Angeles, 29 September 1931; letter from E. Bravo to Alejandro Lubbert, cónsul general de México, San Francisco, 27 September 1931; letters to SRE from members of Liga Protectora Mexicana de Fresno, 27 July 1934 and 25 September 1933. All letters in Archivo de Secretaría de Relaciones Exteriores (SRE), México D.F.

42. Yet this disruption of old communities simultaneously helped to weaken regional barriers and thus acclimatized Mexicans to being part of a national state and identity.

43. Interview with Roberto Castro.

44. Interview with Vicente Trinidad Navarro. Some data suggest that the shift was less sharp: one 1921 report claimed there were 106 women to every 100 men in Jalisco, Guanajuato, and Michoacán. *Resumen del censo general de habitantes de 30 de noviembre de 1921* (México: Talleres Gráficos de la Nación, 1928), quoted in George Sánchez, "Becoming Mexican American: Ethnicity and Acculturation in Chicano Los Angeles, 1900–1943," Ph.D. diss., Stanford University, 1989, 38.

45. Interview with Josefina Arancibia, Madera, California, January 1982.

46. Interviews with Juana Padilla and Juan Gutiérrez-García.

47. Interviews with Juana Padilla; Vicente Trinidad Navarro; Roberto Castro; Guillermo Martínez; Belén Flores, interview, Hanford, California, January 1982; Teresa McKenna, personal communication to author, June 1989. See also Elizabeth Salas, *Soldaderas in the Mexican Military: Myth and History* (Austin: University of Texas Press, 1990).

48. Paul Taylor, "Mexican Women in Los Angeles Industry in 1928," *Aztlán* 11 (Spring 1980), 99–131.

49. Vicki Ruiz, "Obreras y Madres: Labor Activism Among Women and Its Impact on the Family," in *La Mexicana/Chicana*, Renato Rosaldo Lecture Series,

1, 1983–1984 (Tucson: Mexican American Studies and Research Center, University of Arizona, 1985), 22. People were not always married in the church or following legal procedures. Many rural areas were rarely or never visited by priests, and formal marriage was often prohibitively expensive. Nevertheless, relationships were referred to as marriages and were considered as such within the community. One woman framed her marriage license and proudly displayed it over the marital bed.

50. Rosaura Sánchez, "The Chicana Labor Force," in *Essays on la Mujer*, ed. Rosaura Sánchez and Rosa Martínez Cruz (Los Angeles: UCLA, Chicano Studies Research Center Publications, 1977). For a good summation of the literature, see Saldívar, *Chicano Narrative*, 20–23.

51. Interview with Vicente Trinidad Navarro; interviews with Edward Bañales; interview with Guillermo Martínez.

52. In the Imperial Valley, the Miguel Hidalgo Society, a mutual aid society, owned only a typewriter, a piano, and a "moving picture machine." Kathryn Cramp, *Study of the Mexican Population of the Imperial Valley, California* (New York: Committee on Farm and Cannery Migrants, Council of Women for Home Missions, 1926), 14. Young Mexican women associated with the All Nations center in Los Angeles told an investigator in 1929 that they went to movies at least once a week, and some went three times a week. They had all gone enough to have chosen favorite actors and actresses. Rena Blanche Perk, "The Religious and Social Attitudes of the Mexican Girls of the Constituency of the All Nations Foundation in Los Angeles," Master's thesis, University of Southern California, 1929, 79–80. George Sánchez discusses the impact of movies on Mexicans in his "Becoming Mexican-American," 227–29.

53. McWilliams, *North from Mexico*, 226. Another version of this theme can be seen in Gamio, *Mexican Immigration to the United States*, 89.

54. Dr. Miriam Van Waters, referee of Juvenile Court in Los Angeles, said, "The Mexican woman deserts as often as does the man." Interview by Paul S. Taylor, Los Angeles, California, n.d. but circa 1930. Paul S. Taylor Collection, Bancroft Library, University of California, Berkeley. This point was confirmed in many interviews. There are a number of possible reasons for this: it may suggest increasing decision-making for women in private relations; it may suggest a continuation of older patterns in private relations that were hidden because they were proscribed, at least theoretically, by cultural tradition and which have not yet been explored; and it may point to the disruptions of emigration. In any case, the burdens of separations fell heavily on women who, while they could often rely on help from other women and family, usually took care of the children. This did not mean, however, that women were going out on their own: usually they left one man for another or to return to their natal family.

Sexual relations are a tender subject, but an acquaintance related the following story told him by his grandmother, whom he described as traditional, attending church faithfully and cooking tamales for church socials. She was first married to a man whom she divorced as not being a good husband and father. She took a second husband, this time one who was a good husband and father, but she kept her first husband as a lover. They were careful to be discreet, but she made it clear that she was not the only woman in the community who had done this. A great deal more work needs to be done based on the reality of women's lives, not on the ideal which both men and women may ascribe to in discussions or in theory. Mexican women may have had and acted on a less constricted view of sexuality than has been thought.

55. Recent scholarship has decimated the argument that industrial capitalism led to the breakdown of the family. See Bodnar, *The Transplanted.*

56. Griswold del Castillo, *La Familia*, 101; Sánchez, "Becoming Mexican American," 164–205.

57. Bodnar, *The Transplanted*, 57–83; Eric Hobsbawm, *Workers: Worlds of Labour* (New York: Pantheon Books, 1984), 20. Zamora, *The World of the Mexican Worker in Texas.*

58. This was repeated in oral histories. See interviews with Roberto Castro; Edward Bañales; Juana Padilla; Jessie de la Cruz; Edward Cuéllar, interview, Visalia, California, January 1982; Lillie Gasca-Cuéllar, interview, Visalia, California, January 1982; Sabina Cortez, interview, Corcoran, California, January 1982.

59. Perk, "The Religious and Social Attitudes of the Mexican Girls," 83.

60. Interview with Lillie Gasca-Cuéllar.

61. Interview with Sabina Cortez; see also Smith, "The Development of the Mexican People in the Community of Watts," 47–55.

62. Lanigan, "Second Generation Mexicans in Belvedere," 76–77.

63. Griswold del Castillo, *La Familia*, 43; interviews with Sabina Cortez; Edward Bañales; pamphlet by Sociedad Progresista Mexicana, "Aniversario de Oro, 1929–1979" (Bakersfield: La Sociedad Progresista, 29 September 1979), 3; *Fresno Morning Republican*, 14 September 1925; Lea Ybarra and Alex Sarragosa, *Nuestras Raíces: The Mexican Community of the Central San Joaquin Valley* (Fresno: La Raza T.E.A.C.H. Project, California State University, 1980), 18.

64. Kathryn Cramp, Louise Shields, and Charles A. Thomson, "Study of the Mexican Population in the Imperial Valley, California," for the Committee on Farm and Cannery Migrants, Council of Women for Home Missions, New York, 31 March–9 April 1926, typescript, Bancroft Library, University of California, Berkeley, 18; Taylor, *Mexican Labor in the United States: The Imperial Valley,* 64; Devra Anne Weber, "The Organizing of Mexicano Agricultural Workers: Imperial Valley and Los Angeles, 1928–34—An Oral History Approach," *Aztlán* 3, no. 2 (1973), 315–16.

65. Interview with Jessie de la Cruz.

66. Interviews with Edward Bañales; Jessie de la Cruz; María Hernández [pseud.], Corcoran, California, 18 September 1982; Roberto Castro; and Ray Magaña.

67. Interview with Jessie de la Cruz.

68. Interview with Jessie de la Cruz.

69. Those who had picked cotton in Baja California, included the Refugio Hernández family, who picked cotton around Mexicali before crossing the border into the Imperial Valley. Interview with Refugio Hernández, El Centro, California, June 1982.

70. Gov. C. C. Young's report gave the following daily wages for Mexican workers in California in 1928 (without board): laborers, $3.80; agriculture

$3.04; construction $3.90; industry $3.79. Young, *Mexicans in California*, 113–15. Gamio, *Mexican Immigration to the United States*, 39, and see tables of wages in Appendix A.

71. Fuller, "Supply of Agricultural Labor," in La Follette Committee, *Hearings*, 19862; *Annual Report* of the Commission of Immigration and Housing of California, 25 January 1925, 14, typewritten; Francisco Palomares, "Report of the Agricultural Labor Bureau of the San Joaquin Valley for 1927," 1 March 1927, Carey McWilliams Collection, UCLA; interviews with Edward Bañales and Eduardo López.

72. Parker Friselle testimony, United States Congress, House, Committee on Immigration and Naturalization, *Hearings on Seasonal Agricultural Laborers from Mexico*, 69th Congress, 1st session (Washington, D.C.: Government Printing Office, 1926), 5–27.

73. Douglas E. Foley, *From Peones to Politicos: Ethnic Relations in a South Texas Town, 1900–1977* (Austin: Center for Mexican American Studies, University of Texas, 1977), 89. Foley notes that for Mexicans who had been tied to the ranchos of South Texas, migrant life, which began in the 1930s, offered not only more money but freedom from the shackles of indebtedness to ranches and, in their travels, a chance to meet new people and experience new areas. The money they earned gave these workers an economic basis upon which to develop some political power in the areas to which they returned. See also Fuller, "Supply of Agricultural Labor," 19856; interviews with Edward Bañales and Jessie de la Cruz. See also Neil Smelser, *Social Change and Industrial Revolution* (Chicago: University of Chicago Press, 1959); Tamara Haraven, *Family Time and Industrial Time* (Cambridge: Cambridge University Press, 1982), 1–8.

74. Interview with Arnold de la Cruz, Fresno, California, 1 June 1981.

75. Interviews with Edward Bañales.

76. Paul Taylor interview with Martínez family, 1928, Paul Taylor Collection, Bancroft Library; interviews with Edward Bañales, Juana Padilla, Eduardo López, María Hernández.

77. Most were part Purépecha Indian. A gift, in the 1920s, of a Purépecha statue to the local museum thus had added significance, as it was representative not of one individual but of an entire community that had worked and lived in Corcoran for years. Edward Bañales told me simply that most of the people around there were Purépechas. A number of other interviewees were also from the same area of Michoacán and were also part Purépecha Indians. *Corcoran Journal*, October 1937; interviews with Edward Bañales and Roberto Castro.

78. Bars and pool halls were centers for Mexican workers in other areas, such as the Imperial Valley, as well. Taylor, *Mexican Labor in the United States: The Imperial Valley*; Weber, "The Organizing of Mexicano Agricultural Workers"; Rita Arambula-Verde, "The Historical Contributions of the Mexican American in Kern County: A Prospectus for School Curricular Materials Development," Master's thesis, California State College, Bakersfield, July 1976.

79. Roberto Castro, son of contractor Mateo Castro, remembered, "Yo viví en el camp [*sic*] 9 años. Ahí vivían muchas familias. Iban y venían. Por ejemplo, si no había trabajo ahí, iban a la uva por un mes, y volvían. . . . Ahí estaba su casa." ("I lived in the camp for nine years. Many families lived there. They used

to go and come back. For instance, if there was no work there, they'd go pick grapes for a month, then come back. . . . That [the camp] was home to them.") Within a few seasons, migrants began to remain in the camps. The López family would "stay in those labor camps all year round. We were allowed to leave our stuff there because they knew that we would come back . . . and stay there. So we would go up north and pick . . . and then go back to the same labor camp and pick cotton. . . . In winter we'd just stay there at the same camp and then wait until they started planting cotton again and we'd start chopping the cotton. . . . We'd stay on the camp because we had no other place to go. . . . At least half the camp was filled all the time." This helped people to get to know each other. As Jessie de la Cruz said, "There were some that would come each year . . . so we knew them. And there were some new ones every year . . . and later on we'd probably meet them at some other labor camp." Interviews with Jessie de la Cruz; Roberto Castro; Eduardo López.

80. Interview with Lillie Gasca-Cuéllar.

81. Arambula-Verde, "The Historical Contributions of the Mexican American in Kern County," 62–63.

82. Interview with Ray Magaña.

83. The prostitutes too were, in a sense, migrant workers. The consul complained in a letter to an unnamed SRE official that the society's lawyer exploited Mexican workers; the co-director had been jailed four times for reasons that were unclear in Bravo's letter; the secretary (a tailor by trade) had been incarcerated in San Quentin for robbery; and the director had been arrested for violating prohibition laws. The sins Bravo attributes to the society may be ascribed in part to the consuls' feud with them and a parallel economy that existed outside the law and made them subject to arrest as they catered to the demands of the migrant community. Letter from Bravo, probably to official in SRE, n.d. but circa September 1931, SRE, file of Enrique Bravo. This comment was made in the context of a flurry of letters to SRE from Bravo and members of the Mexican community in Fresno.

84. Interview with Jessie de la Cruz. In 1925, 15,000 Mexicans attended the 16 September festival. *Fresno Morning Republican,* 17 September 1925. In 1926, 3,000 attended the 16 September parade in Fresno. The festivities included a bloodless bullfight, a dance, and a barbecue. The event was attended by the mayor, city commissioner, sheriff, and field deputy. *Fresno Morning Republican,* 17 September 1926.

85. *Fresno Morning Republican,* 12 February 1926.

86. As in southern California, these organizations had mutually interlocking memberships and held meetings in one another's halls. The forming meeting of the Comisión Honorífica was held in the salon of the Sociedad Juárez Mutualista Mexicana and was attended by representatives from Sucursal Femenil no. 1; Club Atlético Nacional; Logia no. 81 de la AHA; Asociación Católica Juventud Mexicana; and Sociedades Mutualistas del Valle. From Fresno the Sociedad Morelos Mutualista de Obreros Mexicanos, Surcursal no. 1 of Merced established a new branch in Merced Falls. *La Opinion,* 8 September 1929, 10 and 24 October 1929, cited in Nelson A. Pichardo, "The Role of the Community in Social Protest: Chicano Protest, 1848–1933," Ph.D. diss., University of Michigan, 1990, 124.

The number of organizations in the Valley was never as extensive as in southern California: Los Angeles County boasted over 200 organizations from 1928 to 1933. The organizations give some indication of the Mexican class composition of the town: Bakersfield's Mexican Rotary and athletic clubs reflect the older and more prosperous Mexican middle class; Fresno's prevalence of mutual aid societies suggests the high number of workers in the community; and areas with a large Mexican citizen population tended to have branches of the Comisión Honorífica. Interestingly, several of the organizations were women's branches of other groups; Los Angeles had eighty-seven of these. Pichardo, "The Role of the Community in Social Protest," 131.

87. Pichardo, "The Role of the Community in Social Protest," has compiled an extremely useful list of Mexican voluntary associations in California which he culled from the pages of La Prensa for 1917 to 1923 and from La Opinion for the years from 1927 to 1933. No San Joaquin Valley Mexican organizations are listed for 1917 to 1923, but the following are listed for the years from 1927 to 1933 (in this order).

Fresno: Sociedad Morelos Mutualista de Obreros Mexicanos; Sucursal Femenil no. 2 de Morelos Mutualista de Obreras Mexicanas; Sociedad de Jóvenes de la Primera Iglesia Bautista Mexicana; Comité Patriótica.

Merced: Sucursal no. 1 de la Sociedad Morelos Mutualista de Obreros Mexicanos; Sucursal Femenil no. 3 de la Sociedad Morelos Mutualista de Obreros Mexicanos; Sucursal no. 4 de la Sociedad Morelos Mutualista de Obreros Mexicanos; Sociedad Unión Patriótica Mexicana; Comisión Honorífica Mexicana; Sucursal no. 3 de la Sociedad M.F.

Bakersfield: Sucursal no. 1 de la Sociedad Juárez Mutualista Mexicana; Sucursal Femenil no. 1 de la Sociedad Juárez Mutualista Mexicana; Comisión Honorífica Mexicana; Club Patriótica Mexicana; Club Atlético Nacional; Logia no. 81 de la Alianza Hispano Americana (AHA); Asociación Católica de la Juventud Mexicana; Sociedades Mutualistas del Valle de San Joaquín; Club Rotario Internacional; Comité Mexicano de Benefiencia.

East Bakersfield: (Mexican and black barrio): Comisión Honorífica Mexicana.

DiGiorgio Farms (Arvin): Sucursal no. 2 de la Sociedad Juárez Mutualista Mexicana.

Modesto: Sucursal no. 5 de la Sociedad Morelos Mutualista de Obreros Mexicanos.

Tulare: Comisión Honorífica Mexicana; Logia no. 265 de AHA.

Madera: Sociedad Mutua Anáhuac; Logia no. 191 de AHA.

Visalia: Comisión Honorífica Mexicana; Logia no. 00 de la AHA; Colonia Unida Mexicana de Protección Fúnebre.

Hanford: Comisión Honorífica Mexicana; Logia Femenil no. 9 de UPBMI; Logia Femenil no. 10 de UPBMI; Logia Femenil no. 13 de UPBMI.

88. The Progresistas was a statewide organization, incorporated in 1929 as a fraternal organization. A lodge had been formed in 1929 in El Monte, but it was not until the mid or late 1930s that lodges were formed in San Joaquin Valley cotton towns, as Mexican workers began to settle there. In the 1930s, lodges were formed at Corcoran, Sanger, Arvin, Tulare, Hanford, Anaheim, and on the Tagus

ranch. El Monte and Tagus were also dissolved in the 1930s. From Sociedad Progresista Mexicana, "Aniversario de Oro."

For the spread of mutualistas to the Imperial Valley see Griswold del Castillo, *La Familia*, 43; interview with Sabina Cortez, Corcoran, California, 18 September 1982; interviews with Edward Bañales; Taylor, *Mexican Labor in the United States: The Imperial Valley*, 63; Sociedad Progresista Mexicana, "Aniversario de Oro," 3; *Fresno Morning Republican*, 14 September 1925; Ybarra and Sarragosa, *Nuestras Raíces*, 18.

89. In the fight Bravo had with several Fresno Mexican organizations, he wrote to the SRE that "ni el acusador o acusadores son mexicanos, son 'pochos' sin matrículas o vínculo con nuestro país o Gobierno."("Neither this accuser or the accusers are Mexicans; they are 'pochos,' without documents or any connections with our country or government.") Letter of 29 September 1931, SRE, Bravo file.

90. Roberto Castro explained this: "Y entonces ya el Boswell, el patrón, él dijo a mi papá que mejor él . . . le hiciera el trabajo. Pues . . . mi papá se entendía con el campo y yo buscaba la gente pa'que [colloquial for *porque*] viniera a trabajar." ("And then Boswell, the owner, he told my father that it would be better if . . . he did the work. Thus . . . my father dealt with the camp and I went to look for people to come to work [on the ranch].") Interview with Roberto Castro.

91. In season, the contractor employed five or six assistants: the *sucaro*, who oversaw the fields and checked for cotton workers left behind; the *pesador*, who weighed the cotton; the *compero*, who cleaned the camp; and assistants who hauled water and chopped down trees in nearby areas to provide wood for fuel. Interview with Edward Bañales, January 1982.

92. Families and social networks were crucial to a contractor's success. Contracting itself was often a family affair and depended on a network of relatives. The Vidaurri brothers worked together; Roberto Castro worked for his father; Felix Ybarra was the brother-in-law of fellow contractor Ramón Peña and a relative of the Ortíz family, also contractors. Interviews with Edward Bañales, January 1982; Roberto Castro; and interview with Narciso Vidaurri, Hanford, California, January 1982.

93. Interviews with Edward Bañales, January 1982; Roberto Castro; Narciso Vidaurri.

94. Interview with Edward Bañales, January 1982.

95. Contractors' reputations influenced workers' choice of where to work. Felix Ibarra lured single men by offering "a big tent for gambling . . . dope peddlers, bootleggers, and prostitution." Adolfo Toledo, however, hand-picked families, forbade drinking and gambling, and expected workers to be in bed by nine or ten at night. Interview with Edward Bañales, May 1983; article in *Fresno Morning Republican*, 29 November 1927; McLean, *That Mexican*.

96. Interview with Edward Bañales, May 1983.

97. Interviews with Roberto Castro and Narciso Vidaurri. For an excellent discussion on the contractor, albeit in Texas, see Foley, *From Peones to Politicos*, 80–89. José Bañales also worked with a labor contractor he had known in Mexico.

98. Interview with Edward Bañales, May 1983.

99. Interview with Guillermo Martínez.

100. Interview with Edward Bañales, January 1982.

101. Ibid.

102. Interviews with Edward Bañales, May 1983, and Guillermo Martínez; *Fresno Morning Republican*, 29 November 1927; McLean, *That Mexican.*

103. Interview with Edward Bañales, January 1982.

104. A literary example of the contractor can be seen in Alejandro Morales, *The Brick People* (Houston: Arte Publico Press, 1988), a fictional rendition of a factual story of Mexican workers in the Simons' brick yard in Pasadena, California.

105. Workers kept in touch with each other, through both word of mouth and letter, letting others know when the crops would be ready for harvest, which camp offered the best conditions, and where and when work would be available. Workers communicated so regularly that the manager of the Agricultural Labor Bureau referred to it as the "grapevine telegraph" and growers depended on it to attract workers.

106. Interview with Edward Bañales, January 1982.

107. Interviews with Edward Bañales; Guillermo Martínez; María Hernández (pseud.); Belén Flores.

108. Interviews with Jessie de la Cruz; Lillie Gasca-Cuéllar; Belén Flores.

109. Testimony of Bearman, in California, Cotton Wage Hearing Board, Minutes, 28 and 29 September 1939, reported by Margaret Brolliar and Bessie Bezzerides (typescript), Fresno, 1939, 97 (typescript).

110. Interview with Luis Lima, Brawley, California, June 1982.

111. Interviews with Jessie de la Cruz and Edward Bañales.

112. Interviews with Edward Bañales, January 1982, and Jessie de la Cruz.

113. Tduvije Marín, aged four, went to sleep in a rack while her mother and siblings were working with 170 other pickers in the fields of a ranch near Corcoran; she was smothered by cotton loaded into the rack. The father was picking peas in San Luis Obispo. *Corcoran Journal*, 22 October 1937.

114. Interview with Jessie de la Cruz by Lea Ybarra, Fresno, California, 27 August 1980.

115. Interview with Luis Lima.

116. Gamio, *Life Story of the Mexican Immigrant*, 147.

117. Cady, "Report of Commission on International and Interracial Factors."

118. John M. Hart, *Revolutionary Mexico: The Coming and Process of the Mexican Revolution* (Berkeley: University of California Press, 1989), 8, 126–62; Raat, *Revoltosos*, 65–91.

119. Hart, *Revolutionary Mexico*, 133, 142; Raat, *Revoltosos*, 66–91; Anderson, *Outcasts in Their Own Land*, 50–52; James D. Cockcroft, *Mexico: Class Formation, Capital Accumulation and the State* (New York: Monthly Review Press, 1983), 85–114.

120. Interviews with Guillermo Martínez and Belén Flores.

121. Memories of the stories of Joaquín Murietta, Tiburcio Vásquez, and the treaty of Guadalupe Hidalgo surfaced in many oral histories. See interviews with

Belén Flores; Guillermo Martínez; Leroy Parra, Los Angeles, California, 18 April 1971; Roberto Castro; Edward Cuéllar.

122. Interview with Edward Bañales, January 1982; Gamio, *Life Story of the Mexican Immigrant*, 107.

123. An unsigned circular, quoted in Anderson, *Outcasts in Their Own Land*, 78.

124. Quoted in Anderson, *Outcasts in Their Own Land*, 59.

125. Interview with Edward Bañales, May 1983.

126. Interview with Edward Bañales, May 1983; see also articles in *Fresno Morning Republican*, 29 October 1927.

127. Interview with Eduardo López.

128. California Department of Industrial Relations, "Report to Governor's Council," Industrial Relations Office of the Secretary of State, California State Archives, Sacramento, 27 and 28 May 1930, 6–7. The problem became acute in cotton. Many of the wage claims were against contractors. Growers stopped paying contractors, who in turn did not pay pickers. The powerlessness of contractors as well as workers made it hard to collect on the claims. In 1934, for example, the contractor S. Cortez was unable to pay the three wage claims against him "owing to his inability to collect from the large ranch on which he is employed as a contractor, the ranch being financed by the Finance company." The problem of having pay withheld was not new. In 1926 there were claims by Mexican workers against Russell Giffen at the Arvin Ranch for $1,000 of wages he had withheld from workers. A judge ordered him to pay the workers. "Cotton Conference, April 24, 1930," Industrial Relations, Labor Law Enforcement folder, Office of the Secretary of State, California State Archives; *Fresno Morning Republican*, 29 September 1926.

129. Interviews with Edward Bañales, January 1982, and Jessie de la Cruz.

130. Interview with Jessie de la Cruz by Lea Ybarra.

131. *Fresno Bee*, 8 and 9 February 1932. The clippings were among those sent by Bravo to his superiors in Mexico City. See also letter from Bravo to SRE, 12 February 1932, Enrique Bravo file, SRE.

132. Interviews with Jessie de la Cruz. An interview with her brother, Eduardo López, confirmed this.

133. Abraham Hoffman, *Unwanted Mexican Americans in the Great Depression: Repatriation Pressures, 1929–1939* (Tucson: University of Arizona Press, 1974), xiii.

134. Government agents raided the placita in downtown Los Angeles and gathered workers from their homes. The Mexican consul helped county officials repatriate destitute Mexicans. In August 1931 alone, 350 were returned to Mexico from the central and northern counties of the state. Evidently Mexicans didn't take well to the involvement of consuls. Consul Enríque Bravo of Fresno complained of being insulted by the Mexicans he was working to deport: "Un mexicano que se repatriará me insultó diciendo que, como yo nomás comía y dormía que me importaba poco los paisanos, y creo que mi papel como consulado es dejarme insultar. Estos casos son muy frecuentes actualmente en los consulados especialmente en Los Angeles pues nuestros compatriotas creen que nosotros los consules tenemos dinero y que si no obtienen la repatriación es

porque no se las queremos proporcionar." ("A Mexican who was being repatriated insulted me, saying that I just ate and slept and didn't care for other Mexicans, and I believe that my role as a consul is to let them insult me. These cases are actually frequent in the consul offices, especially in Los Angeles, since our countrymen believe that we have money and that if we don't help them to go home it's because we don't want to.") He had gotten twenty-five requests for free repatriation. Bravo letter to Sr. Dn. Manuel G. Otalora, Oficial Mayor de la Secretaría de Relaciones Públicas, México, D.F., 29 September 1931.

California Department of Industrial Relations, "Report to Governor's Council," August 1931. Growers were concerned about the increase. The *Los Angeles Times* reported that a scarcity of Mexicans for cotton picking "is believed largely due to the new policy of the State Department in curtailing visas for entry of Mexican laborers. . . . [L]ess than one eighth of the normal flow of Mexicans across the border has been admitted." This shortage was felt in Arizona. *Los Angeles Times*, 2 October 1930.

135. Interview with Jessie de la Cruz.

136. Ybarra and Sarragosa, *Nuestras Raíces*, 22.

CHAPTER THREE. "AS THE FAULTING OF THE EARTH . . ."

1. Department of Labor, "Preliminary Report of Commissioner of Conciliation," 11 October 1933, National Archives RG 280 176/635.

2. Taylor and Kerr, "Documentary History of the Strike of Cotton Pickers," 17.

3. Turner, *White Gold*, 94.

4. McWilliams, *Factories in the Field*, 229.

5. Memo from W. N. Cunningham to J. H. Fallin, Assistant Director, United States Employment Service, Farm Labor Division, Los Angeles, 19 July 1933, George Clements Collection, UCLA.

6. Confidential letter from W. N. Cunningham to J. H. Fallin, Assistant Director, United States Employment Service, Farm Labor Division, Los Angeles, 19 July 1933, George Clements Collection, UCLA.

7. Memo from Clements to Arnoll re Mexican employment meeting, 25 July 1933, George Clements Collection, UCLA, box 44.

8. Memo from Clements to Arnoll, 20 July 1933, George Clements Collection, UCLA, box 40.

9. Before the CAWIU came to the Valley 130 Mexicans near Fresno had struck, demanding a minimum wage under the NRA. *San Francisco Examiner*, 3 August 1933, 7, quoted in Porter Chaffee, "History of the Cannery and Agricultural Workers Industrial Union," Federal Writers Project, Oakland, California, n.d. but circa 1930s, 63 (typescript).

10. *Visalia Times Delta*, 14, 17, 18, 19 August 1933; *Western Worker*, 28 August 1933; interview with Pat Chambers, Wilmington, California, 11 May 1982; interview with Roberto Castro, Corcoran, California, January 1982.

11. *Visalia Times Delta*, 29 August 1933.

12. *Visalia Times Delta*, 14, 15 September 1933; interview with Pat Chambers, University of California, Berkeley, Regional Oral History Office, 1 June 1972; interview with Caroline Decker, San Raphael, California, June 1982.

13. Quoted in Taylor and Kerr, "Documentary History of the Strike of Cotton Pickers"; La Follette Committee, *Hearings*, part 54, 19955. Pat Chambers also mentions the effectiveness of the workers' communication. Interview with Pat Chambers, 11 May 1982, Wilmington, California.

For other accounts of the strike see Jamieson, *Labor Unionism in American Agriculture*, 100–105; Federal Writers Project, "A Documentary History of Migratory Farm Labor in California: The California Cotton Pickers Strike, 1933," Oakland, California, 1938 (typescript); Johns, "Field Workers in California Cotton."

14. Interviews with Pat Chambers and Caroline Decker; *Visalia Times Delta*, 19 September 1933.

15. *Western Worker*, 25 September 1933; "Resolution Adopted by Cotton Strike Conference Called by CAWIU in Tulare," Carey McWilliams Collection, UCLA, 17 September 1933; *Times Delta*, 12 September 1933.

16. Wofford B. Camp, "Cotton, Irrigation, and the AAA," an oral history conducted by Willa Baum, 1962–1966, 131; *Pacific Rural Press*, 17 March and 3 June 1933; *Corcoran Journal*, 22 September 1933; *Times Delta*, 3 October 1933; "Bank of America Business Review" (typescript), Bank of America Archives, San Francisco, 3, no. 10, 20 October 1933, 21, and 4, no. 1, 20 January 1934, 15; "The Effects of the Depression on Agricultural Income in California," United States Congress, House, Committee on Labor, *Hearings on H.R.6288, Labor Disputes Act*, 74th Cong., 2d session (Washington, D.C.: Government Printing Office, 1935), 348–52; Marian A. de Ford, "Bloodstained Cotton in California," *The Nation* (30 December 1933), 705–6.

17. Hoffman, *Unwanted Mexican Americans in the Great Depression*, 100; George Clements wrote confidentially that "the loss of 150,000 Mexicans from California in the last three years is causing a marked shortage . . . of agricultural labor to such an extent as to create the bidding against each other by the employers for the remaining Mexicans in the state." Letter from Clements to Mr. Harman, 22 August 1933, George Clements Collection, UCLA.

One grower's suggestion of 75 cents was rejected, he said, because "growers objected to the seventy-five cent wage, saying that if we set seventy-five cents, they would strike for eighty-five cents and so on; it was therefore better to stand on sixty cents." Interview with unnamed grower by Paul Taylor, n.d. but circa 1933, in Paul Taylor Collection, Bancroft Library, University of California, Berkeley.

Wages were also increased in other crops. Due to strikes or the threat of strikes, the average hourly farm labor wage increased in 1933 and 1934 from a depression low of 15 cents to 25 cents an hour. La Follette Committee, *Report*, part 4, 619.

18. *Oakland Tribune*, 3 October 1933; *Visalia Times Delta*, 3 October 1933.

19. *Western Worker*, 9 October 1933.

20. Interview with Caroline Decker. This point was supported by Pat Chambers as well. Interview with Pat Chambers, 1 June 1972. *Visalia Times Delta*, 19 September 1933.

21. *San Francisco Examiner*, 3 August 1933, 7, quoted in Chaffee, "History of the Cannery and Agricultural Workers Industrial Union," 63.

22. Quoted in Daniel, *Bitter Harvest*, 42.

23. From photograph in the Paul Taylor Collection, Bancroft Library, University of California, Berkeley. The sign endorsing the NRA was located directly above the sign designating the burlap tent as the union office.

In a letter dated 19 March 1934, a W. W. Grimms of the Tulare CAWIU sent a letter to H. H. Pike of the USDA which said, in part, "In view of the fact that the President of the United States has advocated collective bargaining between employers and employees granting the workers the right to organize into a union of their own choice for this purpose," and went on to report the intimidation of CAWIU members in Tulare which ran contrary to this aim. Letter, file Misc. M, Records of the National Labor Board, Region IX, Record Group 25, San Francisco Administrative Files, Misc. Correspondence, 1933–1934, box 498, National Archives.

24. In 1933, 47,575 workers were involved in thirty-two strikes. Twenty-five of these strikes, almost four-fifths, were led by the CAWIU. Twenty-one CAWIU-led strikes resulted in partial wage increases; four strikes were lost. AFL-affiliated unions led two strikes, independent unions led two, and three were spontaneous. Jamieson, *Labor Unionism in American Agriculture*, 87.

25. See Weber, "The Organizing of Mexicano Agricultural Workers," 307–47.

26. Interview with Pat Chambers 11 May 1982.

27. Interview with Caroline Decker.

28. From 1878 to 1884, Mexican peasants had rebelled against land expropriations. The widespread revolts focused on areas that would later have a heavy out-migration: Michoacán, Guanajuato, Querétaro, San Luis Potosí, Durango, Sinaloa, Chihuahua, Coahuila, Hidalgo, México, Pueblo, and Morelos. John M. Hart, *Anarchism and the Mexican Working Class, 1860–1931* (Austin: University of Texas Press, 1978), 68–69. The peasantry of Mexico has continued to engage in revolts. See Baird and McCaughan, *Beyond the Border*.

29. By 1865 the mutualistas led a series of industrial strikes at the San Ildefonso and La Colima cotton mills. Their success in these strikes and growing membership led to their incorporation, in 1869, into El Gran Círculo de Obreros de México. By 1875 the organization had twenty-eight branches in twelve states. See Rodolfo Acuña, *Occupied America: The Chicano Struggle for Liberation* (San Francisco: Canfield Press, 1972), 190–92. For a discussion of ideology see Hart, *Anarchism*, 43–59; Anderson, *Outcasts in Their Own Land*, 99–154.

30. For accounts of the Partido Liberal Mexicano and strikes leading up to the revolution, see Anderson, *Outcasts in Their Own Land*, 99–222; Cockcroft, *Intellectual Precursors of the Mexican Revolution*, 134–56; Raat, *Revoltosos*, 65–123; Torres Parés, *La Revolución sin frontera*.

31. Interview with Luis Lima, Brawley, California, June 1982.

32. Romo, *East Los Angeles*, 35, 39, 50. For an excellent discussion of Laguna and the revolts, see William K. Meyers, "La Comarca Lagunera: Work, Protest, and Popular Mobilization in North Central Mexico," in Thomas Benjamin and William McNellie, eds., *Other Mexicos: Essays on Regional Mexican History, 1876–1911* (Albuquerque: University of New Mexico Press, 1984), 243–74.

33. The influence of migrants on conflicts in Mexico is being studied in new works. See Taibo, *Bolshevikis*; Friedrich, *Agrarian Revolt in a Mexican Village*; Torres Parés, *La Revolución sin frontera*; Devra Weber, "Beyond Borders: Overview of Mexican Migrants and Transnational Labor Organization," paper presented at the Annual Meeting of the Organization of American Historians, April 1991.

34. Meyers, "La Comarca Lagunera," 243–74.

35. There are numerous accounts of the revolution. Some of those used here are: Adolfo Gilly, *The Mexican Revolution* (London: Verso Press, 1983); Frank Tannenbaum, *Peace by Revolution: Mexico After 1910* (New York: Columbia University Press, 1933); Cockcroft, *Intellectual Precursors of the Mexican Revolution*; Hart, *Revolutionary Mexico*. For the impact of the revolution on Michoacán, see Luis González y González, *San José de Gracia: Mexican Village in Transition* (Austin: University of Texas Press, 1964); Moreno and Fonseca, *Jaripo*.

36. Gilly, *The Mexican Revolution*.

37. Interview with Belén Flores, Hanford, California, January 1982.

38. See interviews with Juana Padilla, Brawley, California, 25 June 1982; Refugio Hernández, El Centro, California, June 1982; Guillermo Martínez, Los Angeles, California, 21 April 1971; Luis Lima, Brawley, California, June 1982; Edward Bañales, Corcoran, California, January 1982 and May 1983. The importance of the revolution is explained in Gamio's chapter on "The United States as a Base for Revolutionary Activity" in his *Life Story of the Mexican Immigrant*, 29–40.

39. For the relationship among Mexicans, unions, and the left see: Gómez-Quiñones, "The First Steps"; Zamora, *The World of the Mexican Worker in Texas*; Zamora, "Sara Estela Ramírez"; Durón, "Mexican Women and Labor Conflict in Los Angeles." The special issue on labor history and the Chicano of *Aztlán*, 6, no. 2 (Summer 1975), contains a number of articles on Mexican labor activity. Some of the most pertinent to this discussion are Emilio Zamora, "Chicano Socialist Labor Activity in Texas, 1900–1920," 224–26; Luis Leobardo Arroyo, "Notes on Past, Present and Future Directions of Chicano Labor Studies," 137–50, and Víctor B. Nelson-Cisneros, "La clase trabajadora en Tejas, 1920–1940," 239–66. Taibo, *Bolshevikis*; Jesús González Monroy, *Ricardo Flores Magón y su actitud en la Baja California* (México D.F.: Editorial Academia Literaria, 1962); Douglas Monroy, "Anarquismo y comunismo: Mexican Radicalism and the Communist Party in Los Angeles During the 1930s," *Labor History* 24, no. 1 (Winter 1983), 33–59; James Kluger, *The Clifton-Morenci Strike: Labor Difficulty in Arizona 1915–1916* (Tucson: University of Arizona Press, 1970).

40. By May 1931 the following organizations belonged to CUOM: Cooperative "Agricola" CEERES, Los Angeles; Federation of Los Angeles County; Federation of San Diego County; Union of Calle 38; Union of Colonia Carmelitas; Union of Escondido; Union of Jardin; Union of Santa Monica; Union of Southern Pacific Lines, Los Angeles; Union of Sur Santa Ana, Gloryeta; Union of Vista; Union of Watts; Union of Westminster. Letter from CUOM of 1 May 1931 to President Pascual Ortíz Rubio, Archivo de SRE, IV-339-18, CUOM. By 1934, CUCOM claimed fifty affiliated organizations with an estimated membership of between 5,000 and 10,000. (The large range probably reflects the large numbers of members who were migratory and thus not paid members for several months of the year.) Jamieson, *Labor Unionism in American Agriculture*, 76–77, 91; Weber, "The Organizing of Mexicano Agricultural Workers," 326–28.

41. By December 1933 Armando Flores, an official in CUCOM, wrote to C. Eucario León, Secretary General of CROM, thanking him for the money and promising the continuation of funds. In the same month, in a letter to Fernando Torreblanca, the subsecretary of the SRE, Flores mentions the CROM, "de la que somos miembros" ("of which we are members"). Nicholas Avilla and Evelyn Velarde Benson, sister of Guillermo Velarde, CUCOM leader, also said they had worked with the CROM. Archivo de SRE, 21-6-4, file of Ricardo Hill; Archivo de SRE, 27-6-4, personal file of Ricardo Hill; interview with Nick Avilla, Los Angeles, by author and Juan Gómez-Quiñones, April 1971; interview with Evelyn Velarde Benson, Los Angeles, California, 4 May 1971.

42. Juan Gómez-Quiñones, *Sembradores: Ricardo Flores Magón y el Partido Liberal Mexicano* (Los Angeles: UCLA, Chicano Studies Research Center Publications, 1973); Hart, *Anarchism*; Cockcroft, *Intellectual Precursors of the Mexican Revolution*; interview with Guillermo Martínez; interview with Leroy Parra, Los Angeles, California, 18 April 1971; interview with Dorothy Healey, Los Angeles, California, 6 May 1971.

43. William Wilson McEuen, "A Survey of the Mexicans in Los Angeles," Master's thesis, University of Southern California, 19114, 89. Many workers interviewed remembered Ricardo Flores Magón as an important revolutionary leader. Interviews with Roberto Castro; Vicente Trinidad Navarro, Anchihuácuaro, Michoacán, Mexico, 17 July 1987; Eduardo López, Calexico, California, 21 May 1982; Guillermo Martínez; Leroy Parra; Evelyn Velarde Benson.

44. Mexicans who had belonged to the IWW in the United States were later involved in labor and left-wing organizing in Mexico. One, Primo Tapia, became the leader of one of the largest peasant unions to join the Communist party. Francisco M. Rodríguez, later active in the Communist party in Tijuana and author of *Baco y Birján: Una historia sangrante y dolorosa de lo que fue y lo que es Tijuana* (México D.F.: B. Costa-Amic, 1968), had worked with the IWW in Los Angeles and been involved with the San Pedro strike. For a discussion of Primo Tapia see: Friedrich, *Agrarian Revolt in a Mexican Village*; Arnoldo Martínez Verdugo, *Historia del comunismo en México* (México D.F.: Grijalbo, 1985), 77, 78, 80; Taibo, *Boshevikis*, 189–92. Taibo's engaging book in part explores the role United States leftists played in the formation of Mexico's Communist party, although the subject would have been better served had more attention been paid to returning Mexicans.

For cross-border organizing see, in addition to works previously cited, Lowell L. Blaisdell, *The Desert Revolution: Baja, California, 1911* (Madison: University of Wisconsin Press, 1962); Raat, *Revoltosos*; Hart, *Anarchism*; Weber, "Sin Fronteras."

45. González Monroy, *Ricardo Flores Magón*. Pat Chambers, who had been an anarchist before joining the Communist party, remembered that Wobblies had sent organizers to the Imperial Valley in 1912, probably to work with Mexicans there. Interview with Pat Chambers, 11 May 1982.

46. This was, of course, true of Anglos as well. Many anarchists sympathetic to the IWW later joined the Communist party. Some changed their political position. Others remained philosophical anarchists, albeit members of the Communist party. Pat Chambers, a major organizer for the CAWIU, had been influenced by anarchists and, no doubt to the annoyance of some party officials, maintained certain forms of organizing which focused more on workers' organization than top-down directives and decisions reached from democratic centralism. Chambers stressed that "consensus and grass roots organizing supersede individual leaders. Leadership will come to the surface." Chambers's concept of grass-roots organizing meshed well with forms of self-organization within the Mexican communities. It may also have contributed to the fact that only a small number of workers were aware of any centralized union or union structure. Interview with Pat Chambers, 11 May 1982.

47. Guillermo's Velarde's father, Fernando Velarde, had read Marx and knew the PLM leaders Enrique and Ricardo Flores Magón well enough to have hidden them in his house in Watts when they were being hunted by police. The Velarde family, like other Mexican leftists, were internationalists. Velarde's daughter, Evelyn Velarde Benson, remembers meetings at the Placita in downtown Los Angeles when Emma Goldman spoke, and the Velarde children were kept home from school to mourn the execution of Joe Hill. PLM member Josefina Arancibia also recalled the meetings at the Placita. Arancibia remembered Emma Goldman as the Russian woman "with little tiny glasses." Interviews with Evelyn Velarde Benson; with Josefina Arancibia, Madera, California, January 1982. See also Weber, "The Organizing of Mexicano Agricultural Workers."

48. Interviews with Evelyn Velarde Benson; Guillermo Martínez; Nick Avilla.

49. Interview with Dorothy Healey. She also made this point in her autobiography. Dorothy Healey and Maurice Isserman, *Dorothy Healey Remembers: A Life in the American Communist Party* (New York: Oxford University Press, 1990), 45.

50. Stanley Hancock, testimony before the House Committee on Un-American Activities, 1 March 1954, quoted in Healey and Isserman, *Dorothy Healey Remembers*, 46.

51. According to Julio Antonio Mella, by the 1930s it was the CROM, not the PCM, that was more active among Mexicans in the United States. Julio Antonio Mella, "La situación del proletariado Mexicano en los Estados Unidos," originally published in *Defensa Proletaria* 1, no. 5 (20 January 1929); reprinted in *Boletín del CEMOS: Memoria*, 1, no. 6 (February–March 1984), 128–31.

Although there was no formal connection between the groups, the influence of left-wing organizing on both sides of the border was profound. Emigrants from the United States, "los slackers," were influential in the formation of the Communist party of Mexico. Mexicans who worked with the IWW in the United States, such as Guillermo Palacios and Primo Tapia, returned to Mexico to take an active role in the Communist party.

52. There were all-Mexican cells in Colton, San Bernardino, Santa Ana, Cucamonga, Etiwanda, San Bernardino, and Brawley, and probably in Tulare as well. At least one of these received literature, in Spanish, from the Soviet Union and conducted study groups.

53. Interviews with Edward Cuéllar, Visalia, California, January 1982; Leroy Parra; Guillermo Martínez; Dorothy Healey; Caroline Decker. *Visalia Times Delta*, 18 October 1933; *Western Worker*, 6 and 20 November 1933, 9 July and 18 December 1934. Stanley Hancock remembered the Communist party as having twenty to twenty-five Mexican party members recruited from the Imperial Valley. Quoted in Healey and Isserman, *Dorothy Healey Remembers*, 46.

54. Interviews with Eduardo López; Luis Lima; Narciso Vidaurri, Hanford, California, January 1982. One area that deserves more study is the extent to which miners who worked in agriculture directly influenced the form and direction of labor resistance.

55. Torres's memories were inexact. He remembered that the strike occurred in 1917, 1918, or 1919 and involved Japanese, Hindus, Mexicans, and whites. He claims that 10,000 were involved, but this figure seems much too high. At first he said he led it, then he said he was "just a little kid." The possibility that the IWW was involved is raised in part by the period, in which the IWW was the major group organizing in agriculture, and by a statement made by Torres. Torres later became a loyal and hard-driving foreman on the Tagus ranch. Yet when asked about communist organizing on the Tagus ranch in 1933, he replied with anger at what had happened to the leaders at Haymarket, "This movement, it started in 1883 in Chicago. You know what they did with those guys? They burned them alive, they hung them on trees . . . that's the *labor leaders*, on that day. . . . They called it los martyres de Chicago." This sympathetic reference to the Haymarket martyrs suggests that he was at least exposed to the anarchist movement. Interview with Carlos Torres, Tulare, California, January 1982.

56. Interviews with Luis Lima and Eduardo López.

57. Although many Mexicans worked with the TUUL and the CAWIU, other Mexicans opposed the organizations. In the Imperial Valley and El Monte strikes, there were ideological divisions within the Mexican organization: some allied with the left-wing Anglos; others, more conservative, refused these alliances and often worked more closely with the Mexican consuls. In any case, they suggest an ideologically diverse Mexican community and organizations.

58. Interview with Leroy Parra; Dorothy Healey made the same point when I interviewed her.

59. Many remembered discussions about the revolution. Interviews with Lillie Gasca-Cuéllar, Visalia, California, January 1982; Belén Flores.

60. Interview with Guillermo Martínez.

61. Many of the people interviewed used this phrase to describe the intentions of a wide variety of organizations, including the IWW, PLM, CAWIU, Socialist party, and Communist party. Interviews with Carlos Torres, Belén Flores, and Roberto Castro. As a commissioner pointed out, they were open to labor organization and "easy subjects for the Communist organizers." Letter from Charles T. Connell, commissioner of Conciliation, to H. L. Kerwin, director of Conciliation Services, Department of Labor, Washington, D.C., 19 February 1930, National Archives, RG 280, folder "Imperial Valley."

62. Paul Taylor interviews, Paul Taylor Collection, Bancroft Library, University of California, Berkeley.

63. Interview with Roberto Castro.

64. Jamieson, *Labor Unionism in American Agriculture*, 86; letter from Clements to Mr. Harman, 22 August 1933, George Clements Collection, UCLA.

65. Taylor and Kerr, "Documentary History of the Strike of Cotton Pickers," 19955; Arambula-Verde, "The Historical Contributions of the Mexican American"; *La Opinion*, 7 October 1933.

66. Interviews with Belén Flores; Roberto Castro; María Hernández [pseud.], Corcoran, California, 18 September 1982; Edward Bañales; Magdalena Gómez [pseud.], Los Angeles, California, February 1982; Adelaida Sálazar [pseud.], Corcoran, California, January 1982; Lillie Gasca-Cuéllar; Narciso Vidaurri; Pat Chambers, 11 May 1982; Caroline Decker. See also Taylor and Kerr, "Documentary History of the Strike of Cotton Pickers," 19975–77.

67. Federal Emergency Relief Administration, "Report on Cotton Strikers" (typescript), circa 1933, Paul Taylor Collection, Bancroft Library, University of California, Berkeley (hereafter cited as FERA, "Report on Cotton Strikers"), 3. Newspapers and other sources estimated the number of strikers at between 3,000 and 4,000. Taylor writes, "It is difficult to estimate the number of persons at Corcoran camp. Generally it has been placed at about 3,000 but more than likely this figure is too generous, since growers wished to enhance the menace, law officers the importance of their problem, and strikers the success of the strike." Taylor, *Labor on the Land*, 62.

68. Taylor and Kerr, "Documentary History of the Strike of Cotton Pickers," in *On the Ground in the Thirties*, 61.

69. Interview with Roberto Castro; Taylor and Kerr, "Documentary History of the Strike of Cotton Pickers."

70. Taylor interviews with Corcoran strikers, Paul Taylor Collection, Bancroft Library, University of California, Berkeley; *Corcoran Journal*, 13 October 1933.

71. Taylor and Kerr, "Documentary History of the Strike of Cotton Pickers"; interview with Roberto Castro.

72. Daniel, *Bitter Harvest*, 186.

73. For a more general discussion of the problems of organizing in agriculture, see Alexander Morin, *The Organizability of Farm Labor in the United States*, Harvard Studies in Labor in Agriculture (Cambridge: Harvard University Press, 1952).

74. Interview with Caroline Decker.

75. Interview with Pat Chambers, 11 May 1982.

76. Chaffee, "History of the Cannery and Agricultural Workers Industrial Union."

77. Interview with Pat Chambers by Berkeley Oral History Project.

78. Taylor and Kerr, "Documentary History of the Strike of Cotton Pickers," 19975; Pat Chambers made the same observation. Interview with Pat Chambers, 11 May 1982.

79. FERA, "Report on Cotton Strikers," 3.

80. The committee was composed of Roberto Castro, Librado Vidaurri, Alfonso Torres, an unnamed delegate from the Corcoran camp, D. Martínez, and Manuel García. Interview with Roberto Castro.

81. *Western Worker*, 26 November 1933. Interviews with Roberto Castro, Lillie Gasca-Cuéllar, Belén Flores; Paul Taylor interviews, Paul Taylor Collection, Bancroft Library, University of California, Berkeley.

82. *La Opinion*, 8 October 1933; interview with Guillermo Martínez.

83. Interview with Pablo Saludado, Earlimart, California, January 1982.

84. He was arrested on charges of inciting workers to leave the fields and was jailed for eight and a half months before being deported. *Visalia Times Delta*, 18 October 1933; *Western Worker*, 6 November 1933 and 9 July 1934.

85. *Western Worker*, 20 November, 18 December 1933, 9 July 1934.

86. Letter from Charles T. Connell, commissioner of Conciliation, to H. L. Kerwin, director of Conciliation Services, Department of Labor, Washington, D.C., 8 May 1930, National Archives, RG 280, 170/5463, folder "Imperial Valley."

A Department of Labor report on the El Monte strike claimed that four or five hundred of the strikers "carried a Red card." This number seems high and may reflect anti-communist hysteria, but it also suggests a proclivity of strikers to work with the CAWIU. Letter from J. H. Fallin, Assistant to Director, Farm Labor Service, United States Department of Labor, to George E. Tucker, Director, United States Employment Service, Kansas City Missouri, 7 April 1934, National Archives, RG 183, box 17.

Mexican workers' organizational ability used in the cotton fields was later demonstrated in the Imperial Valley, where, according to one Anglo organizer, "they warned us against the stool pigeons, gave us information, and let us know the temperature of the workers in the various fields, just how hot or cold they were toward the strike. The workers we visited would arrange little meetings of the strongest union elements among the workers and out of these we would set up committees. When we would have a meeting at a Mexican home a guard would be posted outside. They would take us out the back way and come out in front three or four doors down. In talking to these splendid rank and file representatives of the working class one was heartened. No sacrifice, in the matter of getting them better living and working conditions, seemed too great to make." Interview with H. Harvey, "How We First Went into the Imperial Valley in 1930," in Chaffee, "History of the Canning and Agricultural Workers Industrial Union," 7; similar descriptions were given by Dorothy Healey, who organized in the Imperial Valley in 1934. Interview with Dorothy Healey. See also Healey and Isserman, *Dorothy Healey Remembers*, 42–50.

87. For an intriguing discussion of these issues, see David Thelan, "Memory and American History," *Journal of American History* 75, no. 4 (March 1989).

88. Interview with Belén Flores.

89. *Western Worker*, 20 November 1933.

90. The gender division of labor helped foster a sense of what Temma Kaplan refers to as female consciousness. As Kaplan argues, this stemmed from an acceptance of the gender division of labor and centered "upon the rights of gender, on social concern, on survival." But in the acceptance of this division of labor, which assigns women "the responsibility for preserving life," women often also "demand the rights that their obligations entail, [and the] collective drive to secure those rights . . . sometimes has revolutionary consequences insofar as it politicizes the networks of everyday life." Temma Kaplan, "Female Consciousness and Collective Action: The Case of Barcelona, 1910–1918," *Signs* (Spring 1982), 545. For a further discussion of how this applied to Mexican women see Weber, "Raíz Fuerte."

91. FERA, "Report on Cotton Strikers."

92. Interviews with Lillie Ruth Anne Dunn by Judith Gannon for California Odyssey Project, California State University, Bakersfield, California, 14 and 16 February 1981.

93. It was decided that because they would be less likely to be attacked, the women, not the men, would enter the fields to confront the strikebreakers. Interview with Belén Flores.

94. *Bakersfield Californian*, 24 October 1933.

95. Interview with Belén Flores.

96. *Corcoran Journal*, 25 October 1933. This is also reported in Paul Taylor, *On the Ground in the Thirties*, 37.

97. Belén Flores was the daughter of a small farmer, possibly a sharecropper, from Ponitlán, Jalisco. Her family moved to Texas in 1919, fleeing the revolution. In Texas the family worked in an El Paso dairy. By 1923 they had moved to the Mexican colonia of Maravilla in East Los Angeles. Her brothers worked in a cement factory, and the family picked crops. She met her husband when they were in Lemore picking crops. By 1933 she and her husband had settled in Hanford, and they migrated within the San Joaquin Valley to work. Their contractor was Narciso Vidaurri.

Her description loses something in being translated from the spoken to the written word, and from Spanish to English. Coming from an oral tradition, she told stories in dialogue with brief commentaries, giving each character distinct tones and cadence which both reflected their personalities and, to a greater extent, her judgment about that person. For strikebreakers she used a high-pitched, nervous, pleading voice. For strikers her voice changed markedly and became sonorous, deep and steady.

[She imitates the strikebreaker's voice in a high-pitched, pleading tone.] "No les peguen. Déjenlos. No les peguen." [Her voice drops to indicate the collective voice of the strikers.] "Que se los lleve el esto. Sí, a nosotros nos están llevando de frío y de hambre, pos que a ellos también. ¡Vendidos! ¡Muertos de hambre!"

[Her voice rises as the strikebreakers continue their plea.] "Pos' [colloquial for *pues*] nosotros vivimos muy lejos, venimos de Los Angeles. Tienes que saber de dónde, que tenemos que tener dinero pa' irnos."

"Sí" [her voice lowers and slows as it again becomes the voice of the strikers], "nosotros también tenemos que comer y también tenemos familia. Pero no somos vendidos!"

("Don't hit them. Leave them [the other strikebreakers] alone. Don't hit them."

"Let them go to hell. If we are going cold and hungry, then they should too. They're cowards, sell-outs, scum."

"Because we live far away, we come from Los Angeles. We have to know from where . . . we need to get money to go back."

"Yes, we also have to eat and we also have family," she says. "But we are not sell-outs!")

Interview with Belén Flores, Hanford, California, January 1982. Note: When she says, "Que se los lleve el esto," *el esto* is a substitution for *el chingon* or some other related expletive. For those of you who do not know the term, a trip to the dictionary will enrich your vocabulary. A. Bryson Gerrard, in *Cassell's Colloquial Spanish: A Handbook of Idiomatic Usage* (New York: Collier Books, 1980), defines the term thus: "Chingar. A very rude, but very common, word in Mexico and thereabouts; you should avoid using it though you can hardly escape hearing it. One is obliged to equate it with 'to fuck,' but I fancy it is rather commoner and, as usual with such words, is most used in senses other than its basic one, mainly revolving around frustration, molestation, and booze."

98. *Los Angeles Times*, 22 October 1933; *Times Delta*, 23 October 1933; *Hanford Journal*, 28 October 1933.

99. *Fresno Bee*, 6 October 1933; *Visalia Times Delta*, 6 and 9 October 1933. L. D. Ellett, chairman of the Committee of 14 in Corcoran, threatened as well that "we are going to rid Kings county of all the strikers and strike agitators." *Hanford Journal*, 7 October 1933; *Visalia Times Delta*, 6 October 1933; *San Francisco Examiner*, 9 October 1933.

100. Forrest Frick was named to head the Kern County Farm Bureau in November 1933. *Bakersfield Californian*, 6 November 1933.

101. Ellett was a gin manager and one-time representative of the Pacific Cotton Seed Corporation. Other members of the committee in Kings County included the Corcoran grower, banker, speculator, and member of the Chamber of Commerce, J. W. Guiberson. Taylor and Kerr, "Documentary History of the Strike of Cotton Pickers," 19966.

102. *Visalia Times Delta*, 9 October 1933.

103. Taylor interview, 22 December 1933, Paul Taylor Collection, Bancroft Library, University of California, Berkeley.

104. Ibid.

105. Taylor and Kerr, "Documentary History of the Strike of Cotton Pickers"; *Western Worker*, 11 December 1933.

106. Taylor and Kerr, "Documentary History of the Strike of Cotton Pickers"; *Western Worker*, 30 October 1933.

107. Interview with Jim White [pseud.], Bakersfield, California, 1 January 1982.

108. Walter Goldschmidt, *As You Sow: Three Studies in the Social Consequences of Agribusiness* (Montclair: Allanheld, Osmun, and Co., 1978), 43, 59, 165, 410.

109. Interview with Dan Reuben, owner of Davis Quality Market, by Walter Goldschmidt, 4 April 1944, Federal Records Center, San Bruno, California; interview with Roberto Castro; Taylor and Kerr, "Documentary History of the Strike of Cotton Pickers"; FERA, "Report on Cotton Strikers," 15 October 1933.

110. Taylor and Kerr, "Documentary History of the Strike of Cotton Pickers," 1974.

111. Paul Taylor interview, 22 December 1933, Paul Taylor Collection, Bancroft Library, University of California, Berkeley.

112. Taylor and Kerr, "Documentary History of the Strike of Cotton Pickers," 19992.

113. *Visalia Times Delta*, 19 and 24 October 1933.

114. Taylor and Kerr, "Documentary History of the Strike of Cotton Pickers," 19974.

115. *San Francisco Examiner*, 9 October 1933; *New York Times*, 22 October 1933.

116. *Visalia Times Delta*, 13 October 1933.

117. *The Californian*, 5 October 1933.

118. Interview with Carlos Torres; Taylor and Kerr, "Documentary History of the Strike of Cotton Pickers"; *Visalia Times Delta*, 11 and 12 October 1933.

119. A black striker was shot in the neck near Pixley, and another was shot near McFarland. See *Fresno Bee*, 11 and 12 October 1933; *San Francisco Chronicle*, 15 October 1933.

120. *Visalia Times Delta*, 11 and 12 October 1933.

121. *Hanford Journal*, 13 October 1933; *Corcoran Journal*, 12 October 1933.

122. In a newspaper interview Bravo pointed out that while Mexicans are "unarmed and in perfect spirit to abide by the law," the farmers are "using force [and are] armed." *Advance Register*, 12 October 1933.

123. Interview with Roberto Castro; Taylor and Kerr, "Documentary History of the Strike of Cotton Pickers," 70.

124. Paul Taylor quoted the *Tulare Times* as saying, "Estimates of the crowd ranged from 1,200 to 4,000 with experienced observers setting it at a little over 1,500." There is no mention of whose estimates they were, but one would imagine that they came from local officials, the police, and the union. Taylor and Kerr, "Documentary History of the Strike of Cotton Pickers," 19956.

125. *Visalia Times Delta*, 12 and 15 October 1933.

126. Letters from Rabbi Irving Reichart to Rolph, 3 October 1933 and 9 October 1933, Irving Reichart File, Bancroft Library.

127. Letter from Reichart to James Rolph, 3 October 1933, Irving Reichart File, Bancroft Library.

128. Quoted in Daniel, *Bitter Harvest*, 195.

129. Typed statement to newspapers, Irving Reichart File, Bancroft Library. The Social Service Department of the Diocese of California resolved the same day

that it "urgently requests the authorities of the State of California to take immediately all steps necessary to maintain peace and order in all strike areas." Letter to Reichart from B. D. Weigle, Secretary of Social Service of the Diocese of California. Weigle attached a copy of the resolution. Irving Reichart File, Bancroft Library.

130. Taylor and Kerr, "Documentary History of the Strike of Cotton Pickers," 19994; *Visalia Times Delta*, 12 October 1933.

131. Certainly the California state government was not completely opposed to federal involvement, but its acquiescence depended on the terms of intervention. Desperate, Timothy Reardon, the director of California's Department of Industrial Relations, proposed a direct federal subsidy whereby pickers would receive 80 cents per 100 pounds: growers would pay 60 cents a hundred pounds, and the federal government would pay the 20 cents difference. The federal government rejected the proposal as an impossible precedent. *Fresno Bee*, 22 October 1933.

132. Stephen Vaughn, *Holding Fast the Inner Lines: Democracy, Nationalism, and the Committee on Public Information* (Chapel Hill: University of North Carolina Press, 1980).

133. George Creel, *Rebel at Large: Recollections of Fifty Crowded Years* (New York: Putnam and Sons, 1947).

134. Roosevelt created the National Labor Board on 5 August 1933 and appointed Wagner its first chairman. See Tomlins, *The State and the Unions*, 109.

135. Letter from George Creel to Parker Friselle, 7 November 1933, in Records of National Labor Board, Region IX, Record Group 25, National Archives (hereafter cited as NLB), box 497.

136. In August, 130 Mexicans who had struck against peach-growers struck again after the negotiations, "contending that they were entitled to a minimum of 27 cents per hour under the National Recovery Act." *San Francisco Examiner*, 3 August 1933, 7, quoted in Chaffee, "History of the Cannery and Agricultural Workers Industrial Union," 63.

137. Letter from Irving Reichart to Lloyd Frick, 10 October 1933, Irving Reichart File, Bancroft Library.

138. Letter from George Creel to Parker Friselle, 7 November 1933, NLB, box 497.

139. Taylor and Kerr, "Documentary History of the Strike of Cotton Pickers," 19998.

140. Ibid.

141. Letter from Reichart to H. C. Merritt, 13 October 1933, Irving Reichart File, Bancroft Library.

142. Taylor and Kerr, "Documentary History of the Strike of Cotton Pickers," 19999.

143. Taylor and Kerr, "Documentary History of the Strike of Cotton Pickers," 19987; *Visalia Times Delta*, 15 and 16 October 1933.

144. Sam Darcy, "Autobiography," Sam Darcy Collection, New York University, 18 (typescript). The complete quote is: "The next morning as I approached the road where we were to gather I saw a line of jalopies reaching as far as the eye could see. They had been organized during the night by Bill

Hammett and his field leaders. And standing next to each car was a group of prospective occupants. But protruding from each car was a long slim object which did not seem to be any familiar automobile accessory. These turned out to be a great variety of shotguns, rifles, and other weapons. I called Bill Hammett and demanded to know what this meant. He insisted that I had issued orders for them the day before. I was astounded until he explained that this was what they all understood by my urging them to 'be prepared for a business-like demonstration.' They were Oklahoman and Mexican peasants who understood each other well but did not understand the exaggerated emphasis of language common to city people."

145. Interview with Pat Chambers, 11 May 1982; *Visalia Times Delta*, 18 October 1933.

146. *Visalia Times Delta*, 19 October 1933.

147. Interview with District Attorney Walter C. Haight, Tulare County, by Paul Taylor, 21 December 1933, Paul Taylor Collection, Bancroft Library, University of California, Berkeley.

148. Telegram from Al Wirin to Frances Perkins, 19 October 1933, National Archives, RG 280, 176/635.

149. *Hanford Journal*, 28 October 1933.

150. *Corcoran Journal*, 20 October 1933.

151. Taylor and Kerr, "Documentary History of the Strike of Cotton Pickers"; interviews with Roberto Castro and Belén Flores; *Western Worker*, 9 July 1934.

152. Letter of Caroline Decker to Frances Perkins, 24 July 1933, and letter of Caroline Decker to Irving Reichart, 16 August 1933, Irving Reichart File, Bancroft Library.

153. *Western Worker*, 28 August 1933; interviews with Caroline Decker and Pat Chambers, 11 May 1982; Taylor and Kerr, "Documentary History of the Strike of Cotton Pickers," 20003.

154. George Creel, "Memo on Strike of Cotton Pickers in San Joaquin Valley," 10 November 1933, Records of NRA, Region IX, Record Group 25, National Archives.

155. *San Francisco Chronicle*, 24 October 1933; *Visalia Times Delta*, 26 October 1933; interviews with Belén Flores, Roberto Castro, Soledad Regelado, Corcoran, California, September 1982.

156. George Creel to Frank P. Walsh, 30 October 1933, and Creel to L. W. Frick, 6 November 1933, in NLB, IX/RG 25, National Archives; Creel, "Memo on Strike of Cotton Pickers."

157. A series of telegrams document the attempts to direct workers to the strike areas. See George Creel to Robert Wagner, 16 October 1933; J. H. Fallin, Assistant Director, United States Farm Labor Service in Los Angeles, to Creel, 27 October 1933; Fallin to Robert J. Smart, Special Agent, U.S. Farm Labor Service, State Free Employment Service, Stockton City, 27 October 1933; Fallin to William N. Cunningham, Special Agent, U.S. Farm Labor Service, Bakersfield, 27 October 1933; Fallin to Willard Marsh, Special Agent, U.S. Farm Labor Service, Bakersfield, 27 October 1933; Fallin to W. A. Granfield, State Employment Service, 27 October 1933; W. A. Granfield to Fallin, 20 October 1933;

Fallin to George Tucker, 31 October 1933; Fallin to George R. Tucker, 3 November 1933; Tucker to Fallin, 4 November 1933. All in RG 183, box 1174, National Archives.

A side note to this is an ongoing battle between Fallin of the Farm Labor Service, on one side, and Palomares of the Agricultural Labor Bureau and W. A. Granfield, chief of the state Employment Service, on the other. Fallin privately accused the ALB and Granfield of taking credit for all their work, while they called Fallin uncooperative and refused to deal with him. As the tension heated up, Fallin wrote his superior that Granfield's desire was to "take over Farm Labor Service efforts in California by manipulation of state politics." Fallin to George Tucker, FLS, 31 October 1933. Tucker admonished Fallin, who had by now incurred the enmity of growers and state agencies, to be more diplomatic and said that "a representative of this service should never make the mistake of assuming he is a representative of the Federal Government and therefore has certain authority within a state which can disregard the wishes of those in authority within that state." Tucker to Fallin, 4 November 1933, RG 183, box 1174, National Archives.

158. *La Opinion*, 27 October 1933; *Berkeley Gazette*, 27 October 1933.

159. Taylor and Kerr, "Documentary History of the Strike of Cotton Pickers," 20013; *Trabajador Agricolo* (Spanish language paper of the CAWIU), 20 December 1933; Taylor interviews, Paul Taylor Collection, Bancroft Library, University of California, Berkeley, 1933. *Trabajador Agricolo* reported that in December Arvin and Delano still held regular meetings; thirty people were organizing in Tulare; thirty-two people paid dues and met in Earlimart; Shafter had one hundred members, Wasco forty, and Porterville three hundred and fifty. If we consider the size of union meetings in non-strike situations, it indicates a healthy number in attendance.

160. *Daily Californian*, 31 October 1933.

161. J. L. Leonard, William J. French, Simon J. Lubin, United States Special Commission on Agricultural Disturbances in the Imperial Valley, "Report to the National Labor Board by Special Commission," Release No. 3325, 11 February 1934, Pelham D. Glassford Collection, Special Collections, UCLA, 7 (typescript).

162. Tomlins, *The State and the Unions*, 147.

163. Interview with Pat Chambers.

164. George Creel, "Memo on Strike of Cotton Pickers."

165. Creel still intended to push for collective bargaining in agriculture. Other agreements were made under the NRA in agriculture. A district representative of the Tulare County NRA arranged a wage for olive packers and added, "This would conform with a minimum scale under NRA even though the actual packing does not come under any code. In the event of any such code application might later be applied," he continued, growers would be safe if they gave that rate. Letter from Floyd Byrnes of Tulare County NRA to Creel, 2 November 1933, RG 25/497, Records of NRA, National Archives.

In March, Creel answered a letter from Lincoln Steffens about the future of federal intervention in agricultural labor disputes:

Matters mentioned in your various letters have been my deepest concern, and still are for that matter. Unhappily, however, Washington is without any clear idea in the matter. The National Labor Bureau has attempted to secure the cooperation of the Department of Agriculture, but Secretary Wallace seems unwilling to join in any authoritative and helpful effort.

As a consequence, the National Labor Board, in passing on the Imperial Valley Report[,] ruled that correction of the situation was not a federal duty, but entirely within the province of the State of California.

I have had Will French and Simon Lubin in consultation with me, and despite an utter lack of authority, I feel that something is going to be worked out.

> Letter from George Creel, State Director,
> National Emergency Council,
> to Lincoln Steffens, 19 March 1934,
> National Archives, RG 25, file Misc. S,
> Records of NRA, National Archives.

CHAPTER FOUR. THE MIXED PROMISE OF THE NEW DEAL

1. Finegold and Skocpol, "Capitalists, Farmers and Workers"; Finegold, "Agriculture, State, Party, and Economic Crisis." For a discussion of farm policy proposals see Theodore Saloutos, *The American Farmer and the New Deal* (Ames: Iowa State University Press, 1982), 20–33; Christina M. Campbell, *The Farm Bureau and the New Deal* (Urbana: University of Illinois Press, 1962); McConnell, *The Decline of Agrarian Democracy*.

2. Finegold and Skocpol, "Capitalists, Farmers and Workers"; Campbell, *The Farm Bureau and the New Deal*.

3. In Corcoran, for example, cotton growers met with county farm advisers, officials of the nationwide American Cotton Cooperative, and representatives of the local California Cotton Cooperative in order to get local growers to sign up. *Corcoran Journal*, 21 December 1933, 26 January 1934.

4. Wofford B. Camp, "Cotton, Irrigation, and the AAA," an oral history conducted by Willa Baum, 133; Turner, *White Gold Comes to California*.

5. *Pacific Rural Press*, 2 December 1933.

6. *Bakersfield Californian*, 7 December 1933.

7. *Pacific Rural Press*, 2 and 9 December 1933; testimony of Sloan, California, Cotton Wage Hearing Board, Minutes, 28 and 29 September 1939, reported by Margaret Brolliar and Bessie Bezzerides (typescript), Fresno, 1939, 27, 30–31; see also Lawrence Jelinek, "California Farm Bureau Federation," Ph.D. diss., University of California, Los Angeles, 1953.

8. In the United States as a whole, cotton prices jumped 50 percent. Basil Rauch, *The History of the New Deal, 1933–1938* (New York: Capricorn Books, 1963), 101.

9. John Pickett, *California Journal of Development* (January 1934), 30. Typescripts, Bank of America Archives, San Francisco, California: *Bank of America Business Review* 3, no. 10 (20 October 1933), 21; *Bank of America Business Review* 3, no. 11 (20 November 1933); *Bank of America Business Review* 3, no. 12 (20 December 1933); *Bank of America Business Review* 4, no. 2 (20 February 1934).

10. *Pacific Rural Press*, 3 June 1933.

11. Ed Woodruff, California Lands Inc., *Annual Report to Board of Directors for 1934 to January 1935*, Bank of America Archives, San Francisco, 18.

12. La Follette Committee, *Report*, 289.

13. Boswell made many acquisitions, among them 6,500 acres he purchased in Kern County in 1933. *Bakersfield Californian*, 17 October 1933.

14. *Fresno Bee*, 24 October 1937.

15. Giffen owned 54,000 acres, most of which were planted in cotton. "California Cotton Rush," *Fortune* (May 1949).

16. *Western Worker*, 7 July 1937; La Follette Committee, *Hearings*, part 51, 18574.

17. Cotton's proportion of California's AAA allotment grew from 50 percent in 1933 to 80 percent in 1938. The payments increased from $416,919 in 1933 to $3,356,361 in 1938. La Follette Committee, *Hearings*, part 51, 8418.

18. La Follette Committee, *Hearings*, exhibit 8418, 18792.

19. Taken from the La Follette Committee, *Hearings*, part 62, "1938 Agricultural Conservation Program: Names and Addresses of Persons Receiving Net Payments in Excess of $1000," 22889–906.

20. Letter to *Lubin Society Newsletter*, Carey McWilliams Collection, Special Collections, UCLA [n.d. but circa 1938].

21. Paul S. Taylor, "Again the Covered Wagon," *Survey Graphic* 24 (July 1935).

22. *Bakersfield Californian*, 21 October 1933; Wofford B. Camp, "Cotton, Irrigation, and the AAA."

23. Arthur M. Schlesinger, *The Age of Roosevelt* (Boston: Houghton Mifflin Co., 1959), 73, 209; Alan Wolfe, *The Limits of Legitimacy: Political Contradictions of Contemporary Capitalism* (New York: Free Press, 1977), 138–75.

24. *Bakersfield Californian*, 9 September 1933.

25. *Bakersfield Californian*, 9 September 1933.

26. *Bakersfield Californian*, 9 September 1933. See Appendix B for full proposal.

27. *Bakersfield Californian*, 9 September 1933.

28. The idea of federal intervention in favor of the smaller growers did not die completely. C. O. Griffin, a real estate man from Lindsay, California, wrote to Harry Hopkins of the Federal Relief Administration. Concerned about the mounting activities of the Tulare County Farm Bureau in developing "vigilante groups" (i.e., the Associated Farmers), and yet opposed to unions, he urged that "farmers should organize a bargaining agency that is fully representative of their several crops and groups with the purpose of meeting representatives of labor with a representative of the NRA or the Government . . . to determine by arbitration and agreement . . . the rights in the matter." Letter from C. O. Griffin to Harry Hopkins, 23 March 1934, National Archives, RG 69, box 34–460, FERA state files, 1935–1936.

29. Minutes, Agricultural Labor Subcommittee, 6 November 1933, in La Follette Committee, *Hearings*, part 55, exhibit 8830, 920235.

30. Ibid.

31. La Follette Committee, *Report*, 576–77; minutes of a meeting of the Agricultural Labor Subcommittee, Los Angeles Chamber of Commerce, 6 November 1933, George Clements Collection, UCLA.

32. La Follette Committee, *Hearings*, part 63, 22961; letter from George Clements to M. R. Benedict, director of the Giannini Library, 1 August 1935.

33. La Follette Committee, *Report*, 573–636.

34. Wofford B. Camp, "Cotton, Irrigation, and the AAA," 250.

35. La Follette Committee, *Hearings*, part 55, 29978.

36. *Daily Worker*, 27 October 1934; *Western Worker*, 1, 8, 16, 20, and 23 August 1934, 15 January 1935, in La Follette Committee, *Hearings*, part 69, exhibit 11800, 25322, and exhibit 11804, 25323–24; La Follette Committee, *Report*, 631–32.

37. Report on the economic structure, State Relief Administration, box 36, file 12, 26; *Business Week* (4 September 1937).

38. La Follette Committee, *Report*, 503.

39. La Follette Committee, *Hearings*, part 50, 18608; La Follette, *Report*, 278–79, 515.

40. La Follette Committee, *Hearings*, part 50, 18225.

41. Tomlins, *The State and the Unions*, 147.

42. Wolfe, *The Limits of Legitimacy*, 158.

43. Tomlins, *The State and the Unions*, 148.

44. Quoted in Stein, *California and the Dust Bowl Migration*, 263.

45. *Bakersfield Californian*, 1, 11 August 1933.

46. *Bakersfield Californian*, 11 August 1933.

47. Daniel, *Bitter Harvest*, 235; J. L. Leonard, Simon J. Lubin, and William J. French, "Report to the National Labor Board by Special Commission," Release No. 3325, 11 February 1934, Pelham D. Glassford Collection, Special Collections, UCLA, 14 (typescript).

48. Merle Weiner, "Cheap Food, Cheap Labor: California Agriculture in the 1930's," *Insurgent Socialist* 8, no. 2–3 (Fall 1978).

49. United States Congress, Senate, Committee on Education and Labor, *Hearings on S.2926, To Create a National Labor Board*, 73d Congress, 2d session (Washington, D.C.: U.S. Government Printing Office, 1934), 1000; United States Department of Agriculture, AAA, *Agricultural Adjustment*, a Report of the Administration of the AAA, May 1933–February 1934 (Washington, D.C.: U.S. Government Printing Office, 1935), 5.

50. Vito Marcantonio was a representative from East Harlem who later worked as a lawyer for the Workers Alliance and was president of the International Labor Defense. See Alan Schaffer, *Vito Marcantonio: Radical in Congress* (Syracuse: Syracuse University Press, 1966); Salvatore LaGumina, *Vito Marcantonio: The People's Politician* (Dubuque, Iowa: Kendall Hall, 1969); Harvey Klehr, *The Heyday of American Communism* (New York: Harper Bros., 1984), 292–95.

51. Schlesinger, *The Age of Roosevelt*, 267.

52. The federal government administered money to the states, and $250 million was paid on a matching basis: one dollar of federal funds was given for every three dollars of state money that had been spent on relief in the preceding three months. Another $250 million was given on the basis of need. California received twice as much federal relief as any other state: by 1935, over 200,000 California families were receiving federal relief.

53. California State Relief Administration, *Review of Activities of the State Relief Administration of California, 1933–1935* (San Francisco: California State Relief Administration, April 1936), 22; Stein, *California and the Dust Bowl Migration*, 142.

54. M. H. Lewis, *State Welfare Survey* (California State Relief Administration, n.d. [1935?]), part 1, unpaginated.

55. Letter from George Clements to R. N. Wilson of the State Chamber of Commerce, 19 June 1935, George Clements Collection, UCLA.

56. Jamieson, *Labor Unionism in American Agriculture*, 41.

57. *California Journal*, 30 November 1934.

58. Memorandum of the meeting of the Farm Labor Committee of the State Advisory Council of the State Employment Service, 19 July 1935, George Clements Collection, UCLA. For the statement that large numbers on relief who would otherwise have migrated were staying in metropolitan areas, see telegram from W. V. Allen, U.S. Farm Placement Service, to Lincoln McConnell, Director, Farm Placement Service, U.S. Dept. of Labor, Washington, D.C., 23 September 1935, George Clements Collection, UCLA, box 64; Jamieson, *Labor Unionism in American Agriculture*, 141; David Ziskind, "The Suspension of Relief in Agricultural Areas," Research Section of the Planning Division for the Labor Relations Division, 20 August 1935, Farm Security Administration Collection, Bancroft Library.

59. Letter from George Clements to R. N. Wilson of the California Chambers of Commerce, 10 June 1935. George Clements Collection, Special Collections, UCLA.

60. Francisco Palomares at Housing Conference of December 1934 in La Follette Committee, *Hearings*, part 62, 22552. In a confidential memo Palomares confessed that he expected things to be "hot" for the cotton harvest. Francisco Palomares, "Confidential Report," 7 August 1935, George Clements Collection, UCLA, box 8.

61. Palomares, "Confidential Report."

62. Letter from Arthur Clark, head of the Special Labor Committee, to Francisco Palomares in La Follette Committee, *Hearings*, exhibit 53, part 8774, 19705–6.

63. Schlesinger, *The Age of Roosevelt*, 274.

64. Ibid., 277.

65. Don Henderson, "Agricultural Workers," *American Federationist* (May 1936).

66. Letter from McLaughlin to Palomares, 5 October 1935, in La Follette Committee, *Hearings*, part 63, 22991.

67. La Follette Committee, *Hearings*, pt. 63, 22964, Report of the Special Labor Committee, 22 November 1935; see also Palomares in 11 March 1936 meeting of the Agricultural Committee, La Follette Committee, *Hearings*, part 63, 22964.

68. Minutes of the Southern Statewide Agricultural Committee, 6 August 1935, Bancroft Library.

69. Letter from Harold Pomeroy to W. V. Allen, Asst. Director, Farm Placement Service, USES, 21 September 1935, George Clements Collection, UCLA;

letter from Arnoll to Clements, 1 September 1935, in La Follette Committee, *Hearings*, part 63, 22987; letter from Clements to Arnoll, 8 October 1935, La Follette Committee, *Hearings*, pt. 53, 19708. Telegram from W. V. Allen, U.S. Farm Placement Service, to Lincoln McConnell, Director of Farm Placement Service, U.S. Dept. of Labor, Washington, D.C., 23 September 1935.

70. Letter from Clements to Arnoll, 8 October 1935, in La Follette Committee, *Hearings*, part 53, 19709.

71. He said that although it was "unfortunately true that farm wages are lower than relief wages . . . I am not making the demand" that wages be raised. *Western Worker*, 19 April 1936.

72. Ibid.

73. Jamieson, *Labor Unionism in American Agriculture*, 117.

74. *People's World*, 26 September 1936.

75. Noral had been active in the 1919 Seattle general strike and had organized in the Northwest. Noral, born in 1890, had apprenticed as a boiler maker, blacksmith, and steamfitter and had worked mines in Utah. In Utah he had helped organize the Mine, Mill and Smelterman's Union and, as a result, was blackballed by mining companies. Noral had belonged to the Socialist party from 1912 to 1919. *Western Worker* claimed he had been a member of the Communist party from 1923 to 1928. In 1929 Noral had gone to the U.S.S.R. and worked there for a year and a half. By 1934, he was the state organizer of the State Federation of Unemployed and Allied Organizations. *Western Worker*, 11 October 1934, 29 June 1936. In a "Confidential Report" (probably by Francisco Palomares), George Clements Collection, Special Collections, UCLA, the author claims that Noral had been organizing in Kern County since May or June 1935.

76. *Fresno Bee*, 13 May 1936; La Follette Committee, *Hearings*, part 51, 18839.

77. Resolution from the Weed Patch Grange in Paul Taylor, "From the Ground Up," in Taylor, *On the Ground in the Thirties*, 191.

78. Figures are unavailable on the Valley locals, but they appear to have been active. The camps also provided a base for the alliance: workers in the Arvin and Shafter Farm Security Administration camps supported active alliance chapters. The numbers are notoriously inexact, with the alliance tending to inflate the numbers. See Stuart Jamieson, "The Origins and Present Structure of Labor Unions in Agricultural and Allied Industries in California," La Follette Committee, *Hearings*, exhibit 9376, part 62, 22531–40; see also Alex Noral's testimony in California, Cotton Wage Hearing Board, Minutes, 58.

79. Stein, *California and the Dust Bowl Migration*, 153.

80. Ibid., 157.

81. Tom Collins, "Kern Migratory Camp Report," 8 February 1936, Records of the Farm Security Administration, Record Group 96, Federal Archives and Regional Center, box 23, San Bruno; Stein, *California and the Dust Bowl Migration*, 140–162.

82. Ibid.

83. Dean Hutchison's testimony in La Follette Committee, *Hearings*, part 62, 22541; Mrs. Violet Bright's testimony, "Report of the Bakersfield Conference on Agricultural Labor: Health, Housing and Relief," 29 October 1938, Bakersfield,

Carey McWilliams Collection, John Steinbeck Committee File, UCLA (type-script).

84. L. T. Mott and F. J. Rugg, "Housing for Migratory Agricultural Labor," Department of Industrial Relations, Division of Immigration and Housing, 31 December 1934.

85. U.S. Senate Committee on Appropriations, *Hearings*, 76th Congress, 1st session, 1939, 192, quoted in Stein, *California and the Dust Bowl Migration*, 157.

86. "Conference on Housing of Migratory Agricultural Laborers," 12 October 1935, Marysville, California, quoted in La Follette Committee, *Hearings*, part 62, 22592.

87. Letter from Clements to Palomares, 21 March 1935, in George Clements Collection, UCLA; *Western Worker*, May 1936; Stein, *California and the Dust Bowl Migration*, 253.

88. The RA faced staunch opposition. Farm organizations were either in-different or opposed to the idea. Business was wary. Labor felt that communalism threatened to create rural sweatshops. And the high cost of the project made it politically unattractive. Stein, *California and the Dust Bowl Migration*, 156.

89. Finegold and Skocpol, "Capitalists, Farmers and Workers."

CHAPTER FIVE. NEW MIGRANTS IN THE FIELDS

1. The Tolan Committee cited the number of migrants who entered California in need of manual employment from June 1935 to December 1940: 1935 (June to Dec.), 57,012; 1936, 97,344; 1937, 105,185; 1938, 85,166; 1939, 79,246; 1940, 80,200. United States Congress, House, Select Committee to Investigate the Interstate Migration of Destitute Citizens, Pursuant to H. Res. 63,491,729 (76th Congress) and H. Res. 16 (77th Congress), *Interstate Migration Report* (Washington, D.C.: U.S. Government Printing Office, 1941) (here-after cited as the Tolan Committee, *Report*), 324.

Francisco Palomares remembered it a bit differently, testifying at the La Follette Committee *Hearings*, "In 1933, 1934, 1935, and 1936 we had mostly locals—what I mean, that lived there. Mexicans usually traveled from crop to crop . . . in 1935 we began to get a migration . . . from Oklahoma. . . . But the big migration was in . . . 1939, 1938, and 1937[—]I think that was the peak."

Louis Kaplan reports that the SRA began receiving reports of the exodus of Mexicans, in part due to fear of deportation. A report of November 1936 reported that 90 percent of the cotton pickers were Anglo-Americans, and that Mexicans "practically passed out of the picture": La Follette Committee, *Hearings*, part 51, 18600; Louis Kaplan, "The Problems of Relief and Agriculture in California," December 1937, Carey McWilliams Collection, UCLA, box 19 (typescript). Sources taken from confidential SRA files, most by relief officials in the fields.

2. McWilliams, *Ill Fares the Land*, 196; Jacqueline Sherman, "The Okla-homans in California During the Depression Decade, 1931–1941," Ph.D. diss., University of California, Los Angeles, 1970, 109; Paul Taylor, "Refugee Labor Migration to California, 1937," *Monthly Labor Review* 47, no. 4 (April 1939);

Paul Taylor, "What Shall We Do with Them?" address before the Commonwealth Club of California, San Francisco, April 1938, Carey McWilliams Collection, Special Collections, UCLA.

3. James Gregory points out that migrants who journeyed to California in the Depression years came from both urban and rural areas and had worked both on farms and in industrial jobs, and that the majority moved to cities (especially Los Angeles) in California, not to rural areas. But migrants, whether urban or rural, went to the places they knew best. Of the 40 percent who headed to the Valley, many had been tenants, sharecroppers, or small farmers. Certainly they had experienced the effects of expanding capital. As Mexicans, many had combined work in a variety of jobs such as construction, oil, and railroads. Many of those who had worked in industrial jobs, whether as construction laborers, miners, lumberjacks, or workers on the oil rigs of the Oklahoma oil fields, had been born and raised on the land. Perhaps more important, their assumptions and aspirations were tied to the land.

A 1939 survey of occupations of 21,238 heads of households found that 57 percent had nonagricultural jobs prior to migration to California. But this does not take into account the number who still had some ties to the land, either through family or in their own hope of farming land again. While there are no figures on those who aspired to farming, it is likely that a larger percentage still had some connection with the land.

U.S. Department of Bureau of Agricultural Economics, Seymour J. Janow, "Volume and Characteristics of Recent Migration to the Far West," United States Congress, House, Select Committee to Investigate the Interstate Migration of Destitute Citizens, Pursuant to H. Res. 63,491,729 (76th Congress) and H. Res. 16 (77th Congress), Hearings (Washington, D.C.: U.S. Government Printing Office, 1940–1941) (hereafter cited as Tolan Committee, Hearings), part 6, table 16a, 2307; Gregory, American Exodus, 17.

4. Interview with Fred Ross, Los Angeles, California, March 1983.

5. McWilliams, Ill Fares the Land, 195.

6. The Tolan Committee, investigating the causes of interstate migration in 1940, listed five causes for migration: an over-reliance on agriculture, mechanization, drought, soil depletion, the Agricultural Adjustment programs, and rural poverty. Tolan Committee, Report, 324. A discussion of the New Deal's impact on Oklahoma also occurs in Dale Harrington, "New Deal Farm Policy and Oklahoma Populism," in Studies in the Transformation of U.S. Agriculture, ed. A. Eugene Havens, Gregory Hooks, Patrick Mooney, and Max Pfeffer (Boulder: Westview Press, 1986), 179–205.

7. McWilliams, Ill Fares the Land, 195.

8. Quoted in Stein, California and the Dust Bowl Migration, 11.

9. McWilliams, Ill Fares the Land, 201; see also Tolan Committee, Report; for Texas information see Paul S. Taylor, "Letters to Blaisdell," in On the Ground in the Thirties, 200.

10. Tulsa Tribune, 14 September 1937, quoted in Sherman, "The Oklahomans in California," 62.

11. Quoted in Dorothea Lange and Paul S. Taylor, An American Exodus (New Haven: Yale University Press, 1969), 68.

12. Taylor, "Again the Covered Wagon," 349.

13. "Social Attitudes Expressed by Migrants" (April 1940) in Carey McWilliams Collection, UCLA, "relief" folder.

14. Lange and Taylor, *An American Exodus*, 47.

15. California, Cotton Wage Hearing Board, Minutes, 15.

16. Tom Collins Reports, United States Department of Agriculture, Agricultural Stabilization and Conservation Commission Papers, Record Group 145, Federal Records Center, San Francisco, 17 October 1936 (hereafter cited as Collins Reports); Sheila Goldring Manes, "Depression Pioneers: The Conclusion of an American Odyssey, Oklahoma to California, 1930–1950—A Reinterpretation," Ph.D. diss., University of California, Los Angeles, 1982, 393.

17. Gregory, *American Exodus*, 22–23.

18. Farm Security Administration Case Histories, United States Department of Agriculture, Farm Security Administration, Region IX papers, Bancroft Library, University of California, Berkeley, carton 2, folder 6, "Louis" (hereafter cited as FSA Case Histories).

19. FSA Case Histories, "Chester." The California State Chamber of Commerce published the comparative wage rates: California State Chamber of Commerce, *Migrants: A National Problem and Its Impact on California*, reprinted in Tolan Committee, *Hearings*, part 6, 2785.

20. Lange and Taylor, *American Exodus*, 47; Enid Baird and Hugh P. Brinton, United States Works Project Administration, Division of Social Research, "Average General Relief Benefits, 1933–1938" (Washington, D.C.: U.S. Government Printing Office, 1940), 18.

21. Quoted in Hurly, "The Migratory Worker," 87.

22. *Corcoran Journal*, 22 May 1936.

23. Interview with Jessie de la Cruz, Fresno, California, 1 June 1981.

24. Collins Reports, 24 October 1936.

25. Mr. J. W., Wasco Field Notes, Walter Goldschmidt's 1941 field notes in Professor Goldschmidt's possession, Department of Anthropology, UCLA. Quoted in Gregory, "The Dust Bowl Migration." See California State Chamber of Commerce, *Migrants: A National Problem and Its Impact on California*, for this argument. For other estimations of wages and average amount picked, see R. L. Adams, "Seasonal Labor Requirements for California Crops," California Agricultural Experiment Station, Berkeley, California, *Bulletin 623* (July 1938), table 2, 8. For testimony by workers on the amount they picked see California, Cotton Wage Hearing Board, Minutes, 49–50, (testimony of Alex Noral) 63. The hearings are peppered with testimony on the amount workers picked. These figures coincide with the information gathered in oral interviews.

26. The manager of the State Free Employment Office also made this point in an article in the *Bakersfield Californian*, 19 September 1934.

Advertisements in the *Corcoran Journal* for 21 February 1936 give some indication of prices: Corn meal, 29 cents for 10 lbs.; coffee, 49 cents for 3 lbs.; sugar, 50 cents for 10 lbs.; pink beans, 35 cents for 10 lbs.; eggs, 33 cents for two dozen; dress, $3.95; seersucker dress, $1.00.

27. *Daily Worker*, 25 February 1931.

28. Paul Taylor, "Migratory Farm Labor in the United States," *Monthly Labor Review* 44, no. 3 (March 1937), 4; United States, Special Committee on Farm Tenancy, *Farm Tenancy: Report of the President's Committee*, prepared under the auspices of the National Resources Committee (Washington, D.C.: U.S. Government Printing Office, 1937), quoted in Manes, "Depression Pioneers," 130.

29. "I Wonder Where We Can Go Now?" *Fortune* (April 1939).

30. Arvin's population in 1944 was 6,200. Cited in Gregory, "The Dust Bowl Migration," 142.

31. *Fresno Bee*, 11 November 1939; *Corcoran Journal*, 26 March 1936, 17 January 1938; Sherman, "The Oklahomans in California," 109–10.

32. For this discussion I rely heavily on Gregory, *American Exodus*, 26–35. Gregory has a rich discussion of family and chain migration among the Southwestern migrants.

33. Quoted in Lange and Taylor, *American Exodus*, 66.

34. Collins Reports, 6, 8 August 1936.

35. Interviews with Rev. Billie J. Pate by Michael Neely for California Odyssey Project, California State University, Bakersfield, California, 5 and 12 March 1981.

36. Quoted in Gregory, "The Dust Bowl Migration," 124.

37. Gerald Haslam, "What About the Okies?" *American History Illustrated* 12, no. 1 (1977).

38. Bertha S. Underhill, California Department of Social Welfare, Division of Child Welfare Services, "A Study of 132 Families in California Cotton Camps with Reference to Availability of Medical Care," California Department of Social Welfare, 1936, California State Archives, Sacramento, California.

39. California State Relief Administration, *Review of Activities*, 133–55. Bennett is the fictitious name used for the Hammetts. Roy Hammett confirmed in his interview with Anne Loftis that the "Bennetts" were his family; *Visalia Times Delta*, 14 August 1933; *Western Worker*, 13 November 1933, 7 March 1935.

40. Interview with Jim White [pseud.], Bakersfield, California, 1 January 1982.

41. Taylor, "Again the Covered Wagon," 351.

42. Tolan Committee, *Report*, 344–45.

43. Interview with Edgar Romaine Crane by Judith Gannon for California Odyssey Project, California State University, Bakersfield, California, 7 April 1981.

44. Collins Reports, 8 and 16 August 1936.

45. Memo from Clements in the files of the Agricultural Department of the Los Angeles Chamber of Commerce, 18 December 1936, in La Follette Committee, *Hearings*, exhibit 8752, part 53, 19696 and 19713–14; see also Tolan Committee, *Report*, 376–79.

46. John Steinbeck, "Their Blood Is Strong" (San Francisco: Simon J. Lubin Society, April 1938), 1–4.

47. Kern County Health Department, Sanitary Division, "Survey of Kern County Migratory Labor Problem," Supplementary Report as of 1 July 1938, Public Health Library, University of California, Berkeley (mimeo.).

48. Interview with Frank Stockton by Walter Goldschmidt, Arvin, California, 3 April 1944, Walter Goldschmidt collection, Federal Records Center, San Bruno, California. Italics mine.

49. Interview with wife of potato farmer by Walter Goldschmidt, Arvin, California, 30 March 1944, Records of Walter R. Goldschmidt, Central Valley Project Studies, 1942–1946, Record Group 83, Bureau of Agricultural Economics, Western Regional Office, Federal Archives and Regional Center, San Bruno, California.

50. *Corcoran Journal*, 12 February 1937.

51. *Corcoran Journal*, 30 April 1937.

52. *Corcoran Journal*, 7 January 1938. He also made the point that crime had increased and attributed the 27 percent rise in crime to the influx of 8,000 transients into Kings County to work in cotton. In 1937 there had been 964 criminal cases, compared to 772 in 1936.

53. *Bakersfield Californian*, 4 August 1933.

54. *Bakersfield Californian*, 1 October 1938.

55. Interview with Jessie de la Cruz.

56. Interview with LaRue McCormick, Los Angeles, California, August 1982.

57. Collins Reports, 12 September 1936.

58. *Fresno Bee*, 14 November 1937.

59. Interview with Fred Ross.

60. *Madera Tribune*, 24 August 1939.

61. A woman in a cooperative was nervous about sending her children to schools "where folks mix the races the way they do in California." Katherine Douglas, "Uncle Sam's Coop for Individualists," *Coast Magazine* (June 1939), 7.

62. Interview with Rev. Billie J. Pate.

63. Quoted in Goldschmidt, *As You Sow*, 68.

64. For a history of these relations see Montejano, *Anglos and Mexicans in the Making of Texas*; see also Juan Gómez-Quiñones, "Plan de San Diego Reviewed," *Aztlán* 1, no. 1 (Spring 1970), 124–32.

65. Interview with Fred Ross.

66. Interview with Jessie de la Cruz.

67. Interview with Belén Flores, Hanford, California, January 1982.

68. Goldschmidt, *As You Sow*, 68.

69. Interview with Fred Ross.

70. Quoted in Haslam, "What About the Okies?" 39.

71. James Bright Wilson, "Religious Leaders, Institutions and Organizations Among Certain Agricultural Workers in the Central Valley of California," Ph.D. diss., University of Southern California, 1944, 316, quoted in Gregory, "The Dust Bowl Migration," 428.

72. FSA Case Histories, "Theodore."

73. Interviews with Lillie Ruth Anne Dunn by Judith Gannon for California Odyssey Project, California State University, Bakersfield, California, 14 and 16 February 1981.

74. James Green, *Grass Roots Socialism: Radical Movements in the Southwest, 1895–1943* (Baton Rouge: Louisiana State University Press, 1978), xiii.

75. Garin Burbank, "Agrarian Socialism in Saskatchewan and Oklahoma: Short-Run Radicalism, Long-Run Conservatism," *Agricultural History* 51, no. 1 (January 1977); Green, *Grass Roots Socialism*, 228–269.

76. Gregory, *American Exodus*, 142, 152; Green, *Grass Roots Socialism*, 368–95.

77. Oscar Ameringer, in *Oklahoma Leader*, 27 April 1942, quoted in Green, *Grass Roots Socialism*, 390.

78. Letter from Floyd Murray to Franklin Delano Roosevelt, 19 March 1937, quoted in Green, *Grass Roots Socialism*, 432.

79. Donald Grubbs, *Cry from the Cotton: The Southern Tenant Farmer Union and the New Deal* (Chapel Hill: University of North Carolina Press, 1971), 177–78, 187–88.

80. James Gregory has eloquently pointed out that the popular and historical image of the Southwestern migration did not conform to the experiences of many migrants. But it did for some. The faces Dorothea Lange captured in photographs were probably at the lower economic edge of the migrant group, those with fewest resources and the most desperate. These were the people who, by definition, would most likely be unable to find work in industry, construction, or the oil fields and who would find employment, if at all, in the agricultural fields. The popular image applied more accurately to farm laborers, then, than it did to other segments of the migrant population. Gregory, "The Dust Bowl Migration," 222–56.

81. Interview with Fred Ross.

82. Quoted in Stein, *California and the Dust Bowl Migration*, 263.

83. Interview with Fred Ross.

84. A letter from the party organizer in Bakersfield reported they had recruited 13 new members and pledged 13 more, and vowed to "build the agricultural union by 800%": *Western Worker*, 22 October 1936, 3.

85. *Bakersfield Californian*, 20 September 1934, 18 September 1935. The Townsend Old Age Pension Plan was named for a Long Beach retired physician who proposed that a 2 percent federal sales tax be levied on businesses and the funds used to pay all citizens over sixty years of age a monthly pension of $200. The only stipulation was that the money had to be spent within the month. Townsend theorized that the circulation of large amounts of money would both provide for the elderly and enlarge the shrinking market for national industries and farms. Robert Glass Cleland, *California in Our Time, 1900–1940* (New York: Alfred A. Knopf, 1947), 229–31.

86. *Bakersfield Californian*, 22 September 1933, 13 October 1934.

87. *Bakersfield Californian*, 12 September 1933.

88. Other socialist locals existed in the Valley, including one in Visalia. *Bakersfield Californian*, 23 August 1933.

89. *Bakersfield Californian*, 19 July 1933.

90. *Bakersfield Californian*, 25 July 1933.

91. *Bakersfield Californian*, 7 September 1933.

92. *Bakersfield Californian*, 16 October 1933.

93. Letter from Sam [*sic*] White, Secretary of California Conference of Agricultural Workers, to Henry G. Wallace, Secretary of Agriculture, 4 May 1936, National Archives, RG 145/Calif. A–H.

94. *Bakersfield Californian*, 19 October 1933.

95. *Bakersfield Californian*, 23 October 1933.

96. *People's World*, 30 April 1937; interview with Jim White. Both Fred Ross and Wyman Hicks believed that Champion had been a Wobbly. See interviews with Fred Ross; with Wyman Hicks, Bolinas, California, January 1983.

97. *Kern County Labor Journal*, 6 August 1937.

98. *Bakersfield Californian*, 24 August, 7, 9 September 1933.

99. Dunn was grateful enough to name one of her children after Kearney. She named her other son after Pat Chambers. Interviews with Lillie Ruth Anne Dunn.

100. Interview with Catherine Sullivan by Judith Gannon for California Odyssey Project, California State University, Bakersfield, California, 27 February 1981.

101. Interview with Jim White.

102. *Western Worker*, 7 March 1935. Hammett appeared to have often worked independently or to have joined other organizations. He worked with the CAWIU, although he never joined the union. Yet by 1935 Hammett had apparently broken with earlier allies. In a 1935 potato strike Hammett was accused of undermining the union by urging workers to accept the offered wages and return to work. In 1936 he organized a local organization of the unemployed called the Employees' Security Alliance, to protest the wages set by the Agricultural Labor Bureau. It rivaled the Communist party's unemployed councils. He worked briefly with Lillian Monroe, who, by 1938, had broken with the Communist party and was organizing a union in competition with UCAPAWA. California, State Relief Administration, *Review of Activities*, 133–55; *Visalia Times Delta*, 14 August 1933; *Western Worker*, 13 November 1933, 7 March 1935.

103. Charles E. Larsen, "The EPIC Campaign of 1934," *Pacific Historical Review* 27 (May 1958), 127–48; Leonard Leader, "Upton Sinclair's EPIC Switch: A Dilemma for American Socialists," *Southern California Quarterly* 62 (Winter 1980), 361–80.

104. Sinclair had declared his candidacy on the Democratic ticket in September 1933, shortly before the cotton strike. No doubt the tensions around the strike and its aftermath had some effect on support for Sinclair.

Leading socialists Samuel White and R. W. Henderson did not support Sinclair and helped write a pamphlet explaining their position. *Bakersfield Californian*, 4 September 1934.

105. *Bakersfield Californian*, 29 September, 5 October 1934.

106. *Bakersfield Californian*, 5 October 1934.

107. Editorial, *Bakersfield Californian*, 27 October 1933.

108. *Bakersfield Californian*, 15 October 1934. For other editorials see 15, 17, 19, 22, 23 October 1934.

109. *Bakersfield Californian*, 17 and 20 October 1934.

110. A letter to the editor complained that at a meeting for Merriam, merchants and employers devised a plan whereby each voter would publicly pledge himself to "save California" by indicating his support for Merriam. The author of the letter wrote, "Nearly all present were employers and merchants. I don't see how they were to manage the cards of their employees and the cards of voters indebted to the merchants. I was embarrassed by the procedure. . . . The high public pressure must be let off some time . . . it left me 'cold.'" *Bakersfield Californian*, 20 October 1934.

111. *Bakersfield Californian*, 26 September 1934. The poll was taken informally, with cards being sent to the newspaper. The poll showed 524 votes, with Sinclair leading with 256, Merriam receiving 165, and the third candidate, Raymond Haight, getting the remaining 103. Haight ran on the Progressive ticket and supported the Townsend plan. Haight leaders in Kern County switched to support Merriam in October. Polls continued to show a strong lead for Sinclair. See also 25, 29 September, 4, 5 October 1934.

112. Gregory, *American Exodus*, 152–53.

113. Robert E. Burke, *Olson's New Deal for California* (Berkeley: University of California Press, 1953), 3–6.

114. McWilliams, *Ill Fares the Land*, 39.

115. Jamieson, *Labor Unionism in American Agriculture*, 122. In 1935, CUCOM led six of the eighteen strikes in the state.

116. Jamieson, *Labor Unionism in American Agriculture*, 124. For more on CUCOM, see ibid., 122–30; Weber, "The Organizing of Mexicano Agricultural Workers." Although Velarde and many of the CUCOMistas were anarchists, they maintained a respectful, if sometimes distant, working relationship with organizers from the Communist party. Interview with LaRue McCormick.

117. Mexicans continued to be active in strikes that surfaced in newspaper reports. In May 1937, for example, 150 cotton choppers, mostly Mexican, went on strike near Tulare. Most were Mexicans from the lower San Joaquin Valley, and they led the American workers there. *Corcoran Journal*, 21 May 1937. In the same month it was reported that in Kern County it was the Mexican and Filipino workers especially who wanted to organize. *Farmer Labor News*, 28 May 1937; *Western Worker*, 31 May 1937.

118. Ruiz, *Cannery Women, Cannery Lives*, 52; *UCAPAWA News* (July, August, September, October, November 1939); *San Francisco News*, 22 July 1940.

119. The elected officers were reported as being Joe Gutiérrez, president, and Reola Hill, secretary. *Western Worker*, 12 August 1935.

120. *People's World*, 12 October 1938.

121. *Fresno Bee*, 26 April 1938; *People's World*, 26 April 1938.

122. How migration affects unionization and the spread of strikes is suggested by several organizational tactics used in the 1980s. The Farmworkers Labor Organizing Committee (FLOC), based in Ohio, sent organizers into home areas of workers before the harvest. Most workers lived in Mexico or the south

valley of Texas. The organizers met with workers, discussed the crop and the amount of pay the union was going to ask for, and asked whether a strike was likely. The Arizona Farmworkers had a large constituency of undocumented Mexican workers and sent organizers into the Mexican states of Querétaro and Nayarit. Some union members formed all-union crews before they left their home bases. These groups worked not only in the Southwest but on the East Coast and in Florida, where there were few successful agricultural unions. One crew, interviewed by the author in Florida, was composed of Arizona Farmworkers members who came from the state of Querétaro. In North Carolina they organized black workers in a peach strike. In Florida they organized crews of fellow undocumented Mexican tomato workers and led and won a strike for higher wages. Thus migrancy contributed to spreading strikes and was a form of incipient unionization across several states. Lasting gains, however, depended on the organization of stable unions and federal support.

123. It should be said that this has tended to remain the answer for agricultural workers. Most individuals and families move into agriculture only as temporary work: the second generation tends to leave. But there still are attempts to change the system.

CHAPTER SIX. NEW DEAL RELIEF POLICIES, LOCAL
ORGANIZING, AND ELECTORAL BATTLES

1. Testimony of Eugene Hayes, State Relief Administration, California, Cotton Wage Hearing Board, Minutes, 146; *Business Week* (4 September 1937).

2. *Corcoran Journal*, 6 August 1937; *UCAPAWA Agricultural and Cannery Union News*, 11 August 1937.

3. See *Pacific Rural Press*, 19 June 1937; Eric H. Thomson, "Why Plan Security for the Migratory Laborer?" paper presented at the California Conference of Social Work, San Jose, California, 12 May 1937, Federal Record Archives, San Francisco. Eric H. Thomson was a regional sociologist for the U.S. Department of Agriculture.

4. *Oakland Tribune*, 10 July 1937.

5. *Western Worker*, 22 July 1937.

6. *Oakland Tribune*, 13 July 1937.

7. Pomeroy testimony in the La Follette Committee, *Hearings*, part 62, 22643.

8. The WPA had received $2,175 million in 1937 and was to be cut to $1,500 million in 1938. Of this relief assistance the SRA would receive $100 million and the WPA $1,325 million. *Oakland Tribune*, 24 June 1937.

9. *Oakland Tribune*, 2 March 1937.

10. The AFL cotton locals were established in Portersville, Delano, Visalia, and Tulare, and charters were pending in Arvin and Fresno. Jamieson, *Labor Unionism in American Agriculture*; *Western Worker*, 13 June, 23 December 1935; Tom Collins notes that the Agricultural Workers Union in the area received a charter from the AFL. They signed up twenty-five charter members, and workers from Delano, Arvin, and other areas were represented (Collins Reports, 5 September 1936, Records of the Farm Security Administration, Fed-

eral Archives and Regional Center, San Bruno, California). See also *Kern County Labor Journal*, 21 August 1936, and *Rural Worker*, September 1936.

By May 1937, Kern County agricultural unions were reported in Bakersfield, Arvin, Delano, McFarland, Wasco, and Shafter, and one was being organized in Tulare. *Farmer Labor News*, 14 May 1937. By June, organizing had drawn increased interest. In June 400 field workers attended an Arvin meeting, out of which came Arvin Farm Local no. 2, with several dozen members. Ninety-eight percent of the members lived in tents. *Farmer Labor News*, 21 May 1937; *Kern County Labor Journal*, 11 and 18 June 1937. Donald Henderson, president of the National Committee for Agricultural Workers, made this point. See Jamieson, *Labor Unionism in American Agriculture*, 135.

11. This group worked under the National Committee for Unity of Agricultural and Rural Workers.

12. *Farmer Labor News*, 12 March 1937, reported that the Stockton Agricultural Union said that its effect was felt especially strongly among Mexican and Filipino workers there.

13. *Western Worker*, 2 August 1937.

14. Jamieson, *Labor Unionism in American Agriculture*, 27.

15. Ibid.

16. The numbers are notoriously inexact, with the alliance tending to inflate them. See Stuart Jamieson, "The Origins and Present Structure of Labor Unions in Agricultural and Allied Industries in California," La Follette Committee, *Hearings*, exhibit 9376, part 62, 22531–40; see also Alex Noral's testimony in California, Cotton Wage Hearing Board, Minutes, 58.

17. In 1936 the WPA recognized the Workers Alliance and its right to organize on federal projects. This, in effect, made it a union for relief workers. *Western Worker*, 18 October 1936.

18. *Western Worker*, 19 August 1937. The actual number of bales was 3,401,000; *Fresno Bee*, 8 September 1937. The month of October was no better. See Bank of America National Trust and Savings Association, "Commodities-Crops-Marketing," reviews by B. H. Critchfield, Bank of America Archives, San Francisco, 16 October 1937, 142. The review noted that the glut improved the prospects of the government crop-control program pending before Congress.

19. *California Cultivator*, 14 August 1937.

20. *Western Worker*, 19 August 1937.

21. *Fresno Bee*, 3 September 1937.

22. *Fresno Bee*, 3 September 1937.

23. Minutes of the Agricultural Labor Bureau Board of Directors, 22 December 1937, in La Follette Committee, *Hearings*, part 51, 18842.

24. CIO Cannery and Agricultural Organizing Committee, 16 October 1937, Carey McWilliams Collection, Special Collections, UCLA (mimeo.); see also *Western Worker*, 28 October 1937; *Labor Herald*, 30 November 1937.

25. CIO Cannery and Agricultural Organizing Committee, 16 October 1937.

26. *Fresno Bee*, 8, 10 October 1937; *Western Worker*, 22 July 1937.

27. Minutes of a meeting of the Agricultural Labor Bureau, 22 December 1937, in La Follette Committee, *Hearings*, part 51, 18842; CIO Cannery and Agricultural Organizing Committee, 16 October 1937; *Western Worker*, 28

October 1937; *Labor Herald*, 2 November 1937. At this point, 50 percent of the growers were reportedly paying $1.00.

28. Minutes of a meeting of the Agricultural Labor Bureau, 22 December 1937, Fresno, California, in La Follette Committee, *Hearings*, part 51, 18842.

29. *Oakland Tribune*, 1 September 1937; *Sacramento Bee*, 4 September 1937.

30. *Oakland Tribune*, 4 September 1937.

31. *Fresno Bee*, 3 September 1937.

32. *Fresno Bee*, 10 September 1937.

33. *Fresno Bee*, 20 September 1937.

34. *Fresno Bee*, 3, 4 September 1937. Frank Merriam had announced: "We are coming to the cotton harvest. . . . We will have work for every able bodied worker in the state and even then there will be a shortage." *Fresno Bee*, 4 September 1937.

35. *Fresno Bee*, 11 September 1937.

36. *Bakersfield Californian*, 2 September 1937.

37. *San Francisco Call Bulletin*, 18 February 1938; *San Francisco Chronicle*, 18 February 1938, 46.

38. *People's World*, 11 March 1938.

39. *Corcoran Journal*, 18 March 1938.

40. Letter from W. D. Hammett to Mrs. Robert McWilliams, 26 March 1938, Federal Writers' Project, Bancroft Library.

41. *Stockton Record*, 7 February 1938.

42. *Stockton Record*, 7 February 1938; *People's World*, 2 February 1938; *Labor Herald*, 24 March 1938.

43. *San Francisco News*, 14 February 1938. Newspapers such as the *San Francisco News* and the *San Francisco Chronicle* ran a series of articles about camp conditions. Stein makes this point in *California and the Dust Bowl Migration*, 69–70.

44. *San Francisco News*, 14 February 1938.

45. Harold Pomeroy, director of the SRA, quoted the grower. "Migratory Laborers in the San Joaquin Valley, July and August, 1937," *State Relief Administration Report*, December 1937, reprinted in La Follette Committee, *Hearings*, exhibit 9578, part 62, 22642–73, quotation on 22662. This is one of several quotations that blame increased migration on the cotton industry.

46. *San Francisco News*, 14 February 1938.

47. La Follette Committee, *Report*, 373.

48. From Stein, *California and the Dust Bowl Migration*, 54. His sources are Tyr V. Johnson and Frederick Arpke, *Interstate Migration and County Finance in California* (no publication information); La Follette Committee, *Hearings*, part 59, 21866–70.

49. Stein, *California and the Dust Bowl Migration*, 56. His sources are from the La Follette Committee, *Hearings*, part 59, 21866–70.

50. Stein, *California and the Dust Bowl Migration*, 56–58.

51. *San Francisco News*, 1 February 1938.

52. Editorial in the *Bakersfield Californian*, 10 February 1938, quoted in Stein, *California and the Dust Bowl Migration*, 78.

53. La Follette Committee, *Hearings*, part 58, 21929–30; Howard Hill, "Fifty Thousand on Federal Farm Relief in California," *Implement Record*, 30 June 1938, 48; *San Francisco Chronicle*, 18 February 1938.

54. The FSA grant was meant to be only temporary, but by 1940 ten grant offices were located in the San Joaquin Valley, one in each county.

55. *Fresno Bee*, 21 March 1938.

56. *Stockton Record*, 23 February 1938; *Oakland Tribune*, 24 February 1938.

57. *People's World*, 21 January 1938.

58. *San Francisco Call Bulletin*, 18 February 1938.

59. *People's World*, 14 March 1938.

60. *San Francisco Herald*, 24 March 1938.

61. *Rural Observer* (April 1938).

62. Letter from W. D. Hammett to McWilliams; letter from Monroe to McWilliams, Federal Writers' Project, Bancroft Library. Lucy Sullivan McWilliams, a Democrat and head of the labor study group of the San Francisco League of Women Voters, coordinated efforts for donations of food, clothes, and money. Information on McWilliams is sketchy. She was the wife of a San Francisco judge of the Superior Court, Robert Lafayette McWilliams. *Who's Who in California*, 619; William Desmond Ryan, ed., *California Register* (Beverly Hills: Social Blue Book of California, 1954), 622. Letters between McWilliams and Lillian Monroe, 1938, Federal Writers' Project, Bancroft Library.

63. Sheridan Downey later staked his career on fighting the 160-acre limitation law and wrote *They Would Rule the Valley* (1947). See Walter Goldschmidt, "Reflections on Arvin and Dinuba," *Newsline of the Rural Sociological Society* 6, no. 5 (September 1978), 10; Goldschmidt, *As You Sow*, 480–82.

64. Burke, *Olson's New Deal for California*, 1–19.

65. Interview with Dorothy Healey by Joel Gardner, University of California, Los Angeles, Oral History Program, 1982, in Department of Special Collections, University Research Library, UCLA, 188.

In her autobiography, Healey says of the Olson administration:

Our relationships with both the federal government, represented by the NLRB and the FSA, and the state government, once Governor Olson came into office, were very warm and helpful. I began to appreciate two things which I had not understood before: first, that the fact that someone works for the government doesn't mean that they've been co-opted and sold out; and second, that it can make an enormous difference what administration is in power and what kind of appointments they make. The contrast between our experience as organizers at the start of the decade and at the end was incredible. I didn't draw any kind of grand theoretical conclusions from this contrast, but it planted a seed, a doubt about the way we had previously interpreted Marx's comments about the state. I began to understand that the "political superstructure" of society could enjoy a measure of independence from the economic base because of the clash of interests between classes and even within the bourgeoisie itself.

At the same time, she realized that:

Whatever political clout we developed depended on our ability to keep up the pressure for significant reform from outside the realm of electoral politics through the CIO, the LNPL [Labor's Non-Partisan League] and other groups. Culbert Olson's successful campaign for the governor's seat in 1938 was a great triumph for this new kind of politics. It would be an exaggeration to say that Olson triumphed only because of our support, but he certainly would have had a harder time without us.

Healey and Isserman, *Dorothy Healey Remembers*, 72–77.

66. Letter from McWilliams to Monroe, 10 May 1938, Federal Writers' Project, Bancroft Library.

67. Walter Stein makes this argument for the CCA. See Stein, *California and the Dust Bowl Migration*, 96–103.

68. *Fresno Bee*, 25 May 1938; *San Francisco News*, 26 May 1938; Stein, *California and the Dust Bowl Migration.*

69. Stein, *California and the Dust Bowl Migration*, 97.

70. Ibid., quoting from "Contributors to California Citizens Association," typewritten list in Simon J. Lubin Papers, carton 12, Bancroft Library, University of California, Berkeley.

71. La Follette Committee, *Hearings*, part 51, 18824.

72. *Corcoran Journal*, 16 September 1938.

73. Rogin and Shover, *Political Change in California*, 119. For other discussions of California state government, see Nash, *State Government and Economic Development*. Among the works which convey a sense of California's turbulent politics during the 1930s are Royce D. Delmatier, Clarence F. McIntosh, and Earl G. Waters, eds., *The Rumble of California Politics, 1848–1970* (New York: Wiley, 1970); Ronald Chinn, "Democratic Party Politics in California, 1920–1956," Ph.D. diss., University of California, Berkeley, 1958; Royce D. Delmatier, "The Rebirth of the Democratic Party in California, 1928–1938," Ph.D. diss., University of California, Berkeley, 1955.

74. Rogin and Shover, *Political Change in California*, 124.

75. In 1934 the Democrats received 31 percent of the vote in the San Joaquin Valley, while the Republicans garnered 38.8 percent and the Commonwealth party 29.3 percent: Rogin and Shover, *Political Change in California*, 132.

76. *Business Week* (September 1938).

77. The population increases in cotton belt counties, 1935–1940, were as follows: Kern, 52,554 (63.6 percent); Kings, 9,783 (38.5 percent); Tulare, 29,710 (38.4 percent); Madera, 6,150 (35.8 percent); Merced, 10,240 (27.9 percent); Fresno, 34,186 (23.7 percent). Commonwealth Club of California, *The Population of California* (San Francisco: Commonwealth Club, 1946), 19–20.

78. *Visalia Times Delta*, 28 September 1938.

79. Burke, *Olson's New Deal for California*, 33.

80. Letter from Philip Bancroft to Hiram Johnson, 11 November 1938, in Hiram Johnson Papers, Bancroft Library, University of California, Berkeley, III, box 23, as quoted in Stein, *California and the Dust Bowl Migration*, 94.

81. Culbert Olson, "Gov. Culbert L. Olson's Statement Before Senate Committee on Education and Labor," San Francisco, 6 December 1939, in La Follette Committee, *Hearings*, part 47, 17250.

82. *People's World*, 15 February 1939.

83. McWilliams, *Factories in the Field*, 325. The book had first appeared as a series of muckraking articles in 1935. McWilliams gives his own account of the impact of the 1939 publication of *Factories in the Field* in his *Ill Fares the Land*, 42–44.

84. John Watson, president of the Associated Farmers, wrote to an FSA administrator that Carey McWilliams's "every action has been so destructive with repeated attempts to build class hatreds between farmers and farm workers that he has been dubbed agricultural pest number one in California." Telegram from John S. Watson, president of California Associated Farmers, to William Alexander, FSA administrator, 5 March 1940, RG 96, National Archives, Washington, D.C.

85. *San Francisco News*, 3 March 1939; *People's World*, 9 March 1939; Burke, *Olson's New Deal for California*, 90–91; Stein, *California and the Dust Bowl Migration*, 121.

86. *Sacramento Bee*, 21 February 1939.

87. *San Francisco Chronicle*, 7 March 1939; *People's World*, 15 February, 9 March 1939; *Los Angeles Examiner*, 15 February 1939.

88. Reflecting a national move toward conservatism, more Republicans had been elected to the California state assembly in 1938. See Burke, *Olson's New Deal for California*, 34.

89. *People's World*, 15 February 1939.

90. Burke, *Olson's New Deal for California*, 88–95; *People's World*, 18 March 1939.

91. Part of the union's position can be explained by a jurisdictional fight that was taking place at the same time. The State, County, and Municipal Workers of America–CIO was busy organizing the SRA staff in Bakersfield, and the more conservative CSEA opposed it. See Burke, *Olson's New Deal for California*, 83; *San Francisco Examiner*, 3 April 1939.

92. Burke, *Olson's New Deal for California*, 92; *People's World*, 19 August 1939.

93. Burke, *Olson's New Deal for California*, 83; *People's World*, 16 June 1939. This fragmentation also had to do with the onslaught of the Nazi-Soviet pact and the Communist party's break with the Roosevelt administration. Healey and Isserman, *Dorothy Healey Remembers*, 77.

CHAPTER SEVEN. END OF A HOPE:
THE STRIKES OF 1938 AND 1939

1. *Corcoran Journal*, 3 June, 15 July 1938.

2. *People's World*, 12 August 1938.

3. The CIO was especially strong among the mine, mill, and smelter workers, the borax workers, the store clerks, and the automobile mechanics of Bakersfield. *People's World*, 30 April 1937; interview with Jim White [pseud.], Bakersfield, California, 1 January 1982.

4. *Western Worker*, 29 April 1938; *Farmer Labor News*, 14 May, 1937, 23, 30 April 1938.

5. *Kern County Labor Journal*, 24 September 1937.

6. AFL-CIO relations were friendly enough that Lyman N. Sisley of the U.S. Department of Labor Conciliation Service wrote to his director that "the CIO group met in the A.F. of L. Labor Temple in Bakersfield, and I particularly noticed the C.I.O. charters hanging on the same wall with the various craft organizations of the A.F. of L." Letter from Lyman N. Sisley to Dr. John R. Steelman, Director of Conciliation Services, Department of Labor, Washington, D.C., 10 October 1938, National Archives, RG 280, 199/2542.

7. *Fresno Bee*, 26 April 1938.

8. As has been noted before, W. B. Camp was in charge of California Lands, a subsidiary of Bank of America, and had worked for the Agricultural Adjustment Administration for California. See Wofford B. Camp, "Cotton, Irrigation, and the AAA," an oral history conducted by Willa Baum, 1962–1966, University of California, Berkeley, Regional Oral History Office.

9. See letter from David Kindead, assistant camp manager at Shafter, to Edward J. Rowell, regional labor administrator of the FSA, 7 October 1938, RG 280, National Archives, Washington, D.C.

10. According to *People's World*, the Camp ranch agreed to pay $1.00: 12 October 1938.

11. *San Francisco News*, 6 October 1938; *Los Angeles Times*, 11 October 1938.

12. *Fresno Bee*, 10 October 1938; La Follette Committee, *Report*, 519; *Berkeley Gazette*, 30 September 1938. For a brief discussion of the United States Employment Service and the Wagner-Peyser Act, see Joseph G. Rayback, *A History of American Labor* (New York: Free Press, 1966), 325–26.

13. Dorothy Healey, a veteran of the 1933 strikes of the CAWIU, was elected vice president of UCAPAWA and went to the area. LaRue McCormick was there with the International Labor Defense. *UCAPAWA News*, n.d. but probably 1938, Carey McWilliams Collection, UCLA; Dorothy Healey interview by Joel Gardner, 1982, University of California, Los Angeles, Oral History Program, University Research Library, UCLA, vol. 1, 177; interview with LaRue McCormick, Los Angeles, California, August 1982.

14. *UCAPAWA News*, n.d. but probably 1938, Carey McWilliams Collection, UCLA.

15. *Fresno Bee*, 26 April 1938; *People's World*, 26 April 1938; Felix Rivera of Bakersfield represented the San Joaquin Valley at the UCAPAWA meeting; interview with Wyman Hicks, Bolinas, California, January 1983.

In the town of Arvin, a magnet for incoming white migrants, most union members were apparently Anglo. This observation comes from various descriptions of strike activity and from lists of arrests during the 1938 strike. In Corcoran, by 1939, out of the 175 members of the local, 75 percent were Anglos, 20 percent Mexicans, and 5 percent blacks. Testimony of Mr. Gregg, secretary of UCAPAWA in Corcoran, California, Cotton Wage Hearing Board, Minutes; letter from Eugene E. Gregg, secretary and treasurer of UCAPAWA in Corcoran to Carey McWilliams, 2 October 1939, Carey McWilliams Collection, UCLA. For this point see Víctor Nelson-Cisneros, "UCAPAWA and Chicanos in California: The Farm Worker Period, 1937–1940," *Aztlán* 7, no. 3 (Fall 1976),

453–78; Ruiz, *Cannery Women, Cannery Lives*, 52; Jamieson, *Labor Unionism in American Agriculture*, 186; *Farmer Labor News*, 28 May 1938; *Western Worker*, 31 May 1938.

16. *People's World*, 26 April 1938.

17. *People's World*, 24 June 1938. Differing accounts by newspapers on conflicting sides of the fight cloud the picture but suggest that there were union headquarters in both the federal camp and the Mexican colony. The *Corcoran Journal* reported that union headquarters were located in the federal camp at Shafter. *Corcoran Journal*, 28 October 1938.

18. *People's World*, 10 October 1938. Allegedly, 2,000 workers were attending nightly meetings during the height of the strike. *People's World* always reported the meetings as being held in the Mexican colony. *People's World*, 12 October 1938.

19. *People's World*, 10, 12 October 1938; *Fresno Bee*, 5 October 1938. By the October strike, growers said the strike was being led by Pete Nichols, "an outsider who came to the Shafter district this season," according to the *Fresno Bee*. According to this source, he had appointed Clyde Champion chairman of the strike committee, along with R. F. Johnson, A. H. Beard, R. E. Fisher, and M. M. Rodríguez. Nichols's name does not come up in other reports.

There is some confusion as to the identity of Rodríguez. Some sources refer to him as M. M. Rodríguez. Others mention a Steve Rodríguez in the same position in the union. If these are the same man, Stephen Rodríguez was active in the Congreso del Pueblo de Habla Española, was one of the Mexican CIO organizers sent to Madera in 1939, and served as UCAPAWA representative to statewide conventions.

20. *People's World*, 12, 21 October 1938; *Fresno Bee*, 8, 10 October 1938. Newspapers' emphasis varied on who and how many went on strike. While the *People's World* crowed that 75 percent of the Arvin campers were on strike, the *Fresno Bee* vaguely reported that occupants of the Shafter FSA camp "are generally working and a large majority have taken no part in the strike" and briefly remarked that two labor camps south of Shafter (at least one of which was in the Mexican colony) were still on strike. *People's World*, 13 October 1938; *Fresno Bee*, 19 October 1938.

21. *UCAPAWA News*, n.d. but probably 1938, Giannini Foundation of Agricultural Economics Library, University of California, Berkeley; *Kern County Labor Journal*, 22 September 1939.

22. *Fresno Bee*, 26 April 1938.

23. *Fresno Bee*, 26 October 1938; *People's World*, 5 November 1938; Progress Report from Lyman N. Sisley, Commissioner of Conciliation, U.S. Department of Labor, 2 November 1938, and Final Report, Lyman N. Sisley, 3 November 1938, National Archives, RG 280, 199/2542. Both reports cited "intimidation and arrests" as the causes of the strike's failure. H. T. Geiger of the Oil Workers International Union, Kern River Local no. 19, concurred that by the time Lyman Sisley arrived, "the strike was about a lost issue as the morale of the strikers had been broken on account of the jailing . . . [and] there was nothing that your Department could have done in the matter at that late date." Letter from H. T. Geiger, Representative of Oil Workers' International Union–

CIO, Kern County Local no. 19, Bakersfield, California, to John R. Steelman, Department of Labor, Washington, D.C., 5 November 1938, National Archives, RG 280, 191/2542.

24. Shafter camp notes, San Francisco Federal Archives, 24 March 1939; letter from A. I. Tucker, Chairman of Labor and Farm Management Committee, Kern County Agricultural Economic Conference Committee, Bakersfield, California, to Dr. N. E. Dodd, Director, Western Division, U.S. Department of Agriculture, Agricultural Adjustment Administration, Washington, D.C., 11 April 1939, complaining about the administrators of the Shafter camp and claiming that they have "sanctioned, if not fostered, the creation of unrest and class hatred among residents of the camps." National Archives, RG 96, 12–129, file AD-124.

25. *Visalia Times Delta*, 24 September 1939; mimeographed newsletter from UCAPAWA, n.d. but circa October 1939.

26. Conference Minutes, UCAPAWA Field Workers Local of San Joaquin Valley, Fresno, 26 November 1939 (typescript), and postcards, petitions, and letters from many organizations including the Shafter Farm Workers Community Council. Both in Carey McWilliams Collection, UCLA.

27. *Visalia Times Delta*, 28 September 1939; *Fresno Bee*, 11 November 1939; *Corcoran Journal*, 26 March 1936, 7 January 1938; *Madera Daily Tribune*, 21 September 1939; La Follette Committee, *Report*, 1511; Sherman, "The Oklahomans in California," 109–10.

28. La Follette Committee, *Report*, 1496, 1497.

29. La Follette Committee, *Report*, 1501–2; Jamieson, *Labor Unionism in American Agriculture*, 169; interview with Gene Luna, Madera, California, January 1982.

30. Stuart Jamieson, "The Origins and Present Structure of Labor Unions in Agricultural and Allied Industries in California," in La Follette Committee, *Hearings*, exhibit 9376, part 62, 22531–40; see also Alex Noral's testimony in California, Cotton Wage Hearing Board, Minutes, 58; for a discussion of the Chowchilla Workers Alliance, see testimony of Charles Fleming, in ibid., 109.

31. McWilliams's testimony, La Follette Committee, *Hearings*, part 51, 18716–17; La Follette Committee, *Hearings*, part 51, 18716; *San Francisco Examiner*, 21 May 1939; *El Cerrito Review*, 9 and 19 June 1939.

32. *Madera Daily Tribune*, 11 May 1939.

33. La Follette Committee, *Hearings*, part 51, 18897, quoted in Jamieson, *Labor Unionism in American Agriculture*, 177.

34. *People's World*, 12 May 1939.

35. Carey McWilliams, "Memorandum on Housing Conditions Among Migratory Workers in California," Division of Immigration and Housing, 29 March 1939, Carey McWilliams Collection, UCLA (typescript); Tolan Committee, *Report*, 446–47.

36. *Madera News*, 25 May 1939.

37. Stein, *California and the Dust Bowl Migration*, 204; see also Gregory, *American Exodus*, 97.

38. For an explication of the rationale and methodology behind the image presented in the photographs of Dorothea Lange and the text of Paul S. Taylor,

see William Stott, *Documentary Expression and Thirties America* (New York: Oxford University Press, 1973).

39. Stein, *California and the Dust Bowl Migration*, 210–15.

40. *The Annalist*, an industry newsletter, quoted in the Simon J. Lubin Society *Newsletter* 2, no. 3 (August 1939).

41. *Fresno Bee*, 11 October 1939; *San Francisco Chronicle*, 21 September 1939.

42. Carey McWilliams on cotton-chopping hearings in Madera, 9 May 1939, Carey McWilliams Collection, UCLA.

43. Ben Hayes testimony, *Minutes*, Cotton Chopping Hearings, California State Relief Administration, Madera, 9 May 1939, 9.

44. Some growers were paid more than the $10,000 maximum benefit payment because total payments covered more than one ranch. Confidential News Service, Simon J. Lubin Society, Carey McWilliams Collection, UCLA, box 10 (typescript). Figures taken from Agricultural Conservation Program for Madera County. Data for the AAA are spotty because the original AAA records appear to have been destroyed, and could not be found in the National Archives, local government archives, or local offices. La Follette Committee, *Hearings*, part 51, 18937–43.

45. La Follette Committee, *Hearings*, part 47, 17, 382.

46. Víctor Nelson-Cisneros points out that the union had spent $18,000 on organizing farm workers from 1 December 1938 to 30 November 1940. Only $6,000 had been collected through dues and initiation fees from field workers. The union relied heavily on donations from such organizations as the National Committee to Aid Agricultural Workers. Nelson-Cisneros, "UCAPAWA and Chicanos in California."

47. Harry Schwartz suggested several of the union's problems: the perennial problem of low wages, which impeded developing a solid financial base; the migratory nature of the work force; the lack of cooperation between racial and ethnic groups; the strength of the Associated Farmers. Harry Schwartz, *Seasonal Farm Labor in the United States* (New York: Columbia University Press, 1945).

48. CIO organizing in the International Longshoremen's and Warehousemen's Union also assisted labor organizing in cotton. In 1937, the ILWU organized Anderson Clayton's Western Compress Company in San Pedro. This organization was credited with helping to bring attention to conditions of Anderson Clayton's compress workers in the Valley, and in 1937 organizing began in the compress plants of the Valley. *The Dispatcher*, 2 March 1984.

49. *People's World*, 13 and 28 April 1938.

50. Jamieson, *Labor Unionism in American Agriculture*, 173.

51. UCAPAWA argued that growers were better able to pay the higher rate because "those who control cotton will make big profits because of the war." *UCAPAWA News*, n.d. but circa Fall 1939, Carey McWilliams Collection, UCLA.

52. The Madera Farm Bureau and Associated Farmers called the hearings illegal. An editorial in the *Madera Daily Tribune* castigated the hearings as "a curtsey by state officials to the radical elements . . . dominated . . . by the com-

munists." They opposed state intervention and argued that "if it is right that the state shall attempt to set the minimum wage in the agricultural fields it is also right it shall have the power to intervene in all fields of labor activity." *Madera Daily Tribune and Morning Mercury*, 26, 30 September 1939; La Follette Committee, *Report*, 510.

53. *CIO Labor Herald*, 28 September 1939.

54. Testimony of Mr. Rhodes in California, Cotton Wage Hearing Board, Minutes, 114; testimony of Mr. York, ibid., 129; testimony of Mr. Berman, ibid., 129.

55. Report of Cotton Wage Hearing Board; La Follette Committee, *Hearings*, part 51, 18968–70; La Follette Committee, *Report*, 1510; *People's World*, 6 October 1939; *Madera Daily Tribune and Morning Mercury*, 6 October 1939.

56. Report of Cotton Wage Hearing Board.

57. *Fresno Bee*, 24 October 1939.

58. *Visalia Times Delta*, 12 and 13 October 1939; *Madera Daily Tribune*, 14 October 1939; *Corcoran Journal*, 26 September 1939.

59. *Fresno Bee*, 14 October 1939.

60. *Madera Daily Tribune*, 11 October 1939; *Visalia Times Delta*, 12 October 1939.

61. Interview with Wyman Hicks; La Follette Committee, *Hearings*, part 51, 18998.

62. Tall and imposing, Jesse McHenry had been involved in agricultural strikes for twenty years. Not much is known of his early life, but he was active in the 1933 cotton strike and was probably a member of the CAWIU. By 1934 he was a registered member of the Bakersfield branch of the Communist party. La Follette Committee, *Hearings*, part 51, 18998.

63. Interview with Gene Luna.

64. *UCAPAWA News*, October 1939, cited in Ruiz, *Cannery Women, Cannery Lives*, 52.

65. For discussions of El Congreso del Pueblo de Habla Española, see Mario García, *Mexican Americans: Leadership, Ideology and Identity, 1930–1960* (New Haven: Yale University Press, 1989), 145–74; Alberto Camarillo, *Chicanos in California: A History of Mexican Americans in California* (San Francisco: Boyd and Fraser, 1984), 58–74; Alberto Camarillo, "The Development of a Pan-Hispanic Civil Rights Movement: The 1939 Congress of Spanish Speaking People," paper presented at the Annual Meeting of the Organization of American Historians, April 1986.

66. It is unclear how directly membership in the Congreso del Pueblo de Habla Española overlapped with membership in unions or how influential the organization may have been in forming unions. Francisco Palomares of the Agricultural Labor Bureau claimed that new *mutualistas* were organizing workers to demand higher wages. There is no corroborating evidence, but he may have been referring either to the Congress or to more classic mutual aid societies. For mutual aid societies to organize would be consistent with their role in labor struggles and in the Imperial Valley, where they had become the structural basis for labor organization. Certainly there were new *mutualistas*. Mexicans, among them veterans of the 1933 cotton strike, were forming branches of the Progre-

sistas in the Valley. In Corcoran, the branch was organized by 1933 strike leaders Roberto Castro and Lino Sánchez. Mr. Jaramillo, California, Cotton Wage Hearing Board, Minutes, 106; *Corcoran Journal*, 30 October 1938; interview with Roberto Castro, Corcoran, California, January 1982; Sociedad Progresista Mexicana, "Aniversario de Oro, 1929–1979" (pamphlet).

67. La Follette Committee, *Hearings*, part 50, 18641; La Follette Committee, *Report*, 1513; *Madera Daily Tribune*, 12 and 14 October 1939; *Fresno Bee*, 12 October 1939.

68. Letter from Stuart Strathman to Harold Pomeroy, 15 October 1939, printed in La Follette Committee, *Hearings*, part 51, 18919.

69. *Madera Daily Tribune*, 16 October 1939; *Visalia Times Delta*, 18 October 1939.

70. La Follette Committee, *Report*, 1515.

71. *Madera Daily Tribune*, 19 October 1939.

72. *Madera Daily Tribune*, 19 October 1939, quoted in La Follette Committee, *Hearings*, part 51, 18655.

73. The fifteen men surrounded him and told him, "Listen here, you nigger, you've been sitting on a platform at your meetings with white women. Do you know what we do to niggers where we come from? We'll string you up like they do in the South." *Madera Daily Tribune*, 18 October 1939; *People's World*, 21 October 1939; testimony by Elmer Joseph in La Follette Committee, *Hearings*, part 51, 18657–58; interview with Wyman Hicks; interview with Gene Luna.

74. Some of this aggression also drew on southern racial sentiments. One grower testified about leader McHenry, "I have respect for the colored race, and all southerners do, but they can't talk to us in any way and get by with it." Testimony of O. L. Baker in La Follette Committee, *Hearings*, part 51, 18682.

75. *Fresno Bee*, 21 October 1939; La Follette Committee, *Hearings*. O. L. Baker testified that "the people had to take it into their own hands," ibid., 18679.

76. *Visalia Times Delta*, 18 October 1939; *Fresno Bee*, 18 October 1939; *People's World*, 21 October 1939.

77. *Fresno Bee*, 21 October 1939; La Follette Committee, *Hearings*, part 62, 18679.

78. *Fresno Bee*, 21 October 1939; *Visalia Times Delta*, 21 October 1939; *UCAPAWA News*, n.d. but circa 1939, Giannini Foundation of Agricultural Economics Library, University of California, Berkeley.

79. *Madera Daily Tribune*, 21 October 1939; *Visalia Times Delta*, 21 October 1939.

80. *Fresno Bee*, 21 and 22 October 1939.

81. *People's World*, 24 October 1939.

82. Interview with Gene Luna.

CHAPTER EIGHT. "DOWN THE VALLEYS
WILD . . .": CONCLUSION

1. Interviews with Caroline Decker, San Raphael, California, June 1982, and Pat Chambers, Wilmington, California, 11 May 1982.

2. La Follette Committee, *Hearings*, part 62, 22,067.

3. My thanks to Charles Noble for his suggestions. In the course of our discussions, he made explicit the relation between class and the state, in which "where workers are well-organized, and/or capital is divided, and where the state exercises considerable control over the economy, workers are more likely to win concessions from the state. When capital is unified, and/or workers poorly organized, and the state is entirely dependent on private capital accumulation, workers tend to get little or nothing. This . . . makes clear that the analysis depends on systemic and structural considerations, not simply the mobilization and countermobilization of interest groups."

4. La Follette Committee, *Report*, 1615.

Bibliography

BOOKS

Acuña, Rodolfo. *Occupied America: The Chicano Struggle for Liberation.* San Francisco: Canfield Press, 1972.

Agee, James, and Walker Evans. *Let Us Now Praise Famous Men: Three Tenant Families.* New York: Houghton Mifflin, 1939.

Anderson, Benedict. *Imagined Communities: Reflections on the Origin and Spread of Nationalism.* Rev. ed. London: Verso Press, 1991.

Anderson, Perry. *Arguments Within English Marxism.* London: Verso Press, 1980.

Anderson, Rodney. *Outcasts in Their Own Land: Mexican Industrial Workers, 1906–1911.* De Kalb: Northern Illinois University Press, 1976.

Anzaldua, Gloria. *Borderlands/La Frontera: The New Mestiza.* San Francisco: Spinsters/Aunt Lute Book Co., 1987.

Auerbach, Jerold S. *Labor and Liberty: The La Follette Committee and the New Deal.* New York: Bobbs-Merrill Co., 1966.

Baird, Peter, and Ed McCaughan. *Beyond the Border: Mexico and the U.S. Today.* New York: North American Congress on Latin America, 1979.

Balderama, Francisco. *In Defense of La Raza: The Los Angeles Mexican Consulate and the Mexican Community, 1929 to 1936.* Tucson: University of Arizona Press, 1982.

Balibar, Etienne, and Immanuel Wallerstein. *Race, Nation, Class: Ambiguous Identities.* London: Verso Press, 1991.

Barrera, Mario, Alberto Camarillo, and Francisco Hernández. *Work, Family, Sex Roles, Language: The National Association of Chicano Studies Selected Papers 1979.* Berkeley: Tonatiuh-Quinto Sol International, 1980.

Benedict, Murray R. *Can We Solve the Farm Problem? An Analysis of Federal Aid to Agriculture.* New York: Twentieth Century Fund, 1955.

Benjamin, Thomas, and William McNellie, eds. *Other Mexicos: Essays on Regional Mexican History, 1876–1911.* Albuquerque: University of New Mexico Press, 1984.

Blackford, Mansel G. *The Politics of Business in California, 1890–1920.* Columbus: Ohio State University, 1977.

Blaisdell, Donald C. *Government and Agriculture: The Growth of Federal Farm Aid.* New York: Farrar and Rinehart, 1940.

Blaisdell, Lowell L. *The Desert Revolution: Baja, California, 1911.* Madison: University of Wisconsin Press, 1962.

Bodnar, John. *The Transplanted: A History of Immigrants in Urban America.* Bloomington: Indiana University Press, 1985.

Bogardus, Emory S. *The Mexican in the United States.* Los Angeles: University of Southern California Press, 1934.

Bonnifield, Paul. *The Dust Bowl: Men, Dirt and the Depression.* Albuquerque: University of New Mexico Press, 1979.

Brackenridge, R. Thomas, and Francisco O. García-Treto. *Iglesia Presbiteriana: A History of Presbyterians and Mexican Americans in the Southwest.* San Antonio: Trinity University Press, 1974.

Braverman, Harry. *Labor and Monopoly Capital: The Denigration of Work in the Twentieth Century.* New York: Monthly Review Press, 1974.

Brock, William R. *Welfare, Democracy, and the New Deal.* New York: Cambridge University Press, 1988.

Brody, David. *Steelworkers in America: The Nonunion Era.* New York: Harper and Row, 1960.

Brown, Richard Maxwell. *Strain of Violence: Historical Studies of American Violence and Vigilantism.* New York: Oxford University Press, 1975.

Burbach, Roger, and Patricia Flynn. *Agribusiness in the Americas.* New York: Monthly Review Press and North American Congress on Latin America, 1980.

Burbank, Garin. *When Farmers Voted Red: The Gospel of Socialism in the Oklahoma Countryside, 1910–1924.* Westport, Conn.: Greenwood Press, 1978.

Burke, Robert E. *Olson's New Deal for California.* Berkeley: University of California Press, 1953.

Buttel, Frederick H., and Howard Newby. *The Rural Sociology of Advanced Societies: Critical Perspectives.* Montclair, N.J.: Allanheld, Osmun, and Co., 1980.

Calderón, Hector, and José D. Saldívar, eds. *Criticism in the Borderlands: Studies in Chicano Literature and Ideology.* Durham: Duke University Press, 1991.

Camarillo, Alberto. *Chicanos in California: A History of Mexican Americans in California.* San Francisco: Boyd and Fraser, 1984.

Campbell, Christina M. *The Farm Bureau and the New Deal.* Urbana: University of Illinois Press, 1962.

Cardoso, Lawrence A. *Mexican Emigration to the United States, 1897–1931.* Tucson: University of Arizona Press, 1971.

Carlson, Oliver. *A Mirror for Californians.* New York: Bobbs-Merrill Co., 1941.

Carnoy, Martin. *The State and Political Theory*. Princeton: Princeton University Press, 1984.

Carreas de Velasco, Mercedes. *Los Mexicanos que devolvío la crisis 1929–1939*. México D.F.: Secretaría de Relaciones Exteriores, 1974.

Castles, Stephen, and Godula Kosack. *Immigrant Workers and Class Structure in Western Europe*. London: Oxford University Press, 1973.

Cerda Silva, Roberto de la. *El movimiento obrero en México*. México D.F.: Instituto de Investigaciones Sociales, UNAM, 1961.

Chambers, Clarke A. *California Farm Organizations*. Berkeley: University of California Press, 1952.

Chandler, Alfred D. *The Visible Hand: The Managerial Revolution in American Business*. Cambridge, Mass.: Belknap Press, 1977.

Charles, Searle F. *Minister of Relief: Harry Hopkins and the Depression*. Syracuse: Syracuse University Press, 1963.

Clark, Marjorie Ruth. *Organized Labor in Mexico*. Chapel Hill: University of North Carolina Press, 1934.

Cleland, Robert Glass. *California in Our Time, 1900–1940*. New York: Alfred A. Knopf, 1947.

Clifford, James, and George Marcus, eds., *Writing Culture: The Poetics and Politics of Ethnography*. Berkeley: University of California Press, 1986.

Coalson, George. *The Development of the Migratory Farm Labor System in Texas, 1900–1954*. San Francisco: R. and E. Research Associates, 1977.

Cockcroft, James D. *Intellectual Precursors of the Mexican Revolution, 1910–1913*. Austin: University of Texas Press, 1977.

———. *Mexico: Class Formation, Capital Accumulation and the State*. New York: Monthly Review Press, 1983.

———. *Outlaws in the Promised Land: Mexican Immigrant Workers and America's Future*. New York: Grove Press, 1981.

Cohen, Lizabeth. *Making a New Deal: Industrial Workers in Chicago, 1919–1939*. Cambridge: Cambridge University Press, 1990.

Commonwealth Club of California. *The Population of California*. San Francisco: Commonwealth Club, 1946.

Craig, Richard. *The Bracero Program: Interest Groups and Foreign Policy*. Austin: University of Texas Press, 1971.

Creel, George. *Rebel at Large: Recollections of Fifty Crowded Years*. New York: Putnam and Sons, 1947.

Daniel, Cletus. *Bitter Harvest: A History of California Farmworkers, 1870–1941*. Ithaca: Cornell University Press, 1981.

Daniel, Pete. *Breaking the Land: The Transformation of Cotton, Tobacco, and Rice Cultures Since 1880*. Chicago: University of Illinois Press, 1985.

Del Castillo, Adelaida R., ed. *Between Borders: Essays on Mexicana/Chicana History*. Encino, Calif.: Floricanto Press, 1990.

Delmatier, Royce D., Clarence F. McIntosh, and Earl G. Waters, eds. *The Rumble of California Politics, 1848–1970*. New York: Wiley, 1970.

Deutsch, Sarah. *No Separate Refuge: Culture, Class and Gender on an Anglo-Hispanic Frontier in the American Southwest, 1880–1940*. New York: Oxford University Press, 1987.

Dobie, J. Frank. *Puro Mexicano*. Dallas: Southern Methodist University Press, 1935.

Drinnon, Richard. *Facing West: The Metaphysics of Indian Hating and Empire Building*. New York: Schocken Books, 1990.

Dubofsky, Melvyn. *We Shall Be All: A History of the I.W.W.* New York: Quadrangle Books, 1969.

DuBois, Ellen Carol, and Vicki Ruiz. *Unequal Sisters: A Multi-Cultural Reader in U.S. Women's History*. New York: Routledge, 1990.

Dunne, John Gregory. *Delano*. New York: Farrar, Straus and Giroux, 1971.

Durand, Jorge. *Los obreros de Río Grande*. Zamora: Colegio de Michoacán, 1986.

Edmunson, Munro S., et al., eds. *Synoptic Studies of Mexican Culture*. New Orleans: Middle American Research Institute, Tulane University, 1957.

Elac, John C. *The Employment of Mexican Workers in United States Agriculture, 1900–1960: Binational Economic Analysis*. San Francisco: R. and E. Research Associates, 1972.

Fábila, Alfonso. *El problema de la emigración de obreros y campesinos mexicanos*. México D.F.: Talleres Gráficos de la Nación, 1929.

Fisher, Lloyd S. *The Harvest Labor Market in California*. Cambridge: Harvard University Press, 1953.

Fite, Gilbert C. *Cotton Fields No More: Southern Agriculture, 1865–1980*. Lexington: University Press of Kentucky, 1984.

Foley, Douglas E. *From Peones to Politicos: Ethnic Relations in a South Texas Town, 1900–1977*. Austin: Center for Mexican American Studies, University of Texas, 1977.

Forbath, William E. *Law and the Shaping of the American Labor Movement*. Cambridge: Harvard University Press, 1989.

Friedrich, Paul. *Agrarian Revolt in a Mexican Village*. Englewood Cliffs, N.J.: Prentice-Hall, 1970.

Fuller, Varden. *Labor Relations in Agriculture*. Berkeley: Institute of Industrial Relations, 1955.

Galambos, Louis. *Competition and Cooperation: The Emergence of a National Trade Association*. Baltimore: Johns Hopkins University Press, 1966.

Galarza, Ernesto. *Barrio Boy*. Notre Dame: University of Notre Dame Press, 1971.

———. *Farmworkers and Agri-Business in California, 1947–1960*. Notre Dame: University of Notre Dame Press, 1977.

———. *Merchants of Labor: The Mexican Bracero Story*. Charlotte, N.C.: McNally and Loftin, 1964.

———. *Spiders in the House and Workers in the Fields*. Notre Dame: University of Notre Dame Press, 1970.

Gamio, Manuel. *The Life Story of the Mexican Immigrant: Autobiographic Documents*. New York: Dover Press, 1971.

———. *Mexican Immigration to the United States: A Study of Human Migration and Adjustment*. New York: Dover Press, 1971.

García, Mario. *Desert Immigrants: The Mexicans of El Paso, 1880–1920*. New Haven: Yale University Press, 1981.

———. *Mexican Americans: Leadership, Ideology and Identity, 1930–1960.* New Haven: Yale University Press, 1989.

Gilly, Adolfo. *The Mexican Revolution.* London: Verso Press, 1983.

Gluck, Sherna, and Daphne Patai, eds. *Women's Words: The Feminist Practice of Oral History.* New York: Routledge, 1991.

Goldschmidt, Walter. *As You Sow: Three Studies in the Social Consequences of Agribusiness.* Montclair, N.J.: Allanheld, Osmun, and Co., 1978.

Gómez-Quiñones, Juan. *Development of the Mexican Working Class North of the Rio Bravo.* Popular Series 2. Los Angeles: UCLA, Chicano Studies Research Center Publications, 1982.

———. *On Culture.* Popular Series 1. Los Angeles: UCLA, Chicano Studies Research Center Publications, 1977.

———. *Sembradores: Ricardo Flores Magón y el Partido Liberal Mexicano.* Los Angeles: UCLA, Chicano Studies Research Center Publications, 1973.

González y González, Luis. *San José de Gracia: Mexican Village in Transition.* Austin: University of Texas Press, 1964.

González Monroy, Jesús. *Ricardo Flores Magón y su actitud en la Baja California.* México D.F.: Editorial Academia Literaria, 1962.

Goodman, David, and Michael Redclift. *From Peasant to Proletarian: Capitalist Development and Agrarian Transitions.* Oxford: Basil Blackwell, 1981.

Goodwyn, Lawrence. *Democratic Promise: The Populist Movement in America.* New York: Oxford University Press, 1976.

Gramsci, Antonio. *Selections from the Prison Notebooks.* New York: International Publishers, 1971.

Grele, Ron. *Envelopes of Sound: The Art of Oral History.* 2d ed. Chicago: Precedent Publishing Co., 1985.

Green, James. *Grass Roots Socialism: Radical Movements in the Southwest, 1895–1943.* Baton Rouge: Louisiana State University Press, 1978.

Gregory, James. *American Exodus: The Dust Bowl Migration and Okie Culture in California.* New York: Oxford University Press, 1989.

Grindle, Merilee S. *Searching for Rural Development: Labor Migration and Employment in Mexico.* Ithaca: Cornell University Press, 1988.

Griswold del Castillo, Richard. *La Familia: Chicano Families in the Urban Southwest, 1848 to the Present.* Notre Dame: University of Notre Dame Press, 1984.

———. *The Los Angeles Barrio, 1850–1890: A Social History.* Berkeley: University of California Press, 1979.

Grubbs, Donald. *Cry from the Cotton: The Southern Tenant Farmers Union and the New Deal.* Chapel Hill: University of North Carolina Press, 1971.

Gutiérrez, David G. *Walls and Mirrors: Mexican Americans, Mexican Immigrants, and the Politics of Ethnicity in the American Southwest.* Berkeley: University of California Press, forthcoming.

Gutiérrez, Ramón. *When Jesus Came, the Corn Mothers Went Away: Marriage, Sexuality and Power in New Mexico, 1500–1846.* Stanford: Stanford University Press, 1991.

Gutman, Herbert. *Work, Culture and Society in Industrial America.* New York: Vintage Books, 1976.

Haraven, Tamara. *Family Time and Industrial Time*. Cambridge: Cambridge University Press, 1982.

Hart, John M. *Anarchism and the Mexican Working Class, 1860–1931*. Austin: University of Texas Press, 1978.

———. *Revolutionary Mexico: The Coming and Process of the Mexican Revolution*. Berkeley: University of California Press, 1989.

Havens, A. Eugene, Gregory Hooks, Patrick Mooney, and Max Pfeffer, eds. *Studies in the Transformation of U.S. Agriculture*. Boulder: Westview Press, 1986.

Healey, Dorothy, and Maurice Isserman. *Dorothy Healey Remembers: A Life in the American Communist Party*. New York: Oxford University Press, 1990.

Hernández, José Amaro. *Mutual Aid for Survival: The Case of the Mexican American*. Malabar, Fla.: Robert E. Krieger Publishing, 1983.

Herrera-Sobek, Maria. *The Mexican Corrido: A Feminist Analysis*. Bloomington: Indiana University Press, 1990.

Hobsbawm, Eric. *Labouring Men: Studies in the History of Labour*. London: Weidenfeld and Nicolson, 1964.

———. *Workers: Worlds of Labour*. New York: Pantheon Books, 1984.

Hoffman, Abraham. *Unwanted Mexican Americans in the Great Depression: Repatriation Pressures, 1929–1939*. Tucson: University of Arizona Press, 1974.

Hofstadter, Richard. *The Progressive Historians: Turner, Beard, Parrington*. New York: Knopf, 1968.

James, Marquis, and Bessie R. James. *Biography of a Bank: The Story of Bank of America*. New York: Harper and Row, 1954.

Jay, Martin. *The Dialectical Imagination: A History of the Frankfurt School and the Institute of Social Research, 1923–1950*. Boston: Little, Brown, 1973.

Key, V. O. *Politics, Parties, and Pressure Groups*. New York: Thomas Crowell, 1942.

Klehr, Harvey. *The Heyday of American Communism*. New York: Harper Bros., 1984.

Kluger, James. *The Clifton-Morenci Strike: Labor Difficulty in Arizona, 1915–1916*. Tucson: University of Arizona Press, 1970.

Kolko, Gabriel. *Main Currents in Modern American History*. New York: Harper and Row, 1976.

Kushner, Sam. *Long Road to Delano*. New York: International Publishers, 1975.

Laclau, Ernesto, and Chantal Mouffe. *Hegemony and Socialist Strategy: Towards a Radical Democratic Politics*. London: Verso, 1985.

LaGumina, Salvatore. *Vito Marcantonio: The People's Politician*. Dubuque, Iowa: Kendall Hall, 1969.

Lange, Dorothea, and Paul S. Taylor. *An American Exodus*. New Haven: Yale University Press, 1969.

Lantis, David, Rodney Steiner, and Arthur Carinen. *California: The Pacific Connection*. Chico: Creekside Press, 1989.

Levenstein, Harvey. *Labor Organization in the United States and Mexico: A History of Their Relations.* Westport, Conn.: Greenwood Press, 1971.

Limerick, Patricia Nelson. *The Legacy of Conquest: The Unbroken Past of the American West.* New York: W. W. Norton, 1987.

Limón, José. *Mexican Ballads, Chicano Poems: History and Influence in Mexican-American Social Poetry.* Berkeley: University of California Press, 1992.

London, Joan. *So Shall Ye Reap.* New York: Thomas Y. Crowell, 1970.

López, José Timoteo. *La historia de la sociedad protección mutua de trabajadores unidos.* New York: Comet Press, 1958.

López Castro, Gustavo. *La casa dividida: Un estudio de caso sobre la migración a Estados Unidos en un pueblo michocano.* Zamora, Michoacán: Colegio de Michoacán, 1986.

Lowenstein, Norman. *Strikes and Strike Tactics in California Agriculture.* Berkeley: University of California Press, 1940.

Lukács, Georg. *History and Class Consciousness: Studies in Marxist Dialectics.* Cambridge, Mass.: MIT Press, 1971.

McConnell, Grant. *The Decline of Agrarian Democracy.* Berkeley: University of California Press, 1953.

McKenna, Teresa. *Parto de Palabra: Studies on Chicano Literature in Process.* Austin: University of Texas Press, in press.

McLean, Robert N. *That Mexican: As He Really Is, North and South of the Rio Grande.* New York: Fleming H. Revell Company, 1928.

McWilliams, Carey. *California: The Great Exception.* New York: A. A. Wyn, 1949.

———. *The Education of Carey McWilliams.* New York: Simon and Schuster, 1978.

———. *Factories in the Field: The Story of Migratory Farm Labor in California.* Orig. published 1939. Hamden: Archon Books, 1969.

———. *Ill Fares the Land: Migrants and Migratory Labor in the United States.* New York: Barnes and Noble, 1967.

———. *North from Mexico: The Spanish-Speaking People of the United States.* Westport, Conn.: Greenwood Press, 1948.

———. *Southern California Country: An Island on the Land.* Santa Barbara: Peregrine Smith, 1979.

Majka, Linda, and Theo Majka. *Farmworkers, Agribusiness and the State.* Philadelphia: Temple University Press, 1982.

Marcuse, Herbert. *Eros and Civilization.* New York: Vintage Books, 1962.

Martínez Cerda, Carlos. *El algodón en la región de Matamoros, Tamaulipas.* Banco Nacional de Crédito Ejidal, 1954.

Martínez Verdugo, Arnoldo. *Historia del comunismo en México.* México D.F.: Grijalbo, 1985.

Marx, Karl. *The Eighteenth Brumaire of Louis Bonaparte.* Moscow: Progress Publishers, 1934.

Massey, Douglas, Rafael Alarcón, Jorge Durand, and Humberto González. *Return to Aztlan: The Social Process of International Migration from Western Mexico.* Berkeley: University of California Press, 1987.

Matthiessen, Peter. *Sal Si Puedes: Cesar Chavez and the New American Revolution*. New York: Delta, 1969.

Milkman, Ruth, ed. *Women, Work, and Protest: A Century of U.S. Women's Labor History*. London: Routledge and Kegan Paul, 1985.

Milton, David. *The Politics of U.S. Labor: From the Great Depression to the New Deal*. New York: Monthly Review Press, 1982.

Mitchell, H. L. *Mean Things Happening in This Land: The Life and Times of H. L. Mitchell, Co-Founder of the Southern Tenants Farmers Union*. Montclair, N.J.: Allanheld, Osmun, and Co., 1979.

Moley, Raymond. *After Seven Years*. New York: Harper and Brothers, 1939.

Montejano, David. *Anglos and Mexicans in the Making of Texas, 1836–1986*. Austin: University of Texas Press, 1987.

Montgomery, David. *The Fall of the House of Labor: The Workplace, the State and American Labor Activism, 1865–1925*. Cambridge: Cambridge University Press, 1987.

———. *Workers' Control in America*. Cambridge: Cambridge University Press, 1979.

Moody, J. Carroll, and Alice Kessler-Harris, eds. *Perspectives on American Labor History: The Problems of Synthesis*. DeKalb: Northern Illinois University Press, 1990.

Morales, Alejandro. *The Brick People*. Houston: Arte Publico Press, 1988.

Moreno, Lilia, and Omar Fonseca. *Jaripo: Pueblo de migrantes*. Jiquilpan, Michoacán: Centro de Estudios de la Revolución Mexicana "Lázaro Cárdenas," 1984.

Morin, Alexander. *The Organizability of Farm Labor in the United States*. Harvard Studies in Labor in Agriculture. Cambridge: Harvard University Press, 1952.

Nash, Gerald. *Creating the West: Historical Interpretations 1890–1990*. Albuquerque: University of New Mexico Press, 1991.

———. *State Government and Economic Development: A History of Administrative Policies in California, 1849–1933*. Berkeley: University of California Press, 1964.

Nelson, Cynthia. *The Waiting Village: Social Change in Rural Mexico*. Boston: Little, Brown, 1971.

Nelson, Eugene. *Huelga: The First Hundred Days of the Great Delano Grape Strike*. Delano: Farm Worker Press, 1966.

Nourse, Edwin G. *Three Years of the Agricultural Adjustment Administration*. Washington, D.C.: Brookings Institution, 1937.

Oxnam, G. Bromley. *The Mexican in Los Angeles*. Los Angeles: Interchurch World Movement of North America, 1920.

Perkins, Frances. *The Roosevelt I Knew*. New York: Viking Press, 1946.

Portelli, Allesandro. *The Death of Luigi Trastulli and Other Stories: Form and Meaning in Oral History*. Albany: State University of New York Press, 1991.

Portes, Alejandro, and Robert Bach. *Latin Journey: Cuban and Mexican Immigrants in the United States*. Berkeley: University of California Press, 1985.

Raat, Dirk. *Revoltosos: Mexico's Rebels and the United States, 1903–1923*. College Station: Texas A and M University Press, 1981.

Rauch, Basil. *The History of the New Deal, 1933–1938*. New York: Capricorn Books, 1963.

Rayback, Joseph G. *A History of American Labor*. New York: Free Press, 1966.

Reisler, Mark. *By the Sweat of Their Brow: Mexican Immigrant Labor in the United States, 1900–1940*. Westport, Conn.: Greenwood Press, 1976.

Richards, Harry I. *Cotton and the A.A.A.* Washington, D.C.: Brookings Institution, 1936.

Ríos Bustamante, Antonio, ed. *Mexican Immigrant Workers in the United States*. Los Angeles: UCLA, Chicano Studies Research Center, 1981.

Rodríguez, Francisco M. *Baco y Birján: Una historia sangrante y dolorosa de lo que fue y lo que es Tijuana*. México D.F.: B. Costa-Amic, 1968.

Rogin, Michael Paul, and John L. Shover. *Political Change in California: Critical Elections and Social Movements, 1890–1966*. Westport, Conn.: Greenwood Press, 1970.

Romo, Ricardo. *East Los Angeles: History of a Barrio*. Austin: University of Texas Press, 1983.

Ruiz, Vicki. *Cannery Women, Cannery Lives: Mexican Women, Unionization and the California Food Processing Industry, 1930–1950*. Albuquerque: University of New Mexico Press, 1987.

Salas, Elizabeth. *Soldaderas in the Mexican Military: Myth and History*. Austin: University of Texas Press, 1990.

Saldívar, Ramón. *Chicano Narrative: The Dialectics of Difference*. Madison: University of Wisconsin Press, 1990.

Saloutos, Theodore. *The American Farmer and the New Deal*. Ames: Iowa State University Press, 1982.

Samuel, Raphael, ed. *People's History and Socialist Theory*. History Workshop Series. London: Routledge and Kegan Paul, 1981.

Sánchez, George J. *Becoming Mexican American: Ethnicity, Culture, and Identity in Chicano Los Angeles, 1900–1945*. New York: Oxford University Press, 1993.

Sánchez, Rosaura, and Rosa Martínez Cruz, eds. *Essays on La Mujer*. Los Angeles: UCLA, Chicano Studies Research Center Publications, 1977.

Santamaría Gómez, Arturo. *La izquierda norteamericana y los trabajadores indocumentados*. N.p.: Universidad Autónoma de Sinaloa, Ediciones de Cultura Popular, S.A., 1988.

Sartre, Jean-Paul. *Search for a Method*. New York: Vintage Books, 1963.

Schaffer, Alan. *Vito Marcantonio: Radical in Congress*. Syracuse: Syracuse University Press, 1966.

Schlesinger, Arthur M. *The Age of Roosevelt*. 3 vols. Boston: Houghton Mifflin, 1957–1960.

Schlissel, Lillian, Vicki L. Ruiz, and Janice Monk, eds. *Western Women: Their Land, Their Lives*. Albuquerque: University of New Mexico Press, 1988.

Schwartz, Harry. *Seasonal Farm Labor in the United States*. New York: Columbia University Press, 1945.

Simpson, Eyler N. *The Ejido*. Chapel Hill: University of North Carolina Press, 1973.

Slotkin, Richard. *The Fatal Environment: The Myth of the Frontier in the Age of Industrialization, 1800–1890*. Middletown: Wesleyan University Press, 1985.

———. *Gunfighter Nation: The Myth of the Frontier in Twentieth Century America*. New York: Atheneum, 1992.

Smelser, Neil. *Social Change and Industrial Revolution*. Chicago: University of Chicago Press, 1959.

Smith, Henry Nash. *Virgin Land: The American West as Symbol and Myth*. New York: Random House, 1950.

Smith, Wallace. *Garden of the Sun: A History of the San Joaquin Valley, 1772–1939*. Los Angeles: Lyman House, 1939.

Sociedad Progresista Mexicana. *Aniversario de Oro, 1929–1979*. Bakersfield: La Sociedad Progresista, 29 September 1979. Personal collection of Luis Sálazar.

Stein, Walter J. *California and the Dust Bowl Migration*. Westport, Conn.: Greenwood Press, 1973.

Steinbeck, John. *The Grapes of Wrath*. New York: Viking Press, 1939.

———. *In Dubious Battle*. New York: Viking Press, 1938.

Sternsher, Bernard. *Rexford Tugwell and the New Deal*. New Brunswick: Rutgers University Press, 1964.

Stott, William. *Documentary Expression and Thirties America*. New York: Oxford University Press, 1973.

Street, James H. *The New Revolution in the Cotton Economy: Mechanization and Its Consequences*. Chapel Hill: University of North Carolina Press, 1957.

Taibo II, Paco Ignacio. *Bolshevikis: Historia narrativa de los orígenes del comunismo en México, 1919–1925*. México D.F.: Editorial Joaquín Mortiz, 1986.

Tannenbaum, Frank. *The Mexican Agrarian Revolution*. New York: Macmillan and Co., 1929.

———. *Peace by Revolution: Mexico After 1910*. New York: Columbia University Press, 1933.

Taylor, Paul S. *Labor on the Land: Collected Writings, 1930–1979*. New York: Arno Press, 1981.

———. *Mexican Labor in the United States: The Imperial Valley*. University of California Publications in Economics 6, no. 1. Berkeley: University of California Press, 1930.

———. *Mexican Labor in the United States: Migration Statistics*. University of California Publications in Economics 12, no. 3. Berkeley: University of California Press, 1934.

———. *On the Ground in the Thirties*. Salt Lake City: Peregrine Books, 1983.

———. *A Spanish-Mexican Peasant Community: Arandas in Jalisco, Mexico*. Ibero-Americana 4. Berkeley: University of California Press, 1933.

Thomas, Benjamin, and William McNellie, eds., *Other Mexicos: Essays on Regional Mexican History, 1876–1911*. Albuquerque: University of New Mexico Press, 1984.

Thompson, Edward P. *The Making of the English Working Class*. New York: Vintage, 1963.

————. *The Poverty of Theory and Other Essays*. New York: Monthly Review Press, 1978.

Tilly, Louise A., and Joan W. Scott. *Women, Work, and Family*. New York: Holt, Rinehart and Winston, 1978.

Tomlins, Christopher L. *The State and the Unions: Labor Relations, Law, and the Organized Labor Movement in America, 1880–1960*. Cambridge: Cambridge University Press, 1985.

Torres Parés, Javier. *La Revolución sin frontera: El partido Liberal Mexicano y las relaciones entre el movimiento obrero de México y el de Estados Unidos, 1900–1923*. México D.F.: Facultad de Filosofía y Letras Universidad Nacional Autónoma de México, 1990.

Turner, Frederick C. *The Dynamic of Mexican Nationalism*. Chapel Hill: University of North Carolina Press, 1968.

————. *The Significance of the Frontier in American History*. Edited by Harold P. Simonson. New York: Frederick Ungar, 1963.

Turner, John. *White Gold Comes to California*. Bakersfield, Calif.: California Planting Cotton Seed Distributors, 1981.

Valdez, Armando, Alberto Camarillo, and Tomás Almaguer. *The State of Chicano Research in Family, Labor and Migration Studies*. Stanford: Stanford Center for Chicano Research, 1983.

Vaughn, Stephen. *Holding Fast the Inner Lines: Democracy, Nationalism, and the Committee on Public Information*. Chapel Hill: University of North Carolina Press, 1980.

Vittoz, Stanley. *New Deal Labor Policy and the American Industrial Economy*. Chapel Hill: University of North Carolina Press, 1987.

White, Margaret Bourke, and Erskine Caldwell. *You Have Seen Their Faces*. New York: Viking Press, 1937.

White, Richard. *"It's Your Misfortune and None of My Own": A New History of the American West*. Norman: University of Oklahoma Press, 1991.

Williams, Raymond. *Problems in Materialism and Culture*. London: Verso Press, 1980.

Witte, Edwin E. *The Development of the Social Security Act*. Madison: University of Wisconsin Press, 1962.

Wolfe, Alan. *The Limits of Legitimacy: Political Contradictions of Contemporary Capitalism*. New York: Free Press, 1977.

Worster, Donald. *The Dust Bowl in the Southern Plains*. New York: Oxford University Press, 1980.

Wright, Erik Olin. *Class, Crisis, and the State*. London: New Left Books, 1978.

Yans-McLaughlin, Virginia. *Family and Community: Italian Immigrants in Buffalo, 1880–1930*. Ithaca: Cornell University Press, 1977.

Ybarra, Lea, and Alex Sarragosa. *Nuestras Raíces: The Mexican Community of the Central San Joaquin Valley*. Fresno: La Raza T.E.A.C.H. Project, California State University, 1980.

Zamora, Emilio. *The World of the Mexican Worker in Texas During the Early 1900s*. College Station: Texas A and M Press, 1993.

Zavella, Patricia. *Women's Work and Chicano Families: Cannery Workers of the Santa Clara Valley*. Ithaca: Cornell University Press, 1987.

ARTICLES AND PERIODICALS

Adams, R. L. "Why Farm Leases Are Changing." *Pacific Rural Press*, 27 August 1933.

Alba-Hernández, Francisco. "Éxodo silencioso: la emigración de trabajadores Mexicanos a Estados Unidos." *Foro Internacional* (October–December 1976).

"Analyzing Labor Requirements for California: Major Seasonal Crop Operations." *Journal of Farm Economics* 37, no. 43 (November 1945).

Arroyo, Luis Leobardo. "Notes on Past, Present and Future Directions of Chicano Labor Studies." *Aztlán* 6, no. 2 (Summer 1975).

Athearn, Leigh. "Unemployment Relief in Labor Disputes: California's Experience." *Social Science Review* 14, no. 4 (December 1940).

Bach, Robert L. "Mexican Immigration and the American State." *International Migration Review* 12, no. 4 (Winter 1978).

Bamford, Edwin F. "The Mexican Casual Problem in the Southwest." *Journal of Applied Sociology* 8 (July–August 1924).

Beckley, Stewart D. "Cotton Adds $47,000,000 to Wealth of West." *California Cotton Journal* (September 1926).

———. "Financing the Cotton Producer." *California Cotton Journal* (January 1926).

Benson, Jackson J., and Anne Loftis. "John Steinbeck and Farm Labor Unionization: The Background of *In Dubious Battle*." *American Literature* 52, no. 2 (May 1980).

Biagi, H. E. Bruno. "The Regulation of Collective Employment Relations in Agriculture in Italy." *International Labour Review* 29, no. 3 (March 1934).

Bloch, Louis. "Facts About Mexican Immigration Before and Since the Quota Restriction Laws." *American Statistical Association* 24 (1929).

Bogardus, Emory S. "Current Problems of Mexican Immigrants." *Sociology and Social Research* 25 (January-February 1940).

———. "The Mexican Immigrant." *Journal of Applied Sociology* 11 (May 1927).

———. "The Mexican Immigrant and the Quota." *Sociology and Social Research* 12 (March–April 1928).

———. "Mexican Repatriates." *Sociology and Social Research* 18 (November–December 1933).

———. "Second Generation Mexicans." *Sociology and Social Research* 13 (January–February 1929).

Bowles, Samuel, and Herbert Gintis. "The Crisis of Liberal Democratic Capitalism." *Politics and Society* 11, no. 1 (1982).

Brody, David. "The Old Labor History and the New." In Daniel J. Leab, ed., *The Labor History Reader*. Chicago: University of Illinois Press, 1985.

Burawoy, Michael. "The Functions and Reproduction of Migrant Labor: Comparative Material from Southern Africa and the United States." *American Journal of Sociology* 81, no. 5 (March 1976).

Burbank, Garin. "Agrarian Socialism in Saskatchewan and Oklahoma: Short-Run Radicalism, Long-Run Conservatism." *Agricultural History* 51, no. 1 (January 1977).

Bustamante, Jorge. "Commodity Migrants: Structural Analysis of Mexican Immigration to the United States." In S. R. Ross, ed., *Views Across the Border: The United States and Mexico*. Albuquerque: University of New Mexico Press, 1978.

———. "The Historical Context of Undocumented Mexican Immigration to the United States." In Antonio Ríos Bustamante, ed., *Mexican Immigrant Workers in the United States*. Los Angeles: UCLA, Chicano Studies Research Center, 1981.

"California's Cotton Rush." *Fortune* (May 1949).

Cardenas, Gilbert. "United States Immigration Policy Towards Mexico: An Historical Perspective." *Chicano Law Review* (Summer 1975).

Chacon, Ramon D. "Labor Unrest and Industrialized Agriculture in California: The Case of the 1933 San Joaquin Valley Cotton Strike." *Social Science Quarterly* 65, no. 2 (June 1984).

Clarke, Simon. "Socialist Humanism and the Critique of Economism." *History Workshop* 8 (Autumn 1979).

Clements, George P. "If Not Mexicans, Who?" *Los Angeles Times*, 11 May 1930.

———. "Immigration Bill, Economic Loss." *Pacific Rural Press*, 17 December 1927.

———. "Should Mexican Immigration Be Restricted?" *California Cotton Journal* (June 1926).

"Collective Labour Agreements in Italian Agriculture: I." *International Labour Review* 14, no. 5 (November 1926).

"Collective Labour Agreements in Italian Agriculture: II." *International Labour Review* 15, no. 3 (January–June 1927).

Commonwealth Club of California. "Land Tenancy in California." *Transactions of the Commonwealth Club* 17, no. 10 (November 1922).

Cornelius, Wayne. "Mexican Immigrants and Southern California: A Summary of Current Knowledge." Working Papers in U.S.-Mexican Studies 36. La Jolla, Calif.: Center for U.S.-Mexican Studies, 1982.

Cramp, Kathryn. *Study of the Mexican Population of the Imperial Valley, California*. New York: Committee on Farm and Cannery Migrants, Council of Women for Home Missions, 1926.

Dahl, Leif. "Agricultural Labor and Social Legislation." *American Federationist* 44, no. 1 (1937).

Daniel, Cletus. "Agricultural Unionism and the New Deal: The California Experience." *Southern California Quarterly* 59, no. 2 (1977).

Davis, James J. "Labor and Social Conditions of Mexicans in California." *Monthly Labor Review* 32, no. 1 (January 1931).

de Ford, Marian A. "Bloodstained Cotton in California." *The Nation* (30 December 1933).

Dorton, Robert E. "Financing Cotton in California." *Burroughs Clearing House* (January 1944).

Douglas, Katherine. "Uncle Sam's Coop for Individualists." *Coast Magazine* (June 1939).

Durón, Clementina. "Mexican Women and Labor Conflict in Los Angeles: The ILGWU Dressmakers' Strike of 1933." *Aztlán* 15, no. 1 (Spring 1984).

"Efforts of Agricultural Labor Bureau Bring Good Results." *California Cotton Journal* (April 1927).

Ferguson, Thomas. "From Normalcy to New Deal: Industrial Structure, Party Competition and American Public Policy in the Great Depression." *International Organization* 38, no. 1 (Winter 1984).

Finegold, Kenneth. "From Agrarianism to Adjustment: The Political Origins of New Deal Agricultural Policy." *Politics and Society* 2, no. 1 (1982).

Fisher, Lloyd. "The Harvest Labor Market in California." *Quarterly Journal of Economics* 65, no. 4 (November 1951).

Frisch, Michael. "American History and the Structures of Collective Memory: A Modest Exercise in Empirical Iconography." *Journal of American History* 75, no. 4 (March 1989).

———. "The Memory of History." *Radical History Review* 25, no. 2 (June 1980).

Fuller, Varden. "Farm Labor." *Annals of the American Academy* (January 1977).

Galarza, Ernest. "Without Benefit of Lobby." *Survey* 66 (May 1931).

García, Mario T. "Americanization and the Mexican Immigrant, 1880–1930." *Journal of Ethnic Studies* 6 (Summer 1978).

———. "The Chicana in American History: The Mexican Women of El Paso, 1880–1920—A Case Study." *Pacific Historical Review* 49 (May 1980).

———. "La Familia: The Mexican Immigrant Family, 1900–1930." In Mario Barrera and Alberto Camarillo, eds., *Work, Family, Sex Roles, Language.* Berkeley: Tonatiuh-Quinto Sol International, 1980.

Gedicks, Al. "The Social Origins of Radicalism Among Finnish Immigrants in Midwest Mining Communities." URPE *Review of Radical Political Economics* 8, no. 3 (Fall 1976).

Geertz, Clifford. "Deep Play: Notes on the Balinese Cockfight." *Daedalus* 101, no. 1 (Winter 1972).

Genovese, Eugene, and Elizabeth Fox Genovese. "The Political Crisis of Social History: A Marxian Perspective." *Journal of Social History* 10 (Winter 1976).

Goldschmidt, Walter. "Large Scale Farming and the Rural Social Structure." *Rural Sociology* 43, no. 3 (1978).

———. "Reflections on Arvin and Dinuba." *Newsline of the Rural Sociological Society* 6, no. 5 (September 1978).

Gómez-Quiñones, Juan. "The First Steps: Chicano Labor Conflict and Organizing, 1900–1920." *Aztlán* 3, no. 1 (Spring 1972).

———. "On Culture." *Revista Chicano-Riqueña* 5 (Spring 1977).

———. "Plan de San Diego Reviewed." *Aztlán* 1, no. 1 (Spring 1970).

Gorz, André. "Immigrant Labour." *New Left Review* 61 (May–June 1970).

Gourevitch, Peter. "Breaking with Orthodoxy: The Politics of Economic Policy Responses to the Depression of the 1930s." *International Organization* 38, no. 1 (Winter 1984).

Guérin, Daniel. "Agriculture et Capitalisme aux Etats-Unis." *Les Temps Modernes* 63 (January–February 1951).

Hareven, Tamara. "Family Time and Industrial Time: Family Work in a Planned Corporate Town, 1900–1924." *Journal of Urban History* 1 (May 1975).

———. "The Laborers of Manchester, New Hampshire, 1912–1922: The Role of Family and Ethnicity in Adjustments to Industrial Life." *Labor History* 16 (Spring 1975).

Harrington, Dale. "New Deal Farm Policy and Oklahoma Populism." In A. Eugene Havens, Gregory Hooks, Patrick Mooney, and Max Pfeffer, eds., *Studies in the Transformation of U.S. Agriculture*. Boulder: Westview Press, 1986.

Haslam, Gerald. "What About the Okies?" *American History Illustrated* 12, no. 1 (1977).

Henderson, Don. "Agricultural Workers." *American Federationist* (May 1936).

Hodson, R. J. "Corcoran Cotton Finds Ready Market." *California Cotton Journal* (November 1925).

"I Wonder Where We Can Go Now?" *Fortune* (April 1939).

Johnson, Richard. "Critique: Edward Thompson, Eugene Genovese, and Socialist-Humanist History." *History Workshop* 6 (Autumn 1978).

Jones, Lamar B. "Labor and Management in California Agriculture, 1864–1964." *Labor History* 11, no. 1 (Winter 1970).

Joy, Al C. "Cotton in the 'Garden of the Sun.'" *California Cotton Journal* (December 1925).

Kaplan, Temma. "Female Consciousness and Collective Action: The Case of Barcelona, 1910–1918." *Signs* (Spring 1982).

Kerr, Clark. "Industrial Relations in Large-Scale Cotton Farming." *Proceedings of the Nineteenth Annual Conference of the Pacific Coast Economic Association*. Palo Alto: Pacific Coast Economic Association, 1940.

Kieffer, Donald L. "Cotton Growers Better Watch the Dealers." *Pacific Rural Press*, 3 March 1928.

Large, David C. "Cotton in the San Joaquin Valley: A Study of Government in Agriculture." *Geographical Review* (July 1957).

Larsen, Charles E. "The EPIC Campaign of 1934." *Pacific Historical Review* 27 (May 1958).

Leader, Leonard. "Upton Sinclair's EPIC Switch: A Dilemma for American Socialists." *Southern California Quarterly* 62 (Winter 1980).

Longmore, T. Wilson, and Homer L. Hitt. "A Demographic Analysis of First and Second Generation Mexican Population of the United States: 1930." *Southwestern Social Science Quarterly* 24 (September 1943).

López, Ron. "The El Monte Berry Strike." *Aztlán* 1, no. 1 (Spring 1970).

McClelland, Keith. "Some Comments on Richard Johnson, 'Edward Thompson, Eugene Genovese, and Socialist-Humanist History.'" *History Workshop* 7 (Spring 1979).

MacDonald, J. S. "Agricultural Organization, Migration and Labour Militancy in Rural Italy." *Economic History Review* (August 1963).

McLachlan, Argyle. "The Cotton Industry of California." *California Cotton Journal* (February 1926).

McLean, Robert. "Getting God Counted Among the Mexicans." *Missionary Review of the World* 46 (May 1923).
――――. "Goodbye Vincente." *Survey* 66 (May 1931).
――――. "Tightening the Mexican Border." *Survey* 65 (April 1930).
McWilliams, Carey. "Getting Rid of the Mexican." *American Mercury* 28 (March 1933).
――――. "They Saved the Crops." *The Inter-American* (August 1943).
――――. "What's Being Done About the Joads?" *New Republic* (20 September 1939).
Mella, Julio Antonio. "La situación del proletariado Mexicano en los Estados Unidos." Originally published in *Defensa Proletaria* 1, no. 5 (20 January 1929; reprinted in *Boletín del CEMOS: Memoria* 1, no. 6 (February–March 1984).
Meyers, William K. "La Comarca Lagunera: Work, Protest, and Popular Mobilization in North Central Mexico." In Thomas Benjamin and William McNellie, eds., *Other Mexicos: Essays on Regional Mexican History, 1876–1911*. Albuquerque: University of New Mexico Press, 1984.
Monds, Jean. "Workers' Control and the Historians: The New Economism." *New Left Review* 97 (May–June 1976).
Monroy, Douglas. "Anarquismo y comunismo: Mexican Radicalism and the Communist Party in Los Angeles During the 1930s." *Labor History* 24, no. 1 (Winter 1983).
――――. "An Essay on Understanding the Work Experience of Mexicans in Southern California, 1900–1939." *Aztlán* 12, no. 1 (Spring 1981).
Moore, E. B. "What Will Best Serve the Cotton Interests of California at This Time?" *California Cotton Journal* (November 1926).
Mumford, E. Philpott. "Early History of Cotton Cultivation in California." *Quarterly of the California Historical Society* (June 1927).
Musoke, Moses S., and Alan L. Olmstead. "The Rise of the Cotton Industry in California: A Comparative Perspective." *Journal of Economic History* 42, no. 2 (June 1982).
Nelson-Cisneros, Víctor. "La clase trabajadora en Tejas, 1920–1940." *Aztlán* 6, no. 2 (Summer 1975).
――――. "UCAPAWA and Chicanos in California: The Farm Worker Period, 1937–1940." *Aztlán* 7, no. 3 (Fall 1976).
Neuberger, Richard. "Who Are the Associated Farmers?" *Survey Graphic* 28, no. 9 (September 1939).
Ormsby, Herbert F. "The Cotton Industry in California." *California Cotton Journal* (February 1927).
――――. "Stabilizing the Cotton Industry in California." *California Cotton Journal* (November 1925).
Palomares, Frank J. "Activity of Agricultural Labor Bureau of the San Joaquin Valley, Inc. Outlined in Annual Report." *Western Cotton Journal and Farm Review* (March 1928).
Pascoe, Pegge. "Western Women at the Cultural Crossroads." In Patricia Nelson Limerick, Clyde Milner, and Charles Rankin, eds., *Trails: Toward a New Western History*. Lawrence: University Press of Kansas, 1991.

Patterson, James T. "The New Deal in the West." *Pacific Historical Review* 38 (August 1969).

Patterson, Tim. "Notes on the Historical Application of Marxist Cultural Theory." *Science and Society* no. 39 (1975.

Penny, Lucretia. "Pea Pickers Child." *Survey Graphic* (July 1935).

Pomeroy, Harold L. "How Cotton Came to California." *Pacific Rural Press*, 13 June 1931.

———. "Pure Cotton in the San Joaquin Valley." *Pacific Rural Press*, 8 February 1930.

Portelli, Allesandro. "The Peculiarities of Oral History." *History Workshop* 12 (Autumn 1981).

Portes, Alejandro. "Toward a Structural Analysis of Illegal (Undocumented) Immigration." *International Migration Review* 12, no. 4 (Winter 1978).

Pratt, Stanley. "Building the Cotton Industry of the San Joaquin Valley on a Sound Conservative Basis." *California Cotton Journal* (November 1925).

Quadagno, Jill. "Welfare Capitalism and the Social Security Act of 1935." *American Sociological Review* 49 (October 1984).

Reich, William. "King Cotton in California." *The Land* (Spring 1950).

Romo, Ricardo."Responses to Mexican Immigration, 1910–1930." *Aztlán* 6, no. 2 (Fall 1975).

———. "Work and Restlessness: Occupational and Spatial Mobility Among Mexicanos in Los Angeles, 1918–1928." *Pacific Historical Review* 46, no. 2 (May 1977).

Ruiz, Vicki. "Obreras y Madres: Labor Activism Among Women and Its Impact on the Family." In *La Mexicana/Chicana*, Renato Resaldo Lecture Series, 1, 1983–1984. Tucson: Mexican American Studies and Research Center, University of Arizona, 1985.

Saloutos, Theodore. "The American Farm Bureau Federation and Farm Policy, 1933–1945." *Southwestern Social Science Quarterly* 28, no. 4 (March 1948).

Sánchez, Lupe, and Jésus Romo. "Organizing Mexican Undocumented Farm Workers on Both Sides of the Border." Working Papers in U.S.-Mexican Studies 27. La Jolla, Calif.: Center for U.S.-Mexican Studies, 1981.

Sánchez, Rosaura. "The Chicana Labor Force." In Rosaura Sánchez and Rosa Martínez Cruz, eds., *Essays on la Mujer*. Los Angeles: UCLA, Chicano Studies Research Center Publications, 1977.

Saragoza, Alex M. "The Conceptualization of the History of the Chicano Family." In Armando Valdez, Alberto Camarillo, and Tomás Almaguer, eds., *The State of Chicano Research in Family, Labor and Migration Studies*. Stanford: Stanford Center for Chicano Research, 1983.

Schenk, A. M. "California Is the New Realm of King Cotton: Story of New Industry Told to Bankers." *California Bank of Italy Bulletin* 6, no. 12 (December 1925).

"The School Follows the Child." *Survey*, 1 September 1931.

Schwartz, Bonnie Fox. "Social Workers and New Deal Politicians in Conflict: California's Branion-Williams Case, 1933–1934." *Pacific Historical Review* (February 1973).

Schwartz, Harry. "Agricultural Labor in the First World War." *Journal of Farm Economics* (February 1942).

———. "Farm Labor Policy." *Journal of Farm Economics* 25, no. 2 (August 1943).

———. "Organizational Problems of Agricultural Labor Unions." *Journal of Farm Economics* 23, no. 2 (May 1941).

———. "Recent Developments Among Farm Labor Unions." *Journal of Farm Economics* 23, no. 4 (November 1941).

Scruggs, Otey. "The Bracero Program Under the F.S.A. 1942–1943." *Labor History* 3, no. 2 (Spring 1962).

———. "Evolution of the Mexican Farm Labor Agreement of 1942." *Agricultural History* 34, no. 3 (July 1960).

———. "The First Mexican Labor Program." *Arizona and the West* 2, no. 4 (Winter 1960).

Silberling, Norman. "The Return of Farm Prosperity." *California Journal of Development* (November 1934).

Skocpol, Theda, and Kenneth Finegold. "State Capacity and Economic Intervention in the Early New Deal." *Political Science Quarterly* 97, no. 2 (Summer 1982).

Smith, Fred C., and T. Lynn Smith. "The Influence of the AAA Cotton Program Upon the Tenant, Cropper, and Laborer." *Rural Sociology* 1, no. 4 (December 1936).

Spaulding, Charles B. "The Mexican Strike at El Monte, California." *Sociology and Social Research* 25 (November–December 1940).

Stein, Walter J. "The 'Okie' as Farm Laborer." *Agricultural History* 49, no. 1 (January 1975).

Stuart, James, and Michael Kearney. "Causes and Effects of Agricultural Labor Migration from the Mixteca of Oaxaca to California." Working Papers in U.S.-Mexican Studies 28. La Jolla, Calif.: Center for U.S.-Mexican Studies, 1981.

Sufrin, Sidney. "Labor Organization in Agricultural America, 1930–1935." *American Journal of Sociology* 43, no. 4 (January 1938).

Taylor, Paul S. "Again the Covered Wagon." *Survey Graphic* 24 (July 1935).

———. "Migratory Farm Labor in the United States." *Monthly Labor Review* 44, no. 3 (March 1937).

———. "More Bars Against Mexicans?" *Survey* (April 1930).

———. "Power Farming and Labor Displacement in the Cotton Belt, 1937." *Monthly Labor Review* 46, nos. 3–4 (March–April 1938).

———. "Refugee Labor Migration to California, 1937." *Monthly Labor Review* 47, no. 4 (April 1939).

Taylor, Paul S., and Clark Kerr. "Documentary History of the Strike of Cotton Pickers in California, 1933." In Paul Taylor, *On the Ground in the Thirties*. Salt Lake City: Peregrine Books, 1983.

———. "Uprisings on the Farms." *Survey Graphic* 24 (January 1935).

Taylor, Paul S., and Tom Vasey. "Contemporary Background of California Farm Labor." *Rural Sociology* 1, no. 4 (December 1936).

———. "Historical Background of California Farm Labor." *Rural Sociology* 1, no. 3 (September 1936).

Tesche, W. C. "Our Dixieland in the San Joaquin." *Pacific Rural Press*, 2 April 1927.

Thelan, David. "Memory and American History." *Journal of American History* 75, no. 4 (March 1989).

Thompson, Charles A. "What of the Bracero? The Forgotten Alternative in Our Immigration Policy." *Survey* 54 (June 1925).

Los Tulares. "Tagus Ranch." *Quarterly Bulletin of the Tulare Historical Society* 125, no. 2 (December 1979).

Vittoz, Stan. "World War I and the Political Accommodation of Transnational Market Forces: The Case of Immigration Restriction." *Politics and Society* 8, no. 1 (1978).

Walker, Helen. "Mexican Immigrants and American Citizenship." *Sociology and Social Research* 13 (December-January 1928–1929).

———. "Mexican Immigrants as Laborers." *Sociology and Social Research* 13 (December-January 1928–1929).

Weber, Devra Anne. "Mexican Women on Strike: Memory, History and Oral Narratives." In Adelaida R. Del Castillo, ed., *Between Borders: Essays on Mexicana/Chicana History*. Encino, Calif.: Floricanto Press, 1990.

———. "The Organizing of Mexicano Agricultural Workers: Imperial Valley and Los Angeles, 1928–34—An Oral History Approach." *Aztlán* 3, no. 2 (Fall 1973).

———. "Raíz Fuerte: Oral History and Mexicana Farmworkers." *Oral History Review* 17, no. 2 (Fall 1989).

Weiner, Merle. "Cheap Food, Cheap Labor: California Agriculture in the 1930's." *Insurgent Sociologist* 8, no. 2–3 (Fall 1978).

White, Richard. "The Winning of the West: The Expansion of the Western Sioux in the Eighteenth and Nineteenth Centuries." *Journal of American History* 65 (September 1978).

Womack, John. "The Mexican Economy During the Revolution, 1910–1920: Historiography and Analysis." *Marxist Perspectives* 1, no. 4 (Winter 1978).

Yonay, Ehud. "King Cotton." *California* (December 1982).

Zamora, Emilio. "Chicano Socialist Labor Activity in Texas, 1900–1920." *Aztlán* 6, no. 2 (Summer 1975).

———. "Sara Estela Ramírez: una rosa roja en el movimiento." In Magdalena Mora and Adelaida R. Del Castillo, eds., *Mexican Women in the United States: Struggles Past and Present*, Occasional Paper 2. Los Angeles: UCLA, Chicano Studies Research Center Publications, 1980.

GOVERNMENT DOCUMENTS AND PUBLICATIONS

Agricultural Extension Service. Kern County, California. "Acala Cotton Efficiency Study: Kern County California" (typescript). 1931. Giannini Foundation of Agricultural Economics Library, University of California, Berkeley.

———. "Annual Summary: Kern County Efficiency Study, Cotton. For the 1938 Crop Year Including a Final Summary of Five Years of Records" (typescript). April 1938. Giannini Foundation of Agricultural Economics Library, University of California, Berkeley.

———. "Fourth Annual Summary of the Cotton Efficiency Study for Kern County: Crop Year 1929" (typescript). Giannini Foundation of Agricultural Economics Library, University of California, Berkeley.

———. "Third Annual Summary: Cost of Production Study on Cotton" (typescript). 19 July 1929. Giannini Foundation of Agricultural Economics Library, University of California, Berkeley.

Agricultural Extension Service. Kern County, California. Tulare County Farm Bureau. "First Annual Summary of Enterprise Efficiency Study—Cotton, for Tulare County, 1930" (typescript). Giannini Foundation of Agricultural Economics Library, University of California, Berkeley.

———. "Second Annual Summary of Enterprise Efficiency Study—Cotton, for Tulare County, 1931" (typescript). Giannini Foundation of Agricultural Economics Library, University of California, Berkeley.

———. "Third Annual Summary of Enterprise Efficiency Study—Cotton, for Tulare County, 1932" (typescript). Giannini Foundation of Agricultural Economics Library, University of California, Berkeley.

Agricultural Labor Bureau of the San Joaquin Valley. "The Agricultural Labor Bureau of the San Joaquin Valley: Its Aims and Purposes" (pamphlet). N.d. but circa 1926. Carey McWilliams Collection, UCLA.

Baird, Enid, and Hugh P. Brinton. United States Works Project Administration. Division of Social Research. "Average General Relief Benefits 1933–1938." Washington, D.C.: U.S. Government Printing Office, 1940.

Bloch, Louis. "Report on the Mexican Labor Situation in the Imperial Valley." Twenty-second Biennial Report of the Bureau of Labor Statistics of the State of California, 1925–1926. Sacramento: California Bureau of Labor Statistics, 1926.

California. Commission of Immigration and Housing. *Annual Report*. Sacramento: California Printing Office, 25 January 1925.

California Conference of Agricultural Workers. Agricultural Labor Research Bureau. "Ameliorative Proposals for Agricultural Labor in California" (typescript). Berkeley, Bureau of Agricultural Economics, U.S. Department of Agriculture, June 1938. Giannini Foundation of Agricultural Economics Library, University of California, Berkeley.

California. Cotton Wage Hearing Board. Minutes, 28 and 29 September 1939, reported by Margaret Brolliar and Bessie Bezzerides (typescript). Fresno, 1939.

———. Report of Cotton Wage Hearing Board (typescript). Los Angeles: Division of Immigration and Housing, Department of Industrial Relations, 1939(?).

California. Department of Industrial Relations. *Report to the Governor's Council*, 27 and 28 May 1939.

California. Department of Industrial Relations. Division of Housing and Sanitation. "Mexican Survey: Cotton, 1930." Typescript of inspections in cotton camps conducted in September and October 1928. California State Archives, Sacramento, California.

California. Department of Industrial Relations. Division of Immigration and Housing. "Housing for Migratory Agricultural Workers." 31 December 1934 (typescript). Carey McWilliams Collection, UCLA.

California. State Relief Administration. Cotton Chopping Hearings. *Minutes.* Reported by Margaret Brolliar and Bessie Bezzerides. 9 May 1939 (typescript). Madera, California.

California. State Relief Administration. *Review of Activities of the State Relief Administration of California, 1933–1935.* San Francisco: California State Relief Administration, April 1936.

California. State Relief Administration. Division of Research and Surveys. "Survey of Agricultural Labor Requirements in California." December 1935. Giannini Foundation of Agricultural Economics Library, University of California, Berkeley.

Farm Security Administration Case Histories. United States Department of Agriculture, Farm Security Administration, Region IX papers, Bancroft Library, University of California, Berkeley, carton 2, folder 6 (typescript).

Federal Emergency Relief Administration. "Report on Cotton Strikers." Circa 1933 (typescript). Paul Taylor Collection, Bancroft Library, University of California, Berkeley.

Fuller, Varden. "The Supply of Agricultural Labor as a Factor in the Evolution of Farm Organization in California." In United States Congress, Senate, Subcommittee of the Committee on Education and Labor, *Hearings of S. Res. 266, Violations of Free Speech and the Rights of Labor,* part 54, exbihit 8762-A, 19777–898. Washington, D.C.: U.S. Government Printing Office, 1939.

Jamieson, Stuart. *Labor Unionism in American Agriculture.* United States Department of Labor, Bureau of Labor Statistics, Bulletin 836. Washington, D.C.: U.S. Government Printing Office, 1945.

Kern County Health Department. Sanitary Division. "Survey of Kern County Migratory Labor Problem," Supplementary Report as of 1 July 1938. Public Health Library, University of California, Berkeley (mimeo.).

Leonard, J. L., Simon J. Lubin, and Will J. French. "Report to the National Labor Board by Special Commission." Release No. 3325, 11 February 1934 (typescript). In Pelham D. Glassford Collection, Special Collections, UCLA.

Lewis, M. H. *State Welfare Survey.* California State Relief Administration, n.d. [1935?] (typescript). Carey McWilliams Collection, UCLA.

Mott, L. T., and F. J. Rugg. "Housing for Migratory Agricultural Workers." California Department of Industrial Relations, Division of Immigration and Housing, 31 December 1934 (typescript). Carey McWilliams Collection, UCLA.

Palomares, Francisco. "Report of the Agricultural Labor Bureau of the San Joaquin Valley for 1927." 1 March 1927 (typescript). Carey McWilliams Collection, UCLA.

Taylor, Paul S., and Clark Kerr. "Documentary History of the Strike of Cotton Pickers in California, 1933." United States Congress. Senate. Subcommittee of the Committee of Education and Labor. *Hearings on S. Res. 266, Vio-*

lations of Free Speech and the Rights of Labor, part 64, exhibit 8764, 19945–20036.

United States. Special Committee on Farm Tenancy. *Farm Tenancy: Report of the President's Committee*, prepared under the auspices of the National Resources Committee. Washington, D.C.: U.S. Government Printing Office, 1937.

United States Congress. House. Committee on Immigration and Naturalization. *Hearings on Seasonal Agricultural Laborers from Mexico*, 69th Congress, 1st session. Washington, D.C.: U.S. Government Printing Office, 1926.

United States Congress. House. Committee on Labor. *Hearings on H.R.6288, Labor Disputes Act*, 74th Cong., 2d session. Washington, D.C.: U.S. Government Printing Office, 1935.

United States Congress. House. Select Committee to Investigate the Interstate Migration of Destitute Citizens, Pursuant to H. Res. 63,491,729 (76th Congress) and H. Res 16 (77th Congress). *Hearings*. Washington, D.C.: U.S. Government Printing Office, 1940–1941.

United States Congress. House. Select Committee to Investigate the Interstate Migration of Destitute Citizens, Pursuant to H. Res 63,491,729 (76th Congress) and H. Res 16 (77th Congress). *Interstate Migration Report*. Washington, D.C.: U.S. Government Printing Office, 1941.

United States Congress. Senate. *Report, Violations of Free Speech and the Rights of Labor*. No. 1150, 77th Congress, 2d Session. Washington, D.C.: U.S. Government Printing Office, 1942.

United States Congress. Senate. Committee on Education and Labor. *Hearings on S.2926, To Create a National Labor Board*, 73d Congress, 2d session. Washington, D.C.: U.S. Government Printing Office, 1934.

United States Congress. Senate. Subcommittee of the Committee on Education and Labor. *Hearings on S. Res. 266, Violations of Free Speech and the Rights of Labor*. Washington, D.C.: U.S. Government Printing Office, 1939.

United States Department of Agriculture. AAA. *Agricultural Adjustment*. A Report of the Administration of the AAA, May 1933–February 1934. Washington, D.C.: U.S. Government Printing Office, 1935.

United States Department of Agriculture. Farm Security Administration. Division of Information. "Synopsis of 'A Study of 6655 Case Histories in California, 1938'" (typescript). San Francisco, n.d. but circa 1939. Farm Security Administration Collection, Bancroft Library, University of California, Berkeley.

United States Department of Commerce. *15th Census of the United States*. Vol. 3, part 3. Washington, D.C.: U.S. Government Printing Office, 1930.

Young, C. C. *Mexicans in California: Report of Governor C. C. Young's Mexican Fact-Finding Committee*. Sacramento: California State Printing Office, 1930.

PAMPHLETS, MANUSCRIPTS, DISSERTATIONS, AND OTHER UNPUBLISHED SOURCES

Adams, R. L. "Seasonal Labor Requirements for California Crops." California Agricultural Experiment Station, Berkeley, California, *Bulletin 623* (July 1938).

Anderson, Bettie Daingerfield. "Survey of Kern County, California" (typescript). New York, Columbia University, 1932.

Arambula-Verde, Rita. "The Historical Contributions of the Mexican American in Kern County: A Prospectus for School Curricular Materials Development." Master's thesis, California State College, Bakersfield, July 1976.

Bakersfield Conference on Agricultural Labor. "Health, Housing and Relief" (typescript). Bakersfield, California, 29 October 1938. Carey McWilliams Collection, Special Collections, UCLA.

Balderama, Francisco. "Study of Mexican Consulates and the Chicano Community." Ph.D. diss., University of California, Los Angeles, 1973.

Cady, Rev. George L. "Report of Commission on International and Interracial Factors in the Problems of Mexicans in the United States." N.d. but circa 1920s. George Clements Collection, Special Collections, UCLA.

Calcot. "The First Fifty Years"(pamphlet). Calcot Ltd., Bakersfield, California, n.d.

———. "History of Calcot, Ltd." (typescript). Calcot Ltd., Bakersfield, California, 1979.

California State Chamber of Commerce. "Migrants: A National Problem and Its Impact on California: Report and Recommendations of the Statewide Committee on the Migrant Problem" (pamphlet). Sacramento: California State Chamber of Commerce, May 1940.

California State Federation of Labor. "State Conference of Agricultural Unions" (pamphlet). Labor Temple, San Francisco, 27 and 28 February 1937. Agricultural Library, University of California, Davis.

Camarillo, Alberto. "The Development of a Pan-Hispanic Civil Rights Movement: The 1939 Congress of Spanish Speaking People." Paper presented at the Annual Meeting of the Organization of American Historians, April 1986.

Castillo, Pedro G. "The Making of a Mexican Barrio: Los Angeles, 1890–1920." Ph.D. diss., University of California, Santa Barbara, 1979.

Chacon, Ramon. "Labor Unrest and Commercialized Agriculture: The Case of the 1933 San Joaquin Valley Cotton Strike" (typescript). Ethnic Studies Department, Humboldt State University, n.d.

Chaffee, Porter. "History of the Cannery and Agricultural Workers Industrial Union" (typescript). Federal Writers Project, Oakland, California, n.d. but circa 1930s.

Chambers, Clark. "A Comparative Study of Farmer Organizations in California During the Depression Years, 1929–1941." Ph.D. diss., University of California, Berkeley, 1950.

Chambers of Commerce of the United States. Proceedings of the Agricultural Conference of Chambers and Associations of Commerce, Fresno, 26–27 March 1926 (typescript). Washington, D.C.: Chambers of Commerce of the United States, Agricultural Service. Giannini Foundation of Agricultural Economics Library, University of California, Berkeley.

Chinn, Ronald. "Democratic Party Politics in California, 1920–1956." Ph.D. diss., University of California, Berkeley, 1958.

Clements, George P. "Mexican Immigration and Its Bearing on California's Agriculture" (typescript). Address to the Lemon Men's Club, Los Angeles, 2

October 1929. George Clements Collection, University of California, Los Angeles.

Cochran, Clay L. "Hired Farm Labor and the Federal Government." Ph.D. diss., University of North Carolina, 1950.

Cramp, Kathryn, Louise Shields, and Charles A. Thomson. "Study of the Mexican Population in the Imperial Valley, California" (typescript). For the Committee on Farm and Cannery Migrants, Council of Women for Home Missions, New York, 31 March–9 April 1926. Bancroft Library, University of California, Berkeley.

Daniel, Cletus Edward. "Labor Radicalism in Pacific Coast Agriculture." Ph.D. diss., University of Washington, 1972.

Darcy, Sam. "Autobiography" (typescript). Sam Darcy Collection, New York University.

Delmatier, Royce D. "The Rebirth of the Democratic Party in California, 1928–1938." Ph.D. diss., University of California, Berkeley, 1955.

Douglas, Helen Walker. "The Conflict of Cultures in First Generation Mexicans in Santa Ana, California." Master's thesis, University of Southern California, 1928.

Dunn, Cecil, and Phillip Neff. "The Arvin Area of Kern County: An Economic Survey of the Southern San Joaquin Valley in Relation to Land Use and the Size and Distribution of Income" (typescript). 1947. Giannini Foundation of Agricultural Economics Library, University of California, Berkeley.

Edson, George T. "Mexican Labor in the California Imperial Valley" (typescript). Federal Writers Project, Oakland, California, 1927.

Federal Writers Project. "The Contract Labor System in California" (typescript). Oakland, California, no date but circa 1930s.

———. "A Documentary History of Migratory Farm Labor in California: The California Cotton Pickers Strike, 1933" (typescript). Ed. Raymond P. Barry. Oakland, California, 1938.

———. "A Documentary History of Migratory Labor in California: Labor in California Cotton Fields" (typescript). Ed. Raymond P. Barry. Oakland, California, 1938.

———. "Finance Control of California Agriculture." Oakland, California (?), n.d. but circa 1930s.

———. "Organization Efforts of Mexican Agricultural Workers" (typescript). Oakland, California, n.d. but circa 1930s.

———. "Unionization of Agricultural Labor In California" (typescript). Oakland, California, n.d. but circa 1930s.

———. "Unionization of Migratory Labor in California" (typescript). Oakland, California, n.d. but circa 1930s.

Fellows, Lloyd Walker. "Economic Aspects of the Mexican Rural Population in California with Special Emphasis on the Need for Mexican Labor in Agriculture." Master's thesis, University of Southern California, n.d. but circa 1920s.

Fincher, Judith. "Argentine Kansas: The Evolution of a Mexican American Community." Ph.D. diss., University of Kansas, 1975.

Finegold, Kenneth. "Agriculture, State, Party, and Economic Crisis: American

Farm Policy and the Great Depression." Paper presented at the Conference on the Political Economy of Food and Agriculture in Advanced Industrial Societies, Rural Sociological Society Annual Meeting, Guelph, Ontario, 19–23 August 1981.

Finegold, Kenneth, and Theda Skocpol. "Capitalists, Farmers and Workers in the New Deal—The Ironies of Government Intervention: A Comparison of the Agricultural Adjustment Act and the National Industrial Recovery Act." Paper presented at Session on Class Coalitions and Institutions in American Politics, Conference Group on the Political Economy of Advanced Industrial Societies, at the Annual Meeting of the American Political Science Association, Washington, D.C., 31 August 1980.

González, Deena J. "The Spanish-Mexican Women of Santa Fe: Patterns of Their Resistance and Accommodation, 1820–1880." Ph.D. diss., University of California, Berkeley, 1985.

Gregory, James Noble. "The Dust Bowl Migration and the Emergence of an Okie Subculture in California, 1930–1950." Ph.D. diss., University of California, Berkeley, 1983.

Guerin-Gonzáles, Camille. "Cycles of Immigration and Repatriation: Mexican Farm Workers in California Industrial Agriculture, 1900–1940." Ph.D. diss., University of California, Riverside, 1985.

Holt, James S. "The Farm Labor Market in the Eighties" (typescript). Paper presented at the Sixteenth Annual Meeting of the National Council of Agricultural Employers, San Antonio, Texas, 29–31 January 1980.

———. "Introduction to the Seasonal Farm Labor Problem" (typescript). N.d.

Hornsby, Henry Claude. "The Agricultural Adjustment Act and Cotton." Master's thesis, University of California, Los Angeles, 1966.

Hymer, Evangeline. "A Study of the Social Attitudes of Adult Mexican Immigrants in Los Angeles and Vicinity." Master's thesis, University of Southern California, 1923.

Jelinek, Lawrence. "California Farm Bureau Federation." Ph.D. diss., University of California, Los Angeles, 1953.

Johns, Bryan T. "Field Workers in California Cotton." Master's thesis, University of California, Berkeley, 1948.

John Steinbeck Committee to Aid Agricultural Organization. "Report of the Bakersfield Conference on Agricultural Labor: Health, Housing and Relief, October 29, 1938" (typescript). Carey McWilliams Collection, Special Collections, UCLA.

Jones, Victor. "1939 Legislative Problems, No. 4: Transients and Migrants" (typescript). Bureau of Public Administration, University of California, Berkeley, 27 February 1939. Giannini Foundation of Agricultural Economics Library, University of California, Berkeley.

Kaplan, Louis. "The Problems of Relief and Agriculture in California: December 1937" (typescript). Carey McWilliams Collection, Special Collections, UCLA, box 19.

Lanigan, Mary Cecilia. "Second Generation Mexicans in Belvedere." Master's thesis, University of Southern California, 1932.

Lewis, M. H. "State Welfare Survey." N.d.

Loescher, E. F. "Important Farm Labor Developments During 1937" (typescript). Paper presented to the Statewide Agricultural Committee Meeting by Ray Humphreys of Madera, 28 October 1937.

Los Angeles Chamber of Commerce. "Data Compiled on Mexican Labor" (typescript). 1927. George Clements Collection, Special Collections, UCLA.

―――. Minutes of Meeting of the Agricultural Labor Subcommittee of the Los Angeles Chamber of Commerce, 6 November 1933. George Clements Collection, Special Collections, UCLA.

―――. "Proposed Labor Clearing House Plan" (typescript). Offered by the Agricultural Department, Los Angeles Chamber of Commerce. Revised March 1931. George Clements Collection, Special Collections, UCLA.

Lubin, Simon J. "Can the Radicals Capture the Farms of California?" (typescript). Paper presented to the Commonwealth Club, 23 March 1934. Carey McWilliams Collection, Special Collections, UCLA.

McEuen, William Wilson. "A Survey of the Mexican in Los Angeles 1910–1914." Master's thesis, University of Southern California, 1914.

McWilliams, Carey. "Memorandum on Housing Conditions Among Migratory Workers in California" (typescript). 29 March 1939. Carey McWilliams Collection, Special Collections, UCLA.

―――. "Report Submitted to the President's Farm Tenancy Committee at Its Hearing in San Francisco, January 12, 1937" (pamphlet). San Francisco: Simon J. Lubin Society, 1937. Carey McWilliams Collection, Special Collections, UCLA.

Manes, Sheila Goldring. "Depression Pioneers: The Conclusion of an American Odyssey, Oklahoma to California, 1930–1950—A Reinterpretation." Ph.D. diss., University of California, Los Angeles, 1982.

Monroy, Douglas G. "Mexicanos in Los Angeles, 1930–1941: An Ethnic Group in Relation to Class Forces." Ph.D. diss., University of California, Los Angeles, 1978.

Ochoa, Alvaro S. "Arrieros, braceros, y migrantes de Oeste Michoacán (1849–1911)" (typescript). El Colegio de Michoacán, n.d.

Ortegón, Samuel M. "The Religious Status of the Mexican Population in Los Angeles." Master's thesis, University of Southern California, 1932.

Perk, Rena Blanche. "The Religious and Social Attitudes of the Mexican Girls of the Constituency of the All Nations Foundation in Los Angeles." Master's thesis, University of Southern California, 1929.

Perry, Raymond P., ed. "The California Cotton Pickers Strike—1933." Federal Writers Project, Oakland, California, 1938.

―――. "Labor in California Cotton Fields." Federal Writers Project, Oakland, California, 1938.

Philbrick, H. R. "Legislative Investigative Report." UCLA, Special Collections, 28 December 1938.

Pichardo, Nelson A. "The Role of the Community in Social Protest: Chicano Protest, 1848–1933." Ph.D. diss., University of Michigan, 1990.

Recow, Louis. "The Orange County Citrus Strikes of 1935–1936: The Forgotten People in Revolt." Ph.D. diss., University of Southern California, 1972.

Rios Bustamante, Antonio. "The California Cotton Industry and the Background of the 1933 Cotton Strike in the San Joaquin Valley" (typescript). 1973.

Romo, Ricardo. "Mexican Workers in the City: Los Angeles, 1915–1930." Ph.D. diss., University of California, Los Angeles, 1975.

Rose, Margaret. "Women in the United Farmworkers: A Study of Chicana and Mexicana Participation in a Labor Union, 1950–1980," Ph.D. diss., University of California, Los Angeles, 1988.

Sánchez, George. "Becoming Mexican American: Ethnicity and Acculturation in Chicano Los Angeles, 1900–1943." Ph.D. diss., Stanford University, 1989.

———. "Go After the Women: Americanization and the Mexican Immigrant Woman, 1915–1929." Working paper, series 6. Stanford: Stanford Center for Chicano Research, n.d.

Sherman, Jacqueline. "The Oklahomans in California During the Depression Decade, 1931–1941." Ph.D. diss., University of California, Los Angeles, 1970.

Smith, Clara Gertrude. "The Development of the Mexican People in the Community of Watts, California." Master's thesis, University of Southern California, 1933.

Smith, Robert C. " 'Los ausentes siempre presentes': The Imagining, Making and Politics of a Transnational Community Between New York City and Ticuani, Puebla." Columbia University, Institute for Latin American and Iberian Studies, Working Papers on Latin America (typescript). October 1992.

Sociedad Progresista Mexicana. "Aniversario de Oro, 1929–1979" (pamphlet). Bakersfield: La Sociedad Progresista, 29 September 1979.

Solow, Herbert. "Union Smashing in Sacramento: The Truth About the Criminal Syndicalism Trial" (pamphlet). New York: National Sacramento Appeal Committee, August 1935.

Steinbeck, John. "Their Blood Is Strong" (pamphlet). San Francisco: Simon J. Lubin Society, April 1938.

"Survey of the Mexican Labor Problem in California." 1928. Agricultural Economics Library, University of California, Davis.

Tangari, Beverly. "Federal Legislation as an Expression of United States Policy Toward Agricultural Labor, 1914–1954." Ph.D. diss, University of California, Berkeley, 1967.

Taylor, Paul S. "What Shall We Do with Them?" (typescript). Address before the Commonwealth Club of California, San Francisco, 15 April 1938. Carey McWilliams Collection, Special Collections, UCLA.

Thompson, Alvin H. "Aspects of the Social History of California Agriculture." Ph.D. diss., University of California, Berkeley, 1953.

Thomson, Eric H. "Why Plan Security for the Migratory Laborer?" (typescript). Paper presented at the California Conference of Social Work, San Jose, California, 12 May 1937, Federal Records Archives, San Francisco.

Underhill, Bertha S. California Department of Social Welfare, Division of Child Welfare Services. "A Study of 132 Families in California Cotton Camps with Reference to Availability of Medical Care." California Department of Social Welfare, 1936. California State Archives, Sacramento, California.

United Cannery, Agricultural, Packing and Allied Workers of America (UCAPAWA). "La Historia de la UCAPAWA." Escuela de Obreros Betabeleros, Denver, Colorado, April 1940. Giannini Foundation of Agricultural Economics Library, University of California, Berkeley.

———. "Official Proceedings" (pamphlet). First National Convention of United Cannery, Agricultural, Packing and Allied Workers of America, Fraternal Hall, Denver, Colorado, 9–12 July 1937. Washington, D.C.: UCAPAWA. Giannini Foundation of Agricultural Economics Library, University of California, Berkeley.

———. "UCAPAWA Yearbook" (pamphlet). Second National Convention of United Cannery, Agricultural, Packing and Allied Workers of America, San Francisco, California, December 1938. Giannini Foundation of Agricultural Economics Library, University of California, Berkeley.

University of California. College of Agriculture. "Cotton in the San Joaquin Valley." Circular no. 192, February 1918. Giannini Foundation of Agricultural Economics Library, University of California, Berkeley.

Weber, Devra. "Beyond Borders: Overview of Mexican Migrants and Transnational Labor Organization" (typescript). Paper presented at the Annual Meeting of the Organization of American Historians, April 1991.

Wilson, James Bright. "Religious Leaders, Institutions and Organizations Among Certain Agricultural Workers in the Central Valley of California." Ph.D. diss., University of Southern California, 1944.

Woodruff, Ed, for California Lands Inc. "Annual Report to Board of Directors for 1934 to January 1935" (typescript). Bank of America Archives, San Francisco.

Zimrick, Steven John. "The Changing Organization of Agriculture in the Southern San Joaquin Valley, California." Ph.D. diss., Louisiana State University, August 1976.

Ziskind, David. "The Suspension of Relief in Agricultural Areas." Research Section of the Planning Division for the Labor Relations Division, 20 August 1935, Farm Security Administration Collection, Bancroft Library.

COLLECTIONS AND ARCHIVES

Archivo de Secretaría de Relaciones Exteriores. México D.F.

Bank of America Archives. San Francisco, California.

California State Archives. Sacramento, California.

Clements, George, Collection. Special Collections, University of California, Los Angeles.

Collins, Tom, Reports. United States Department of Agriculture. Agricultural Stabilization and Conservation Commission Papers, Record Group 145, Federal Records Center, San Francisco, 17 October 1936 (typescript).

Darcy, Sam, Collection. New York University, New York, New York.

Federal Archives and Regional Center. Records of the Farm Security Administration, Record Group 96. San Bruno, California.

Galarza, Ernesto, Collection. Stanford University.

Giannini Foundation of Agricultural Economics Library, University of California, Berkeley.
Glassford, Pelham D., Collection. Special Collections, University of California, Los Angeles.
Goldschmidt, Walter, Records. Central Valley Project Studies, 1942–1946, Record Group 83, Bureau of Agricultural Economics. Western Regional Office, Federal Archives and Regional Center, San Bruno, California.
Johnson, Hiram, Papers. Bancroft Library, University of California, Berkeley.
McWilliams, Carey, Collection. Special Collections, University of California, Los Angeles.
National Archives. Washington, D.C.
Reichart, Irving, File. Bancroft Library, University of California, Berkeley.
Taylor, Paul, Collection. Bancroft Library, University of California, Berkeley.

NEWSPAPERS AND JOURNALS

Agricultural and Cannery Union News
American Federationist
Bakersfield Californian
Berkeley Gazette
California Cotton Journal
California Cultivator
California Journal of Development
The Californian
CIO Labor Herald
Corcoran Journal
Daily Californian
Daily Worker
The Dispatcher
El Cerrito Review
Farmer-Labor News
Fresno Bee
Fresno Morning Republican
Hanford Journal
Kern County Labor Journal
Labor Herald
Los Angeles Examiner
Los Angeles Times
La Lucha Obrera
Madera Tribune
New York Daily Worker
New York Times
Oakland Tribune
La Opinion
Pacific Rural Press
People's World

Rural Observer
Rural Worker
Sacramento Bee
San Francisco Call Bulletin
San Francisco Chronicle
San Francisco Examiner
Stockton Record
Trabajador Agrícolo
Tulare Advance Register
UCAPAWA News
Visalia Times Delta
Western Cotton Journal and Farm Review
Western Worker

INTERVIEWS

INTERVIEWS BY AUTHOR

Arancibia, Josefina. Madera, California, January 1982.
Avilla, Nick. By author with Juan Gómez-Quiñones. Los Angeles, April 1971.
Bañales, Edward. Corcoran, California, January 1982 and May 1983.
Benson, Evelyn Velarde. Los Angeles, California, 4 May 1971.
Callejo, José [pseud.]. Visalia, California, 23 January 1982.
Castro, Roberto. Corcoran, California, January 1982.
Chambers, Pat. Wilmington, California, 11 May 1982.
Cortez, Sabina. Corcoran, California, 18 September 1982.
Cuéllar, Edward. Visalia, California, January 1982.
Decker, Caroline. San Raphael, California, June 1982.
de la Cruz, Arnold. Fresno, California, 1 June 1981.
de la Cruz, Jessie. Fresno, California, 1 June 1981.
Flores, Belén. Hanford, California, January 1982.
Gasca-Cuéllar, Lillie. Visalia, California, January 1982.
Gómez, Magdalena [pseud.]. Los Angeles, California, February 1982.
Gutiérrez-García, Juan. Ancihuácuaro, Michoacán, Mexico, 18 July 1987.
Healey, Dorothy. Los Angeles, California, 6 May 1971.
Hernández, Maria [pseud.]. Corcoran, California, 18 September 1982.
Hernández, Refugio. El Centro, California, June 1982.
Hicks, Wyman. Bolinas, California, January 1983.
Lima, Luis. Brawley, California, June 1982.
López, Eduardo. Calexico, California, 21 May 1982.
Luna, Gene. Madera, California, January 1982.
Luna, Grace. Madera, California, January 1982.
McCormick, LaRue. Los Angeles, California, August 1982.
Magaña, Ray. Corcoran, California, 17 September 1982.
Martínez, Guillermo. Los Angeles, California, 21 April 1971.
Pacheco, Macario. Brawley, California, 25 June 1982.
Padilla, Juana. Brawley, California, 25 June 1982.

Parra, Leroy. Los Angeles, California, 18 April 1971.
Regelado, Soledad. Corcoran, California, September 1982.
Ross, Fred. Los Angeles, California, March 1983.
Sálazar, Adelaida [pseud.]. Corcoran, California, January 1982.
Sálazar, Luis. Hanford, California, January 1982; Tulare, California, March 1982.
Saludado, Pablo. Earlimart, California, January 1982.
Spector, Frank. Los Angeles, California, 15 April 1971.
Torres, Carlos. Tulare, California, January 1982.
Trinidad Navarro, Vicente. Ancihuácuaro, Michoacán, Mexico, 17 July 1987.
Vidaurri, Narciso. Hanford, California, January 1982.
White, Jim [pseud.]. Bakersfield, California, 1 January 1982.

OTHER INTERVIEWS

Camp, Wofford B. "Cotton, Irrigation, and the AAA," an oral history conducted by Willa Baum, 1962–1966. University of California, Berkeley, Regional Oral History Office.
Chambers, Pat. University of California, Berkeley, Regional Oral History Office, 1 June 1972.
Crane, Edgar Romaine. Interview by Judith Gannon for California Odyssey Project, California State University, Bakersfield, California, 7 April 1981.
de la Cruz, Jessie. Interview by Lea Ybarra, Fresno, California, 27 August 1980.
Dunn, Lillie Ruth Anne. Interviews by Judith Gannon for California Odyssey Project, California State University, Bakersfield, California, 14 and 16 February 1981.
Dunn, Lillian. Interview by Anne Loftis, Bakersfield, California, 10 June 1983. Anne Loftis's personal collection.
Hammett, Roy. Interview by Anne Loftis, Fresno, California, 12 September 1982. Anne Loftis's personal collection.
Hammett, Wilson. Interview by Anne Loftis, Fresno, California, 12 September 1982. Anne Loftis's personal collection.
Healey, Dorothy. Interview by Joel Gardner, 1982. University of California, Los Angeles, Oral History Program. Department of Special Collections, University Research Library, UCLA.
Pate, Rev. Billie J. Interviews by Michael Neely for California Odyssey Project, California State University, Bakersfield, California, 5 and 12 March 1981.
Reuben, Dan. Interview by Walter Goldschmidt, 4 April 1944. Federal Records Center, San Bruno, California.
Stockton, Frank. Interview by Walter Goldschmidt, Arvin, California, 3 April 1944. Walter Goldschmidt collection, Federal Records Center, San Bruno, California.
Sullivan, Catherine. Interview by Judith Gannon for California Odyssey Project, California State University, Bakersfield, California, 27 February 1981.

Index